Sociocultural Explorations of Science Education

Volume 21

The series is unique in focusing on the publication of scholarly works that employ social and cultural perspectives as foundations for research and other scholarly activities in the three fields implied in its title: science education, education, and social studies of science.

The aim of the series is to promote transdisciplinary approaches to scholarship in science education that address important topics in the science education including the teaching and learning of science, social studies of science, public understanding of science, science/technology and human values, science and literacy, ecojustice and science, indigenous studies and science and the role of materiality in science and science education. Cultural Studies of Science Education, the book series explicitly aims at establishing such bridges and at building new communities at the interface of currently distinct discourses. In this way, the current almost exclusive focus on science education on school learning would be expanded becoming instead a focus on science education as a cultural, cross-age, cross-class, and cross-disciplinary phenomenon.

The book series is conceived as a parallel to the journal Cultural Studies of Science Education, opening up avenues for publishing works that do not fit into the limited amount of space and topics that can be covered within the same text. Book proposals for this series may be submitted to the Publishing Editor: Claudia Acuna E-mail: Claudia.Acuna@springer.com

Katerina Plakitsi • Sylvie Barma
Editors

Sociocultural Approaches to STEM Education

An ISCAR International Collective Issue

Springer

Editors
Katerina Plakitsi
Department of Early Childhood Education
School of Education
University of Ioannina
Ioannina, Greece

Sylvie Barma
Department of Teaching and Learning Studies
Laval University
Quebec, QC, Canada

ISSN 2731-0248 ISSN 2731-0256 (electronic)
Sociocultural Explorations of Science Education
ISBN 978-3-031-44376-3 ISBN 978-3-031-44377-0 (eBook)
https://doi.org/10.1007/978-3-031-44377-0

This Springer imprint is published by the registered company Springer Nature Switzerland AG
The registered company address is: Gewerbestrasse 11, 6330 Cham, Switzerland

Paper in this product is recyclable.

Preface

This book contributes to scholarship exploring sociocultural approaches to Science Technology Engineering and Mathematics (STEM) Education. It offers a new interpreting theoretical framework coming from Cultural Historical Psychology and Cultural Historical Activity Theory (CHAT). The authors highlight some elements of the sociocultural context that mediate learning about STEM or via STEM. This book is edited by the founders (Katerina Plakitsi and Sylvie Barma) of the relevant Thematic Section within the International Society of Cultural Historical and Activity Research (ISCAR). The term socio-cultural stems historically from the previous regional conferences of ISCAR. After a series of regional conferences, ISCAR society was formally constituted in June 2002 and in 2005 held the first International ISCAR conference in Seville, Spain. During the third ISCAR conference in Rome (2011), the STEM thematic section was founded by Plakitsi and Barma. The aim of the above ISCAR special section was, among others, to develop a collective identity among members of the ISCAR community interested in STEM Education, situated in formal as well as informal settings.

ISCAR represents many researching groups. Initially, we found that a group of scholars in Madrid developed ideas relevant to the global south and they called themselves (in Spanish) the "sociocultural psychologists". In parallel, a serious critique of the various currents had been emerged around, mainly because they all missed "history" and this was part of coining the term Cultural Historical Activity Theory. In contemporary scholarship, CHAT has been transferred broadly to the academic world and beyond its origins. This theory has been used extensively in the public health and care sector and in STEM Education. It has also been applied to the legal sector, to the study of homelessness and addiction, to social movements, to the exploration of social inequities and to many other forms of activity. The old term "Sociocultural psychology" is a very broad term, much wider than CHAT, but has always included CHAT.

This book connects ISCAR STEM with CHAT and brings together scientists, educators and psychologists to explore new common fields of research. The initial idea to give voice to authors contributing to this book came up during the ISCAR

2017 international conference as the two editors wanted to give voice to a new angle in STEM Education under a cultural historical perspective. The authors take a specific stance to address STEM Education to engage in human development to act for peace, sustainable development and public understanding. The chapters cover a wide range of topics and situations: both inside educational institutions (from the early childhood till university, including in-service training for actors of education), and in all societal "informal" settings where learning, culture and social interactions occur (for example museums, science centers, environmental parks, families, kindergarten). The original idea behind this book aims to develop an understanding of scientific concepts within a sociocultural framework and support professional development of educators and researchers.

Overall, the book addresses the complexity of science teaching and learning in formal and informal settings in the process of fostering transformative thought and logical reasoning within a multidimensional context.

The book is divided into four parts. The first part of the book gathers three texts, respectively by Rodrigues et al., Ratnam and Chaïklin. These three contributions address the core values at the roots of STEM education.

Rodrigues, Camillo and Mattos broaden our view of STEM teaching-learning process adopting a socio-historical cultural perspective, fruitful for new types of solutions and concrete initiatives relevant to science teaching. Considering the existing criticism, they discuss how cultural-historical activity theory might provide a critical lens to examine the history and development of Science, Technology, Engineering, and Mathematics (STEM) education. The authors enlighten STEM initiatives, which contradictorily form a vision for the future as a new transformative activity. Rodrigues and his Brazilian colleagues propose a critical reflection of the epistemology and branches of science education bringing to light the limits of idealizing STEM education in a neoliberal social context. It seems naïve to think that science teaching can be free of contradictions when schools and educational agents don't consider sensitive socioscientific topics, hands-on activities, problem-based learning to foster profound changes in education.

Tara Ratman focuses on Indian education that has remained largely "retrospective" despite the aspirational "prospective" goals set for it in policy statements. The largely passive role to which students are subjected, regarding their interests and voice, renders academic learning irrelevant to their present and future needs. Science, Technology, Engineering, Art, Math (STEAM) presents a way to address this issue by promoting creative and critical thinking adopting an inquiry-based and hands-on learning. Such competencies can prepare students to become self-evolving learners with an ability to co-construct knowledge that adapts to new societal conditions and take into account the ongoing evolution of the culture of schooling. Ratnam focuses on how Science, Technology, Engineering, Arts, Math (STEAM) can become a way to challenge the passive role that students are subjected to by promoting creative and critical thinking as well as inquiry-based and hands-on learning. A long-standing traditional culture of schooling in India too often has been an obstacle to link the micro classroom interactions to the larger institutional and societal dynamics in which they are embedded, so STEAM education gets beyond a dominant form of teaching activity.

Chaïklin introduces the idea of radical-local teaching and learning, and the importance of theoretical thinking for school children's development and discusses their relevance for science education. This new approach to working with theoretical thinking in relation to subject-matter concepts is viewed to propose a more situated curriculum to the students. Seth Chaiklin raises a question related to the origins of Cultural Historical Activity Theory by arguing that the formation of conceptual ideas has not received the proper interest in contemporary research and deserves further examination and consideration. One aim of this chapter is to introduce some of these concepts, including the idea of personality development, and the role of theoretical thinking for school children's development, and discuss their relevance for science education. A second aim is to introduce the idea of radical-local teaching and learning, which both continues this line of thinking, but elaborates a more situated curricular perspective.

The second part brings together the work of researchers interested in developmental psychology and childhood, with a special focus on using Activity theory and Cultural-historical research approach to unite these two opposing approaches to the study of children. Fleer, Remountaki et al. as well as Kolokouri et al. focus on the teaching of STEM through the affordances of CHAT by means of the use of play.

Fleer presents the results of a study where teacher theoretical thinking is the leading activity and fosters new conceptions of STEM practice in the classroom. Dialectical relations between child motives and teacher's actions resulted in a relational model of how to bring STEM concepts into play. Teacher transformation is conceptualized not as a simple problem of practice change but as a maturing of theoretical thinking by early childhood teachers through solving the problem of play and STEM concept formation. Marylin Fleer raised the fact that the long-standing literature into the professional learning of early childhood teachers brings forward the central problem of teacher development. What is not well understood is the psychological content of that development? To answer the question of what is in development, this chapter reports on the outcomes of a 2-year educational experiment and a 2-year follow-up interview of two teachers (digitally recorded practices, interviews, weekly reflections) engaged in STEM teaching. The major conclusion reveals that teacher theoretical thinking was the leading activity which brought forward new conceptions of a STEM practice, and the dialectical relations between child motives and teacher motivated actions resulted in a relational model of how-to bring STEM concepts into play.

Remountaki et al. explore how scientific play can create conditions for pre-schoolers to form the concept of dissolution during everyday educational reality. Their study informs everyday educational practice and science pedagogy by providing new insights into learning and development in science unpacking children's intellectual, social, emotional, and enactive needs. The study informs everyday educational practice by providing a pedagogical framework based on the coherence, consistency, and balance between teaching-learning goals in science and play practices.

Kolokouri and Plakitsi base their chapter on the theoretical framework of Cultural Historical Activity Theory and how scientific thinking can be developed mobilizing conceptual tools to understand dialogue and networks of interacting activity systems. Within this frame, SCOPES (Systems of activity, Contradictions, Outcomes, Praxis, Expansive learning, Science education) is used as a methodological tool so as to integrate scientific knowledge through participatory methods. This third text uses play as well as teaching scientific concepts which are constructed using cartoons. A Science Curriculum for the early years has been developed, which consists of two parts. In the first part, a popular cartoon was used in order to teach floating and sinking. Then, elements from History of Science were incorporated in a narrative about light and colors. Finally, the narrative was turned to an animation in the program Scratch.

The third part is characterized by the importance of mediating tools, so science education opens up to a diversity of techno-creative approaches. STEM education advocates for an interdisciplinary perspective in which techno-creative activities can provide a context supporting development of scientific competencies. Techno-creative activities engages the learners in teams working in ill-defined creative problem-solving activities supported by a diversity of technologies. The activity systems proposed within the STEM educational activities challenges the learners in the pursuit of a solution in which the expansive transformation is accomplished through engaging the learners in reconceptualizing a problem by the mediation of technological artifacts which should be used in a way which is not priorly known by the participants of the activity.

Romero and Barma document how engaging in solving an ill-defined problem during an international educational robotics challenge requires collaboration under a distributed and international approach. The situational learning of the students in a challenge of solving complex problems helped to develop an understanding of the difficulties related to both technological and the necessary coordination efforts. The authors discuss how the complexity of an international synchronous robotics' challenge, involving problem solving and the development of twenty-first century competencies, highlights the fragility of the interconnections between the different communication tools essential for the coordination and the success of this international competition. The teams are engaged in collaborating through a narrative structure based on a natural disaster, a tsunami. The robots had to be remotely programmed to coordinate and operate the smart city's protection systems, to secure four strategic points, engaging in a distributed collaborative problem-solving activity.

Topoliati and Plakitsi focus on the creative approach of Science-Technology-Engineering-Arts-Mathematics (STEAM Education) by preschool students in the context of their participation in innovative educational robotics projects. In this context, a formal, informal and non-formal type of education is applied in which the natural, social and cultural environment is utilized as a primary source of knowledge. The project stimulates creativity, curiosity and critical thinking as well as communication skills and collaboration between children through play and exploration. STEAM is used as a powerful mediating tool aiming at enhancing sustainable development. Following the expansive cycle, children act as scientists and engage in active dissemination of goals of sustainable development throughout their curriculum. Their text brings to light how childhood studies have focused on the study of children anchored in historical time and settings; such approaches are more commonly found within anthropological and sociological traditions, especially those that focus on situated and localized practice with children.

Kornelaki and Plakitsi use Cultural Historical Activity Theory (CHAT) as a theoretical framework for the design and analysis of a science education program with an emphasis on active and interactive learning processes. The use of CHAT unites science, culture and society mobilizing an educational program as a mediating tool to document how museums reveal a powerful setting to develop the scientific method. Grounded in the present, the past and the future, the Archaeological Museum of Ioannina contributes to the interest of children in science and offer a broader perspective of science teaching in formal and informal settings. In the museum's learning community, students practice scientific method processes, interact with the exhibits of the museum as well as with their classmates, instructor and teachers, actively participating in the actions of the educational program "Thunderbolt Hunt" which are developed in a playful way.

Finally, in Part IV, science teachers and their peers as well as students challenge dominant forms of science teaching activity to expand their practice and the conditions in which students learn. Tensions, contradictions play a key role in the transformation of activity systems as they hybridize and share or not a common running object.

Silva et al. present an ongoing research project focusing on expanding the concept of co-teaching, considering the case of the teaching practice (practicum) in the context of the Physics Teacher Education Program. They present the analysis of a co-taught physics class about thermal machines, to a group of Youth and Adult Education class, by one student-teacher (ST) and his mentor (M), an in-service physics teacher, in a Brazilian public High School. They used the Pajek software and considered the verbal interactions of the subjects and identified who spoke to whom for the construction of the network. Based on CHAT, they make us understand that the co-teaching accomplishment is due not just to the good partnership between the student teacher and the mentor, but also to the high school student's acknowledgment, acceptance and legitimization of the student-teacher as a teacher. Thus, the expanded co-teaching brings up the concept of the emergent teacher on its dynamics.

Barma and Voyer present how two science teachers and a pedagogical counselor collaborated during 7 years to co-design and exchange technical and instructional artifacts to meet new curricular demands requiring the integration of technological design to science teaching. The colleagues gave new meanings to conflicting motives in a complex learning setting by means of learning actions. Inspired by developmental work research and ethnomethodology, the chapter traces back 7 years doing CHAT research. Results present how the expansive resolution of conflicts of motives triggered transformative actions that resolved the inner contradictions identified in their activity systems. Boundary crossing was possible as new meaning, new roles and division of labor lead to the expansion of their practice. The authors identify the sharing of personal expertise as a potential germ cell to expand the boundaries of activity systems as science teachers face conflicts of motives in the context of a partnership with a school board and a research team.

The last chapter by Freiman and Lingley pinpoint the importance of tensions as students accept that failure is a key concept in design thinking. The authors explore several STEM-bound (Science, Technology, Engineering, Mathematics) trends that influence the K-12 education system in New Brunswick, Canada, over the past two decades. They present a case study on the method of design thinking class while analyzing the micro-tensions from students' prototyping in a school makerspace. This can be some lessons from an unfinished work. The lenses of CHAT help analyze a micro-level of the student's activity in a Maker space. At every point in the process of creating artifact by means of a 3D printer or a software like Thinkercad, micro-tensions arise, and are revealed as opportunities for students engaging in resolving issues that created those tensions. Overall, this part enhances how science teachers benefit from working creatively and collaboratively taking under consideration the internal and external conflicts and contradictions intervened in their science teaching.

The two editors open a dialogue about sociocultural approaches to STEM Education in an international field, and they welcome open discussions. The current situation of global crisis emerges the necessity of a new STEM education firstly as basic research to find a common core of the different scientific disciplines of Science, Technology, Engineering and Mathematics. So far, the scholastic landscape seems to follow different parallel pathways focusing on one or the other discipline without a unified model that must be educationally grounded. The editors welcome scholars and practitioners who co-develop longitudinal studies on important concept categories or theories like systems of activity, contradictions, outcomes, praxis and expansive learning. The editors foresee that this kind of academia work can bridge the four different disciplines of STEM towards a new hybrid discipline to address the multidimensional global problems concerning peace, one health, global understanding and sustainable development.

Ioannina, Greece Katerina Plakitsi

Quebec, QC, Canada Sylvie Barma
Summer 2023

Introduction: Toward a Zone of Proximal Development in Stem Education

This book as a whole is powerful evidence for the argument that "STEM education, despite its universality aspirations, is a historical and contextual movement still under construction" (Rodrigues, Camillo & Mattos, Chap. 1, in this volume). The impact and viability of such a movement depend largely on our ability to identify historically grounded key dimensions along which development should be examined, envisioned and practically enacted. In this introduction, I propose two key dimensions. When interconnected to form a four-field, these dimensions allow us to construct a collective zone of proximal development for STEM education.

From What to Why

In 2012, Bruce Alberts, Editor-in-Chief of *Science*, wrote an editorial titled "Failure of skin-deep learning." Alberts pointed out that the most meaningful learning takes place when students are challenged to address an issue in depth, which can only be done for a relatively small number of topics in any school year. However, the standard practices of curriculum design, textbook production and classroom teaching tend to promote a superficial coverage of a field, leaving little room for in-depth learning. The curricula, textbooks and instructional practices are skin-deep.

> The factoid-filled textbooks that most young U.S. students are assigned for biology class make science seem like gibberish—an unending list of dry, meaningless names and relationships to be memorized. Take, for example, my 12-year-old grandson's life science textbook. Approved by the State of California, it is filled with elaborate drawings and covers an astonishingly broad range of biology. But the text is largely incomprehensible for its student audience. . . When my grandson and his classmates successfully complete that book and the class based on it, it is clear that they will know nothing of the kind of biology that inspires passion in the souls of the scientists working in the labs around me. . . . (Alberts, 2012, p. 1263)

There is surprisingly little research on this "depth vs. breadth" problem in science education. One major study stands out, namely that of Schwartz and co-authors (2009). Using a sample of 8310 students in introductory biology, chemistry or physics courses in 55 randomly chosen US colleges and universities, these authors found that students who reported covering at least one major topic in depth, for a month or longer, in high school were earning higher grades in college science than did students who reported no coverage in depth. Students reporting breadth in their high school course, covering all major topics, did not appear to have any advantage in chemistry or physics and a significant disadvantage in biology.

Learning in depth – asking and answering "why?" questions – breaks the boundaries between natural, social and human sciences. As Ratnam (Chap. 2 in this volume) points out, this requires "embracing the arts-and-humanities-infused STEAM approach." Natural phenomena have to be reconceptualized in their cultural embeddedness and with their societal consequences. Correspondingly, humans, their artifacts and their societies need to be reconceptualized as often destructive but nonetheless inseparable and foundationally dependent parts of nature. This direction of development means that science learning must tackle the critical challenges that threaten life on our planet; *fateful objects*, as we have called them (Engeström & Sannino, 2021).

From Encapsulation to Partnerships

About 35 years ago, Lauren Resnick characterized the encapsulation of schools as follows.

> The process of schooling seems to encourage the idea that the "game of school" is to learn symbolic rules of various kinds, that there is not supposed to be much continuity between what one knows outside school and what one learns in school. There is growing evidence, then, that not only may schooling not contribute in a direct and obvious way to performance outside school, but also that knowledge acquired outside school is not always used to support in-school learning. Schooling is coming to look increasingly isolated from the rest of what we do. (Resnick, 1987, p. 15)

In an encapsulated classroom, the school text becomes the object of the activity instead of being an instrument for understanding the world. When the text becomes the object, the instrumental resources of the activity are impoverished – students are left "on their own devices." Resnick (1987, p. 13) pointed out that in school, the greatest premium is placed upon "pure thought" activities: "what individuals can do without the external support of books and notes, calculators, or other complex instruments." This instrumental impoverishment produces what Resnick (1987, p. 18) calls "the symbol-detached-from-referent thinking."

In the decades that have passed after Resnick's analysis, the encapsulation has become increasingly dysfunctional due to two major societal transformations. The first one is the spread and pervasive influence of digital media that capture students' attention and offer them unprecedented scopes of entertainment, information acquisition and communication. The second one is the societal turmoil that ranges from racism to the global climate crisis and calls for entirely new depth and breadth of both conceptual understanding and activist involvement.

Going beyond encapsulation – de-encapsulation – may be conceived of in terms of widening spheres or layers of learning. For example, in an examination of the expanding scope of science learning, I suggested five such layers: (1) bringing elements of societal practices into classroom instruction, (2) developing the entire school as an activity system and community of learning, (3) fostering science learning in activity systems outside the school, (4) working with indigenous and other communities as funds of knowledge and alternative epistemologies, and (5) collaborating with social movements as dynamic contexts of activist science learning (Engeström, 2017).

Moving outward along the dimension of de-encapsulation requires partnerships between schools and other activity systems, and eventually construction of heterogenous educational coalitions. Such multi-activity coalitions entail a new phase in activity-theoretical research, a phase we call *fourth-generation activity theory* (Engeström & Sannino, 2021).

Zone of Proximal Development of Stem Education

Moving out of skin-deep learning and encapsulation are two key dimensions along which schools need to construct their collective zones of proximal development. This dual challenge may be depicted with the help of Fig. 1.

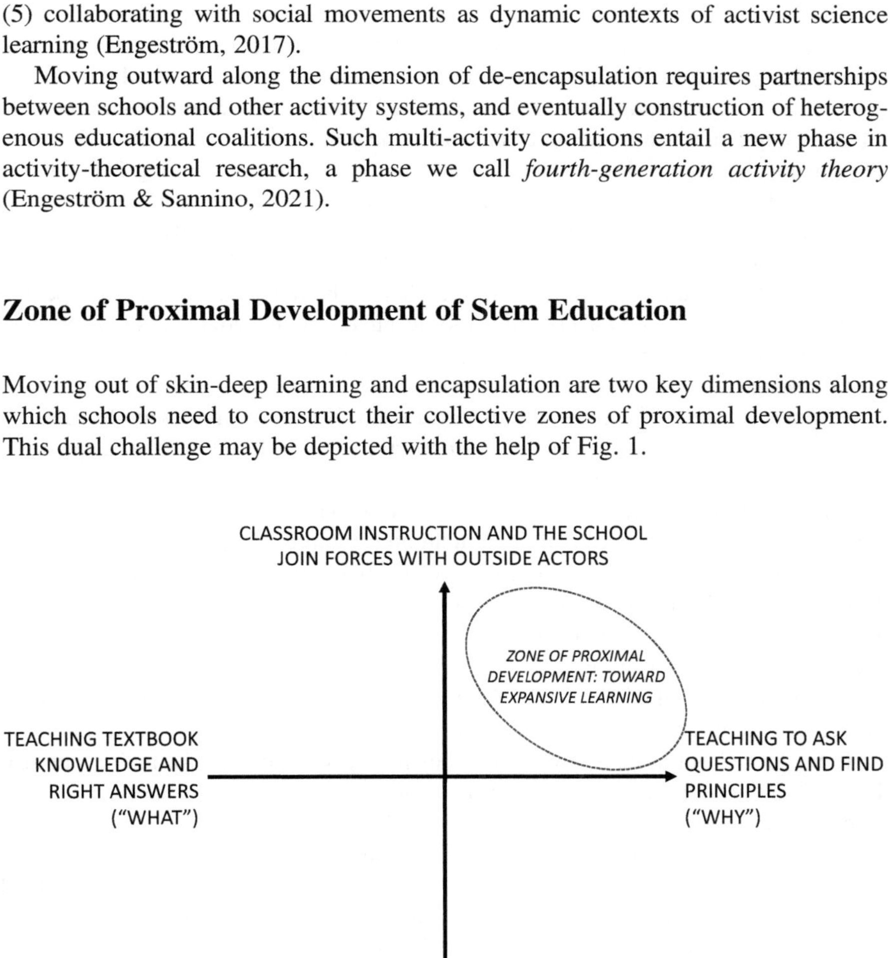

Fig. 1 Zone of proximal development of STEM education

The zone of proximal development is not a fixed destination. It is a gray area of debates, experiments and struggles. There are multiple alternative and complementary paths into and across the zone. The zone is generated in practical actions and in interplay with other fields depicted in Fig. 1. In other words, the zone of proximal development does not emerge out of pure visions. It emerges out of contradictions in existing forms of STEM education and efforts at resolving them.

From What to Why in Chapters of This Book

Davydov's (1990) work on the formation of theoretical concepts and theoretical thinking in instruction is foundational for the movement from what to why in STEM education. This is demonstrated by Chaiklin (Chap. 3, in this volume) with the help of an example of teaching about electromagnetic phenomena. Importantly, Chaiklin also opens up a connection to the de-encapsulation dimension, pointing out that "the main extension in the radical-local approach is to consider the relation or interaction between the content of teaching and the learner's lifeworld explicitly and centrally."

Fleer (Chap. 4, in this volume) shows that theoretical thinking is not only a challenge to the students but also to the teachers: "the theoretical problem appeared to create new developmental conditions for teachers, and the educational experiment brought changes in the dominating motives of the teachers from practice to theoretical thinking."

Concept formation is at the core of the movement from what to why. In cultural-historical activity theory, concept formation is not reduced to the acquisition of verbal definitions and formulas. Concepts grow out of enacted activities, not just school studies narrowly understood but also out of play and work. They are represented in multiple modalities, including pictures and bodily movements (Engeström et al., 2012). Pursuing this perspective, Remountaki, Fragkiadaki and Ravanis (Chap. 5, in this volume) explore "how scientific play created the conditions for the formation of the concept of dissolution by preschool children within early childhood educational settings."

Kolokouri and Plakitsi (Chap. 6, in this volume) used cartoons to develop students' scientific thinking in the early grades. As a mode of representation, cartoons turned out to be a powerful means not only in concept formation but also in building the connection of school instruction with everyday life – another example of connecting the two dimensions depicted in Fig. 1.

The movement from what to why is often understood narrowly as a shift from transmission of descriptive factoids to the acquisition of well-defined theoretical concepts. Romero and Barma (Chap. 7, in this volume) argue that what is increasingly needed is collaborative solving of ill-defined and open-ended problems, in this case saving the city from a tsunami. Problems of this kind can foster systemic and theoretical thinking while at the same time opening up the school curriculum to real-life challenges.

From Encapsulation to Partnerships in Chapters of This Book

Topoliati, Plakitsi and Stylilanidou (Chap. 8, in this volume) describe a project of early years' science education with the help of robotics. The project is a powerful example of opening up the school to create partnerships with activity systems outside. "In this context, the school 'is adopted' by the Ephorate of Antiquities of Ioannina and the 'Little Creative Scientists' participate in a holistic program of substantial acquaintance with the post-Byzantine monuments of Klimatia, the Archaeological and the Byzantine Museum of Ioannina, as well as the Castle of Ioannina. In particular, students gather information through outdoor activities about the post-Byzantine park... Then, the young students engage in a dialogue with the archaeologists comparing the 'past' with the 'present'."

In a similar vein, Kornelaki and Plakitsi (Chap. 9, in this volume) examine the cultivation of the scientific method and students' initiation in the scientific way of thinking in the Archaeological Museum of Ioannina. Importantly, the young participants not only conducted experiments suggested by the instructors – they also designed and implemented their own alternative experiments "both before and after the instructor challenged them to perform the planned experiments."

Barma and Voyer (Chap. 11, in this volume) show that opening schools for partnerships is a lengthy process of boundary crossing, in this case spanning a period of seven years. Such a process involves facing and resolving conflicts. "As the participants of the research projects learned to collaborate and shared their experiences, boundary zones were created and reflected how the sharing of the three collaborators' expertise led to the production of new teaching artefacts integrating technological design to science education."

Besides being encapsulated in space, traditional instruction is also encapsulated in time.

Focusing on learning in a school Makerspace, Freiman and Lingley (Chap. 12, in this volume) open up another angle on the struggles involved in the temporal aspect of de-encapsulation. "What happens when students do not actually achieve their intended goals and leave their work unfinished as they make the abrupt realization that their class has ended?" Using the innovative notion of micro-tensions as an analytical instrument, the authors reveal rich dialectics of virtuality versus materiality and process versus product in the work of the students. "To summarize our findings, we noticed that design thinking activities can engage students through a complex process of finding the significance of the incomplete."

Bringing Together the Two Dimensions

The two dimensions of expansive transformation of STEM education – breaking away from skin-deep learning and breaking away from encapsulation – have traditionally been separate from one another. Scholars and practitioners promoting the first dimension have been preoccupied with the cognitive dynamics of learning and instruction, often ignoring the societal arenas and struggles in which schools operate and students navigate. On the other hand, scholars and practitioners promoting the second dimension have been preoccupied with the societal and cultural emancipation of students, often paying scant attention to what actually goes on – and what could change – in classroom instruction. The question is: Can these two dimensions come together and jointly generate a zone of proximal development for school instruction?

The chapters of this book include significant attempts at bringing together the two dimensions. Yet, we are only beginning to grasp what this challenge involves. In Fig. 1, I characterize the meeting point of the two dimensions as expansive learning (see also Silva et al., Chap. 10, in this volume). Expansive learning means that students and teachers are involved in the formation of theoretical concepts that not only allow them to understand predetermined curricular subject matter knowledge but actually explain contested societal issues and enable transformative action in the world. It means that young people's activism for social and climate justice is empowered by the formation and use of dialectical and historical thinking mediated by systemic models (Engeström, 2022).

Helsinki, Finland Yrjö Engeström

References

Alberts, B. (2012). Failure of skin-deep learning. *Science, 338*(6112), 1263–1263.

Engeström, Y. (2017). Expanding the scope of science education: An activity-theoretical perspective. In In K. Hahl, K. Juuti, J. Lampiselkä, J. Lavonen & A. Uitto (Eds.), *Cognitive and affective aspects in science education research* (pp. 357–370). Springer.

Engeström, Y. (2022). Moving into the zone of proximal development of school instruction. In C. Le Hénaff, M. Le Paven, F. M. Prot & H-L. Go (Eds.), *Comprendre et transformer la forme scolaire: Contributions de la théorie de l'action conjointe en didactique.* Presses Universitaires de Rennes.

Engeström, Y., Nummijoki, J., & Sannino, A. (2012). Embodied germ cell at work: Building an expansive concept of physical mobility in home care. *Mind, Culture, and Activity, 19*(3), 287–309.

Engeström, Y., & Sannino, A. (2021). From mediated actions to heterogenous coalitions: four generations of activity-theoretical studies of work and learning. *Mind, Culture, and Activity, 28*(1), 4–23.

Resnick, L. B. (1987). Learning in school and out. *Educational Researcher, 16*(9), 13–20.

Schwartz, M. S., Sadler, P. M., Sonnert, G., & Tai, R. H. (2009). Depth versus breadth: How content coverage in high school science courses relates to later success in college science coursework. *Science Education, 93*(5), 798–826.

Contents

About the Editors

Katerina Plakitsi is a member of the Governing Council of the University of Ioannina. She is a full professor of Science Education with two bachelor's degrees in Physics and Pedagogy, a master's diploma, and a PhD in Science Education. Her main researching interests are Science Education, Formal and Informal Science Education, and Cultural Historical Activity Theory applied in Science Education. She has written many books in Greek and English while she has published in different international academic journals. She is scientific coordinator in many European Projects and supervises some PhD scholars in Science Education. She has written many schools environmental science textbooks, and she coordinated a science curriculum reform in Greece for the contemporary Education. She co-founded the interdisciplinary master's program "Environmental Sciences and Education for Sustainability" in collaboration with the Faculty of Medicine and the Department of Biological Applications and Technology at the University of Ioannina. She is also the co-founder of the ISCAR-STEM Thematic Section and the principal investigator of the @formal and informal science education group (@fise group). She is the editor-in-chief of the international bilingual journal *Science Education: Research and Praxis*. She had been Head of the Early Childhood Department at the University of Ioannina. Katerina Plakitsi is President of the International Society for Cultural Historical Activity and Research (https://www.iscar.org/).

Sylvie Barma is a full professor of Science Education at the Faculty of Education at Laval University since 2008. After teaching high school science for 20 years and contributing to the development of the new Quebec Science and Technology curriculum, she obtained a Ph. D. in Science Education.

Part I
A Chat Perspective on Transformative Activity in Science Education

Chapter 1
STEM and Its Roots and Branches: Critical Reflections from Cultural-Historical Activity Theory

André Machado Rodrigues, Juliano Camillo, and Cristiano Rodrigues de Mattos

1.1 The Meaning of STEM Education

Education is a field that has been closely linked to political, economic, and scientific trends. Over the years, it has been possible to identify educational policies based on frameworks that, in most cases, seem to be more like fads. This is noticeable in the face of the volatility with which certain trends take the vanguard of educational innovation movements. Such frameworks weaken in the face of the educational process complexity and respond only to part of the problems found in schools. It seems to be a recurrent educational phenomenon, with the creation of new trends shaped by and within the economic and educational policies of their times. Regarding Science, Technology, Engineering, and Mathematics (STEM)[1] education, Li et al. (2020) indicate that the number of research and publications has been growing since 2010, which might suggest that it has obtained recognition as a relevant topic. In this context, it is reasonable to interrogate whether STEM education is a fad or holds the potential for overcoming the current school challenges.

[1] Even though we acknowledge that Science, Technology, Engineering, Arts, and Mathematics (STEAM) education is a slightly different movement, all the considerations developed in this chapter might fit it as well. Thus, hereinafter, we will only refer to it as STEM.

A. M. Rodrigues (✉) · C. R. de Mattos
Department of Applied Physics, Institute of Physics, University of São Paulo, São Paulo, SP, Brazil
e-mail: rodrigues.am@usp.br; crmattos@usp.br

J. Camillo
Department of Teaching and Cultural Practices, School of Education, State University of Campinas, Campinas, SP, Brazil
e-mail: jcamillo@unicamp.br

K. Plakitsi, S. Barma (eds.), *Sociocultural Approaches to STEM Education*, Sociocultural Explorations of Science Education 21,
https://doi.org/10.1007/978-3-031-44377-0_1

Originally from the United States with recent ramifications in the United Kingdom and the rest of Europe, STEM education has started to reach Latin America (Lorenzin, 2019). The acronym's origin goes back to the 1990s, and its spread was supported by institutions like the National Academy of Sciences (NAS) and the National Academy of Engineering, and the National Science Foundation (NFS) (Martín-Páez et al., 2019). Although the goal of this chapter is not to extensively examine the genesis of STEM education, it is important to bear in mind that, to a large extent, an educational initiative like STEM education is oriented by issues of its own context, despite its apparent neutrality and universality.

STEM education has been framed as a *teaching innovation* that can tackle current educational problems. Furthermore, STEM education stems from a certain consensus that the current school is far from providing the needed skills for workers and citizens who encounter a world of fast social and technological changes. As framed by S. Selcen Guzey et al., "more graduates with STEM degrees will be needed in order to maintain America's competitive position in this global economy" (2014, p. 271). This underlying argument regarding the need for a STEM workforce to guarantee technological advances and, therefore, social and economic development, is common in STEM education literature (Donovan et al., 2014).

The literature does not provide any clear definition or conceptualization of STEM education. However, it is possible to relate STEM education to an entanglement of already established science and mathematics education traditions such as project and problem-based learning, interdisciplinarity, hands-on activities, etc. On the one hand, the blurry boundaries of the conceptualization of STEM education make it easy to aggregate new initiatives, projects, and goals since it is open enough to dismiss any deep theoretical commitments and may be responsible for leveraging its growth. On the other hand, it might lead to the pernicious practice of uncritically embracing instrumentalist perspectives on scientific development and salvationist views of STEM (see Zeidler, 2016).

The development of a critical STEM conceptualization is not a matter of preciousness, but rather the development of a structural understanding concerning possible new solutions offered to historic educational problems. In this chapter, we present a critique of STEM education from two analytical tenets grounded on cultural-historical activity theory (CHAT): (i) object-oriented activity; (ii) history and development.

The term STEM is used in the literature to handle at least two educational issues. Typically, STEM refers to an ill-defined field of knowledge, or a group of not necessarily integrated disciplines. For instance, it appears in several studies related to students' interests, career choices, workforce skills, etc. It also appears in many educational policies that refer to these science, math, technology, and related areas but strategically establish no ties to any specific instructional approach or integration among disciplines. While in office, US President Obama mentioned, on several occasions, the relevance of STEM education to the national future. According to him: "It's in these pursuits that talents are discovered and passions are lit, and the future scientists, engineers, inventors, and entrepreneurs are born" (Obama, 2010).

It is worth noting that the term STEM seems strongly associated with local institutional and linguistic aspects. For example, in Brazil, such use of the term STEM is rare since there is already a specific term to express that idea without alluding to the acronym. This use of the term STEM does not entail any specific pedagogical approach or interdisciplinary connection across the related disciplines.

However, in this chapter, we focus on the term STEM used to propose new teaching approaches, envision a new educational practice, and integrate two or more STEM disciplines to some extent. In this sense, it seems fair to state that STEM education is a way to address perceived problems in the current science and mathematics education. Recently, William S. McComas and Stephen R. Burgin (2020) have pointed out that some initiatives have included an I, for *integration* (ISTEM), to emphasize its integrative feature. Despite that, for this chapter, we are not making any further distinctions. We are considering that ISTEM, as a curricular or teaching approach, is part of the object of this analysis. Hereafter, for the sake of simplicity, we will refer only to the term STEM.

The criticism of STEM education came from different areas and pointed to different problems and limitations. For example, Dana L. Zeidler indicated that STEM education might lack a commitment to broad socioscientific issues, particularly values, ethics, and human decisions. According to him, "[...] failing to recognize or willfully ignoring the primacy of sociocultural frameworks will inevitably result in a deficit educational model of STEM initiatives." (2016, p. 12). Similarly, McComas and Burgins are skeptical of STEM, particularly its engineering and technological aspects, considered possibilities to motivate students to learn and use scientific and mathematical concepts, values, principles, and methods. They concluded the critique by underlining:

> We hope that our readers do not see our position as a plan designed to block progress. However, we have witnessed other purported revolutions in science education, that have made little substantive impact on the field, yet garnered much initial support and cost much time and money (McComas & Burgin, 2020, p. 825).

Furthermore, scholars indicate that the STEM education slogan might end up including many teaching initiatives that are far from being innovative or effective (McComas & Burgin, 2020). Radu Bogdan Toma and Antonio García-Carmona (2021) highlight at least three major points of criticism of the current state of STEM education: the claim for its originality when confronted with approaches of science and mathematical education such as Science, Technology, and Society (STS); its effectiveness that are generally neglected in the STEM education reports; and the new markets created under the STEM education slogan.

Considering the existing critiques, we shall discuss how CHAT might help reframe some key underlying problems in STEM education. It should also provide a critical lens to examine the history, development, and future of STEM education and the current schooling process.

1.2 Outlining the Analysis from Cultural-Historical Activity Theory

Lev Vygotsky (1997, p. 240) reminds us that "any concrete phenomenon is completely inexhaustible and infinite in its separate features. We must always search within the phenomenon what makes it a scientific fact". From this perspective, our analysis does not aim to provide a detailed description of STEM education. On the contrary, we rely on cultural-historical activity theory to grasp essential contradictions present at the core of the STEM education phenomenon.

As presented by Yrjö Engeström (2015), CHAT has its roots in Vygotsky's works and has been developed within many fields: psychology, anthropology, and sociology. It is a multidisciplinary research corpus that aims to understand human development from a concrete and historical perspective. From that point of view, human development is a whole process through which humans make up themselves by transforming their reality (which includes knowing, doing, interacting, etc.), i.e., actively through/and in reality (Stetsenko, 2019). According to Wolff-Michael Roth (2020), "activity theory is intended to explain change, learning, and development as an immanent feature of a system, rather than in terms of externally produced cause-effect relations" (p. 20).

With the diffusion of Vygotsky's works in the West in the 1980s, science education research also started incorporating the Vygotskian perspective to focus on conceptual learning, verbal interactions, and the role of social relations in teaching and learning science, for instance. When entering science education research, Vygotsky's works encountered a field permeated by several psychological approaches, including behaviorism and cognitivism (Roth et al., 2009). Roth et al. (2009) have pointed out that, despite the growing interest in this perspective, the potential of CHAT has not yet been fully realized. They argue that the difficulty of adopting this theoretical and methodological framework arises from its philosophical fundamentals. Notably, the dialectical materialist ontology faces difficult acceptance into the non-dialectical thinking underlining a significant part of Western research tradition, which usually addresses science education under formal logic. Annalisa Sannino and Engeström reinforce this perception by mentioning that "activity theory is not an easy approach to adopt and apply. It is built on the philosophical and methodological foundation of Marxist dialectics, a way of thinking alien to academics socialized into the standards of positivism" (2018, p. 44). As both studies imply, the challenge seems persistent through the years.

1.2.1 The Method and Rationale of CHAT

From our point of view, CHAT provides a complex system of tools/concepts to grasp the genesis and the development of the human phenomenon in its *wholeness* and *concreteness*. Furthermore, in the terms we have formulated elsewhere, CHAT provides ways to develop a method as a "rationality within an ontological system" (Rodrigues et al., 2014, p. 584). It means that more than a specific research design, a set of procedures, or an employable category, the *method* we embrace should help us include formulating the problem and the object under investigation.

That said, we direct our efforts toward developing within science education research a method of its own, whose movement is expressed in the dialectical production of the object and the way of knowing it. Vygotsky emphasizes that "every basically new approach to scientific problems leads inevitably to new methods and ways of research" (Vygotsky, 1997, p. 44). Science education (or, more specifically, STEM education) is not an entirely new field of research. However, its own scientific approach to a series of phenomena is still to be realized.

Following Evald Ilyenkov, we assume that "to comprehend a phenomenon means to establish its place and role in the concrete system of interacting phenomena, in which it is necessarily realized, and to find out precisely those traits which make it possible for the phenomenon to play this role in the whole." (1982, p. 177) Within this perspective, concreteness expresses that all objects are "integral [...], multivariously divided within itself, rich in determinations, and historically formed. It is like, not a separate isolated atom, but a living organism, a socioeconomic structure, or similar formation" (Ilyenkov as cited in Bakhurst, 1991, p. 138).

In this direction, it is essential to explore the object of the activity and its historical dimension, understanding its development and identifying tensions historically accumulated. Anna Stetsenko explores the possibility of taking human activities as a totality, an ethic-ontoepistemological system, which allows us to emphasize "the nexus of people changing the world and themselves being changed in this very process – as poles of one and the same (as 'duo in uno'), bi-directional and recursive co-constitution of people and the world in a continual and ceaseless communal praxis of self- and world-realization" (2019, p. 8).

Consistently with the notion of the method presented, we start from the tensions inherent in science education and its object, which are formed from a history and a vision of the future, to address the particular case of STEM education. Then, exploring STEM education through CHAT means understanding what object this activity is aimed at and its historical development to identify the contradictions to which it is subjected and what tensions are produced.

1.3 Object-Oriented Activity

From our perspective, and considering the analysis we aim to develop here, the object-oriented nature of human activity is the first fundamental premise to be brought to the discussion. To understand the emergence and development of any human activity, recognizing its collective, collaborative and complex character, is necessary to examine to what activity is oriented. Engeström (2009), pointing CHAT as a theory of object-driven activity, argues that:

> Objects are concerns; they are generators and foci of attention, motivation, effort, and meaning. Through their activities, people constantly change and create new objects. The new objects are often not intentional products of a single activity but unintended consequences of multiple activities (Engeström, 2009, p. 304).

Object, in this sense, does not exist as an isolated entity that simply exercises its influence upon the subject. On the contrary, the formation and development of the object must be seen as an active process of elaboration and transformation, which contrasts with a myriad of perspectives that assume subjects as mere passive entities (Stetsenko, 2020). For instance, Engeström and Frank Blackler (2005) discuss the centrality of object development and contradiction to organization work, whether dealing with complex diseases, such as diabetes or tracking reuse or recycling processes.

In a similar vein, Cristiano R. Mattos and Luciano B. Tavares (2014) shows how medical staff sees a child with cancer in an oncological center as a patient who needs specific treatments. Meanwhile, this very same child is a son or daughter in a family and a student at the hospital school. The patient-children-student, i.e., the embodiment of a complex object, is the key to understanding the development and articulation among activities throughout the complex oncological treatment system – family-school-hospital. Moreover, while science tutors were keen to provide a fresh teaching approach, putting to service all the recommended science teaching guidelines, students occasionally refused to engage in this new learning approach demanding more traditional and test-oriented teaching. According to the authors, this opened up one opportunity to discuss with students, and sometimes with parents and medical staff, the general purpose of science education in a school within a hospital, which reveals the complex relationship of all activities patient-children-student is involved in. The authors concluded that taking into account the dramatic situation lived by the students, schooling assumed one last link with their previous life, a window of possibilities to come back as soon as possible to their regular schools and lives. Students' engagement in the schooling process unveils the contradictory nature of object formation and development of this particular activity.

The connection between STEM education and its outcomes does not take place directly. Instead, any possible causality is rich, complex, and full of contradictions. It is partially due to the several agents in the activity, sometimes treated as stakeholders of the educational activity. Additionally, it is also due to the synthetic form of educators' planning, teaching, and assessing that encompasses, in one way or another, all the connections as mentioned earlier. In other words, all the desires, goals, and hopes that are projected upon STEM education and are loosely formulated, will be, after all, funneled to educators' practice and students' schooling and daily life. Moreover, as discussed by Engeström:

> [...] the collective, artifact-mediated and object-oriented activity system, seen in its network relations to other activity systems, is taken as the prime unit of analysis. Goal-directed individual and group actions, as well as automatic operations, are relatively independent but subordinate units of analysis, eventually understandable only when interpreted against the background of entire activity systems. Activity systems realize and reproduce themselves by generating actions and operations (Engeström, 2001, p. 136).

In this vein, it is fair to ask: *What is the object of STEM education?* Although we acknowledge some diffuse efforts to produce a definition of STEM education, its object remains overlooked. In many cases, the concern of how to better integrate STEM is presented, albeit not addressing the problem of: *Is the integration towards what?*

One of the most resilient issues while formulating science education goals, which goes back to the 1960s science education reforms era, is the recurrent tension between science education as a pathway to scientific careers *versus* science education for a non-specialist, such as the general workforce or citizenship. Although this issue has been a conundrum for science educators, researchers, and policymakers, literature on STEM education usually frames the matter in a far unproblematic way. Increasing demand for the STEM workforce is established by such literature indicating that, on the one hand, the scientific and technological growing economic competitiveness impacts manufacturing and consumption[2] (Bybee, 2013). On the other hand, there has been a significant decrease in students' interest in STEM topics and careers. Again, a similar scenario goes back to the *era of the projects*, when a myriad of national and, to some extent, international science and mathematics education projects were based on the same justifications. Moreover, an underlying premise that pervades and sustains this whole argument is the connection between STEM education, scientific development, and, far at the end of the line, economic development – in the case of the U.S.; one can argue that there is the need for maintaining economic hegemony and cultural and economic imperialism.

This long line of connections might lead STEM education, and recently STEM literacy – in which the meaning is also excessively imprecise – to a welcomed uncritical appreciation and trust in science, social awareness, and notions of citizenship. This causal line has been taken as an antidote to deal with *fake news* and *post-truth*, themes that have been recently acknowledged as a source of institutional credibility loss and the financial defunding of many scientific endeavors. It is precisely this fuzzy view that has fueled many scientific outreach programs. Evidently, there is no straight answer on how teachers should approach distrust sentiment about science that might affect students and sometimes is misunderstood as scientific skepticism.

Contradictions embedded in STEM education activity are aggravated insofar as all these interconnections between goals and intentionalities are ignored or considered non-problematic. It is often assumed that the STEM goals – education for citizenship and the formation of a twenty-first-century skilled workforce – are automatically and unequivocally aligned with the STEM pipeline, whereas it is rarely the case.

The discussions commonly occur in teaching methodology and approaches are grounded on deeper agreements and disagreements on broad educational goals. To

[2]The way the problem is framed seems to crosscut different approaches in STEM and STEAM education. However, the argument of economic competitiveness seems to be tied to recent United States context.

illustrate that, we outline a discussion from three different positions in the literature on teaching approaches to global warming. The divergences are apparently drawn from the teaching approach, yet indeed stem from how these three studies elaborate the goals of science education related to global warming.

Virgnie Albe and Marie-José Gombert (2012) reported a study on the global warming debate as a citizens' conference. This type of debate in the classroom demands that students step into contradictory positions and explore different kinds of arguments about global warming. They elaborate on the science education goals as such: "Science teaching in this perspective should train future citizens to cope with social problems associated with science and technology" (2012, p. 660).

Ralph Levinson (2012), commenting on the Albe and Gombert (2012) teaching approach, argues that "[...] it is worth noting that the Brazilian statement recognizes school students as not only 'future citizens' but also citizens of today, a small but significant recognition of the rights of school students." (Levinson, 2012, p. 693) In general lines, he advocates that instead of adopting and defending the position of others for democratic deliberation in a consensus conference, it would be fruitful to address what happens when the students make decisions and how they enact change. He appeals to a much more activist approach in which the students enact change. According to him: "[...] there are possibilities of change through pragmatic and purposive courses of action in schools" (Levinson, 2012, p. 699).

Catherine Gautier (2012) also commented on the Albe and Gombert (2012) teaching approach, questioning that controversial debates about global warming in the classroom might perpetuate the sense that there is still a scientific controversy about global warming. Gautier mentions that she used debates in the late 1990s and early 2000s. At the same time, the scientific community was still unsure and IPCC Assessment Reports presented high uncertainty about the role of humankind in global warming. However, nowadays, the teaching effort (including debates) should focus on cutting-edge areas such as carbon capture and sequestration as well as geoengineering solutions once global warming is no longer a scientific controversy. She also underlines that: "One question is what level of scientific literacy is necessary to act appropriately as an informed citizen" and suggests that there may exist a "level of scientific literacy required for participation in our technically-driven society," and "for a democracy to flourish citizens need to attain at least a general understanding of the issues they are called upon to decide". (Gautier, 2012, p. 687)

Despite the subtleness, it is possible to find some epistemological and pedagogical differences of the three teaching approaches presented in the discussion. However, the difference leads us to the meaning given to 'citizenship,' 'future citizens,' and 'informed citizen.' Sylvie Barma et al. (2015) present a case that also illustrates the ways in which such contradictions are embedded into the school setting and within teachers' practice. It showcases how the science education object works as a raw material from which teaching activity is shaped. Furthermore, the disputes that apparently revolve around teaching approaches also include conflicts and contradictions about how educators design teaching objectives.

Drawing on CHAT, we advocated that it is not possible to produce the object for STEM education activity *in vitro* and then later offer schools and educational agents an object that is clean, unproblematic, and free of contractions. On the contrary, there is no other way than to forge the object within the living activity. The risk for STEM educational activity is to isolate its object from the activity itself. What consequently subtracts any strength to evolve as a force for change in school and becomes just another set of abstract rules – an empty shell. The *object* is no mere juxtaposition of interests that put the agents to make local compromises and connections. From our perspective, the challenge is not to have a clear definition or proper set of rules; the challenge lies deeper in the elaboration of the object for any STEM education activity in a manner that moves away from the oversimplified and aggravated version of what we already have in science and mathematics education.

Moreover, from the CHAT perspective, the object of the activity captures the structure of the activity – toward what actions/operations are oriented to achieve specific results. It synthesizes the historical "process of people simultaneously co-creating themselves and the world, as a nexus of these dynamic currents within what is a unified, communal, historical praxis stretching across generations and connecting all people in one seamless (though filled with contradictions) 'fabric' of being and becoming" (Stetsenko, 2020, p. 10). Ultimately, it leads us to include the dynamic dimension of the activity and to examine the matter of history and development in CHAT.

1.4 Historicity and Development: Past, Future, Innovation, and Technology in STEM Education

Engeström pointed out *historicity* as a principle for understanding human activity. According to him, "activity systems take shape and get transformed over lengthy periods of time. Their problems and potentials can only be understood against their own history" (2001, p. 136). For example, Andre M. Rodrigues et al. (2014) have included such an aspect in the research practice while considering the teaching-learning process as a dynamic process. They tracked the teacher's changes in approach throughout a sequence of lessons for different groups of students. Their results indicated that the activities of the teacher are permanently informed and sustained by the feedback of previous interactions, forcing the teacher to reinterpret his initial planning. In other words, the teacher's activity cannot be understood isolated from its own flow.

Moreover, being/becoming a teacher can only be understood with a multitude of activities, inside and outside of school. However, a typical pitfall is a sort of historical oversimplification, which reduces historicity to simply looking back to past events, without considering the genesis and the development of the phenomena under scrutiny. Ultimately it kills the movement of things and the possibility of studying their dynamics. Despite the label of "the history of science," some educational approaches to science and technology may often end up with a pile of frozen historical facts without any resemblance with an actual historical approach.

Moreover, there is the temptation to approach a descriptive level of the dynamics without ever touching the movement's causes. From our perspective, that is why the historical aspect of the activity should not be disassociated from the very concept of development. It is a way to address, as an articulated effort, the movement and transformation of an activity.

Stetsenko (2020) points out that a reductionist perspective also produces a pernicious understanding of concepts like history and development. According to her:

> "'Equipped' with the notion of knowledge as an outcome of internal processes confined within individual solitary minds (or brains), many psychologists inevitably reconstruct history as being a succession of value-, culture-, and politics-free advances (resulting in "naked" facts) carried out in a sociocultural and sociopolitical vacuum" (Stetsenko, 2020, p. 5).

To illustrate the analytical path that considers historicity and development, we shall examine the concepts of *innovation* and *technology* within STEM education.

The notion of innovation that stamps many STEM initiatives play a pivotal role in disseminating the STEM approach. Indeed, innovation is a buzzword that holds the intention of and hope for change, improvement, and ultimately rupture with the current education. Nevertheless, it masks the roots and connections between STEM education and the existing school. Although innovation frequently disguises itself as rupture, it just ignores the resilient, undesired, and unruptured connections with the school at best.

If there is any strength to be extracted from the notion of educational innovation, it stems from a rather contrary rationale. Instead of throwing STEM education into outer space as abstract innovation, it would be worthwhile to insert it back into the schooling dynamic. It implies putting STEM education activity – its actions, intentions, and reflections, within schooling as a whole. We emphasize that STEM education emerges from a particular context and moves toward a particular object while it develops. The typical analysis presented in the STEM initiatives that take it as a novel stand-alone approach, undermines any possibility of understanding what STEM education is or what it could be.

Although it is possible to recognize it as a fragmented attempt, for at least part of teachers and researchers, the task related to STEM education has somehow shifted toward establishing connections and finding its roots. The challenge is to root STEM education in the science, engineering, technological, and mathematics respective teaching traditions and knowledge development.

A recent discussion about the nature of STEM was initiated within the research tradition of the nature of science (NOS). Wonyong Park et al. (2020) acknowledge that the issues around the nature of STEM remain open, despite the recent growth in STEM initiatives. For Candice Quinn et al., as STEM is not a discipline, it would be impossible to identify such a thing as the nature of STEM. They advocate that, after all, "the nature of STEM does not exist but manifests as the NOE [nature of engineering]." (2020, p. 895) One might disagree with the methodological and theoretical approaches to figuring out the nature of STEM. However, to a large extent, it is an attempt to connect STEM education to an already structured debate in science education. One might think that it is a vain effort to understand new and revolutionary concepts through an old lens. However, at the same time, it is in the meaning-making process in which people (students, teachers, policymakers, and scholars) try to make sense of STEM education by connecting and contrasting it with a broader historical process.

Examining the history of science education, it is not hard to find another example of innovation in the late 1950s. This set of science teaching reforms took place in the post-war U.S. and spread to Latin America and many other places like Australia and New Zealand (Openshaw & Walshaw, 2019). It emerges as a vigorous innovation that could cope with the science teaching flaws and school difficulties in meeting societal needs. A 1937s study shows a consistent decline in students' interest in physics courses since the end of the nineteenth century. Later, the reforms received attention and money fueled by the general sentiment that science and math education was in its worse shape, the growing competitiveness due to the Cold War, and the need for consolidation of the American hegemonic power (Donahue, 1993).

Moreover, especially on how the Physical Science Study Committee (PSSC) envisaged the teaching, Roger Openshaw, and Margaret Walshaw (2019, p. 58) highlight that in "their view, placing more emphasis on 'doing physics' rather than 'learning about' physics, was not simply the best, but the only way to present the subject to a new generation of space-age learners". We are not suggesting a point-by-point comparison between the reforms in the 1950s and STEM education. However, it is necessary to acknowledge the scenario. Curious enough, a decade later, the very same innovation was sentenced as a cause of narrowing the science curriculum. It somehow makes some STEM apologists look like old wine in a new bottle. Although it is not fate, some STEM initiatives could repeat similar mistakes and deadlocks without even noticing.

We might find technology apologists and utopists on the other side of the spectrum. Ames (2019) has developed a critical study on *One Laptop per Child* program. Starting in the middle 2000s, it was a program funded mainly by big tech companies and promoted via the World Economic Forum. According to her, educators, policymakers, and technology enthusiasts failed to learn from past failures regarding its utopian goals. The program aimed to solve educational problems in so-called developing countries by handing out laptops to children in various contexts and conditions. Another example that, to some extent, shares similar technological utopism can be found in the project *The Hole in the Wall*. In the later 1990s, the project conducted by Sugata Mitra (2003) was embedded in a wave of enthusiasm for the role computers could play in education and the potential of technology to cope with challenges in twenty-first-century education for marginalized children.

Both projects exemplify the types of technological utopism that pervade educational initiatives, while their results are barely submitted to scrutiny. If some trends advocate the replacement of schools by some technology, others see the educational task as an actualization of its technological content. Today's children will be out of school in a decade, and the technological scenario will be radically different. The estimations and predictions on emerging technologies guide the current educational initiatives (Childress, 2016). Unfortunately, the innovative educational anima often uncritically embraces all sorts of technological utopism, ignoring past experiences and offering a future guided by purely technical development.

A critical point that needs further examination is the vision for the future offered by STEM education, which is often linked with anxiety, uncertainty, and, at best, with the future work market in which technology is critical to competitiveness. The general justification for STEM education is sometimes grounded on adaptation and passivity, in which technological utopianism is converted to technological fatalism. Paradoxically, many STEM initiatives are driven by the notion of students as active learners while offering a relatively closed vision of the future that is a merely aggravated version of today.

1.5 Conclusions

From our perspective, it seems critical to highlight that STEM education, despite its universality aspirations, is a historical and contextual movement still under construction. The STEM education that emerged within the American context in the 1990s with the support from institutions like the National Academy of Sciences (NAS) and the National Science Foundation (NFS) is, to a great extent, grounded on tenets of market-driven skills, national economic competitiveness, and technological utopianism. It has been gaining attention across the world. However, it spreads under an ill-structured and loose conceptualization of STEM education, which congregates a variety of teaching initiatives. The general umbrella of STEM education attempts to tackle students' lack of interest and skills for the always-changing world without any structural change in the current school.

On the one hand, it precisely makes STEM initiatives so attractive since they can fit within the already existing school with minor changes and adjustments. However, on the other hand, STEM education achievements will fall short since it is trying to tackle structural issues only from the perspective of small and local initiatives. In this vein, CHAT has the potential to enable teachers and researchers to examine the connections and change possibilities between the small local initiatives and the rather complex schooling process.

As discussed in this chapter, the stage in which STEM education finds itself comprises two different aspects that might be worth a critical examination from educators and researchers. On the one hand, the lack of a clear conceptualization of STEM education might infuse enthusiasm and give the feeling of being an open grassroots movement. Different stakeholders come together to interact, organize and propose what seems like new initiatives in education. Nevertheless, on the other hand, this very same characteristic may help keep the tenets and goals of STEM education unexamined or uncritically embraced. Under such an umbrella, we can find various concepts such as innovation, technology, future, real problems, socioscientific issues, motivation, and economic development – to mention a few that conforms to the STEM education concept constellation. We advocate that

CHAT can contribute to making sense of concepts and their relations underlying STEM education initiatives.

In research and practice, both science and mathematics education has been identifying, reframing, and battling most of the problems already mentioned in the STEM literature. Moreover, the approaches and strategies such as the STS approach, interdisciplinarity, hands-on activities, work with projects, problem-based learning, etc., have been developed in a long-term process with extensive documentation of its gains and challenges. Thus, we think whether or not STEM education can produce meaningful changes is a remaining critical issue to be examined by everyone involved.

While commentating on the two-page recommendation document produced by the American Association for the Advancement of Science (AAAS), Zeidler (2016) provides an image for reflection:

> While I can appreciate the attempt at trying to make science more interesting, I believe that superficial recommendations like rearranging the furniture is tantamount to rearranging the deck chairs on the Titanic. At best, you might wind up with a better seat to witness the unfolding disaster (Zeidler, 2016, p. 14).

We expect this chapter to provide some provoking points that can be critically examined through the CHAT lens. From our perspective, the critical examination of STEM education should be sought to overcome the superficialities and ephemeral changes in education.

References

Albe, V., & Gombert, M.-J. (2012). Students' communication, argumentation and knowledge in a citizens' conference on global warming. *Cultural Studies of Science Education, 7*(3), 659–681. https://doi.org/10.1007/s11422-012-9407-1

Ames, M. G. (2019). *The charisma machine: The life, death, and legacy of one laptop per child*. MIT Press.

Bakhurst, D. (1991). *Consciousness and revolution in Soviet philosophy. From the Bolsheviks to Evald Ilyenkov*. Cambridge University Press. https://doi.org/10.1017/CBO9780511608940

Barma, S., Lacasse, M., & Massé-Morneau, J. (2015). Engaging discussion about climate change in a Quebec secondary school: A challenge for science teachers. *Learning, Culture and Social Interaction, 4*, 28–36. https://doi.org/10.1016/j.lcsi.2014.07.004

Bybee, R. W. (2013). *The case for STEM education: Challenges and opportunities*. National Science Teachers Association.

Childress, M. (2016). Utopian futures for learning technologies. In *The Wiley handbook of learning technology* (pp. 557–570). Wiley. https://doi.org/10.1002/9781118736494.ch29

Donahue, D. M. (1993). Serving students, science, or society? The secondary school physics curriculum in the United States, 1930-65. *History of Education Quarterly, 33*(3), 321. https://doi.org/10.2307/368196

Donovan, B. M., Mateos, D. M., Osborne, J. F., & Bisaccio, D. J. (2014). Revising the economic imperative for US STEM education. *PLoS Biology, 12*(1), e1001760. https://doi.org/10.1371/journal.pbio.1001760

Engeström, Y. (2001). Expansive learning at work: Toward an activity theoretical reconceptualization. *Journal of Education and Work, 14*(1), 133–156. https://doi.org/10.1080/13639080020028747

Engeström, Y. (2009). The future of activity theory: A rough draft. In A. Sannino, H. Daniels, & K. D. Gutierrez (Eds.), *Learning and expanding with activity theory* (pp. 303–328). Cambridge University Press. https://doi.org/10.1017/CBO9780511809989.020

Engeström, Y. (2015). *Learning by expanding: An activity-theoretical approach to developmental research.* Cambridge University Press.

Engeström, Y., & Blackler, F. (2005). On the life of the object. *Organization, 12*(3), 307–330. https://doi.org/10.1177/1350508405051268

Gautier, C. (2012). A new type of debate for global warming and scientific literacy. *Cultural Studies of Science Education, 7*(3), 683–691. https://doi.org/10.1007/s11422-012-9417-z

Guzey, S. S., Harwell, M., & Moore, T. (2014). Development of an instrument to assess attitudes toward science, technology, engineering, and mathematics (STEM). *School Science and Mathematics, 114*(6), 271–279. https://doi.org/10.1111/ssm.12077

Levinson, R. (2012). A perspective on knowing about global warming and a critical comment about schools and curriculum in relation to socio-scientific issues. *Cultural Studies of Science Education, 7*(3), 693–701. https://doi.org/10.1007/s11422-012-9418-y

Li, Y., Wang, K., Xiao, Y., & Froyd, J. E. (2020). Research and trends in STEM education: A systematic review of journal publications. *International Journal of STEM Education, 7*(1), 11. https://doi.org/10.1186/s40594-020-00207-6

Lorenzin, M. P. (2019). *Sistemas de Atividade, tensões e transformações em movimento na construção de um currículo orientado pela abordagem STEAM* (Dissertação de Mestrado, Ensino de Ciências (Física, Química e Biologia)). Universidade de São Paulo, São Paulo. Retrieved from 10.11606/D.81.2019.tde-10122019-155229.

Martín-Páez, T., Aguilera, D., Perales-Palacios, F. J., & Vílchez-González, J. M. (2019). What are we talking about when we talk about STEM education? A review of literature. *Science Education, 103*(4), 799–822. https://doi.org/10.1002/sce.21522

Mattos, C. R., & Tavares, L. B. (2014). The multiple senses of science teaching at a hospital school. In A. Tiberghien & E. Kyza (Eds.), *E-book proceedings of the ESERA 2013 conference: Science education research for evidence-based teaching and coherence in learning. Part strand 3* (pp. 470–479). European Science Education Research Association.

McComas, W. F., & Burgin, S. R. (2020). A critique of "STEM" education. *Science & Education, 29*(4), 805–829. https://doi.org/10.1007/s11191-020-00138-2

Mitra, S. (2003). Minimally invasive education: A progress report on the "hole-in-the-wall" experiments: Colloquium. *British Journal of Educational Technology, 34*(3), 367–371. https://doi.org/10.1111/1467-8535.00333

Obama Plays Cheerleader For STEM. (2010, November 2). *Education Week.* Retrieved from https://www.edweek.org/teaching-learning/obama-plays-cheerleader-for-stem/2010/11

Openshaw, R., & Walshaw, M. (2019). The rise and fall of P.S.S.C. Physics. In R. Openshaw & M. Walshaw (Eds.), *Transnational synergies in school mathematics and science debates* (pp. 55–76). Springer. https://doi.org/10.1007/978-3-030-28269-1_4

Park, W., Wu, J.-Y., & Erduran, S. (2020). The nature of STEM disciplines in the science education standards documents from the USA, Korea and Taiwan. *Science & Education, 29*(4), 899–927. https://doi.org/10.1007/s11191-020-00139-1

Quinn, C. M., Reid, J. W., & Gardner, G. E. (2020). S + T + M = E as a convergent model for the nature of STEM. *Science & Education, 29*(4), 881–898. https://doi.org/10.1007/s11191-020-00130-w

Rodrigues, A. M., Camillo, J., & Mattos, C. R. (2014). Quasi-appropriation of dialectical materialism: a critical reading of Marxism in Vygotskian approaches to cultural studies in science education. *Cultural Studies of Science Education, 9*(3), 583–589. https://doi.org/10.1007/s11422-014-9570-7

Roth, W.-M. (2020). Activity theory in mathematics education. In S. Lerman (Ed.), *Encyclopedia of mathematics education* (pp. 20–23). Springer International Publishing. https://doi.org/10.1007/978-3-030-15789-0_4

Roth, W.-M., Lee, Y.-J., & Hsu, P.-L. (2009). A tool for changing the world: Possibilities of cultural-historical activity theory to reinvigorate science education. *Studies in Science Education, 45*(2), 131–167. https://doi.org/10.1080/03057260903142269

Sannino, A., & Engeström, Y. (2018). Cultural-historical activity theory: Founding insights and new challenges. *Cultural-Historical Psychology, 14*(3), 43–56. https://doi.org/10.17759/chp.2018140304

Stetsenko, A. (2019). Radical-transformative agency: Continuities and contrasts with relational agency and implications for education. *Frontiers in Education, 4*, 148. https://doi.org/10.3389/feduc.2019.00148

Stetsenko, A. (2020). Hope, political imagination, and agency in Marxism and beyond: Explicating the transformative worldview and ethico-ontoepistemology. *Educational Philosophy and Theory, 52*(7), 726–737. https://doi.org/10.1080/00131857.2019.1654373

Toma, R. B., & García-Carmona, A. (2021). «De STEM nos gusta todo menos STEM». Análisis crítico de una tendencia educativa de moda. Enseñanza de las Ciencias. *Revista de investigación y experiencias didácticas, 39*(1), 65–80. https://doi.org/10.5565/rev/ensciencias.3093

Vygotsky, L. S. (1997). *The collected works of L.S. Vygotsky. Vol. 3: Problems of the theory and history of psychology* (M. J. Hall, Trans.). Springer-Verlag.

Zeidler, D. L. (2016). STEM education: A deficit framework for the twenty first century? A sociocultural socioscientific response. *Cultural Studies of Science Education, 11*(1), 11–26. https://doi.org/10.1007/s11422-014-9578-z

André Machado Rodrigues is an Assistant Professor in the Department of Applied Physics at the Institute of Physics of the University of São Paulo, Brazil. His research focuses on science teacher education and scientific concept formation within the cultural-historical activity theory framework.

Juliano Camillo is an Assistant Professor at the State University of Campinas, Brazil. He is interested in developing a philosophy of science education, concentrating his efforts on understanding the relations between human development and science education.

Cristiano Rodrigues de Mattos has a PhD in Physics from the University of São Paulo investigating artificial cognitive systems. Currently, he is an Associate Professor at the Institute of Physics at the University of São Paulo and the leader of the Research Group in Science and Complexity Education (ECCo). He works on philosophical and psychological fundaments of teaching and learning processes, with emphasis on science education research to establish theoretical-methodological foundations based on the Cultural-Historical Activity Theory from a Freirean perspective. Through this framework, he investigated topics related to the teaching and learning processes of scientific and quotidian concepts, models of dialogic interaction, situated cognition, interdisciplinarity, and complexity, and developed practical educational activities aiming science as an instrument to develop citizenship and democratic education for social and economic equity.

Chapter 2
STEAM Education to Unleash Students' Creativity and Knowledge-Building Capacity: An Indian Perspective

Tara Ratnam

2.1 STEAM to Enrich STEM

> *Reading about the history of scientific thought made me see other competing ideas out there and what the effect of following various ideological pathways has been – and I think this is particularly relevant to science education because it makes one realize that, while there might be some extent to which physics (for instance) can be said to be objectively true, what you choose to do with that physics is very subjective – and knowing what people have chosen to do with such knowledge in the past and with what consequences is helpful in making a more educated choice of what to do with it* (Rohit Prasanna, a humanist and senior scientist at Swift Solar).

This book focuses on the sociocultural approach to STEM education, to explore how it can contribute to addressing the crises experienced in our times. Therefore, I would like to clarify at the outset my preference for STEAM over STEM and the potential STEAM holds for a more robust response to the challenges of a rapidly evolving world. The arts can transform STEM education by highlighting creativity, innovation and problem solving as core practices (e.g., Hardiman & JohnBull, 2019), because these are the very goals that STEM teaching and learning wants to achieve.

STEM education gained importance in the early 2000s in the United States of America to equip students for a knowledge-based economy driven by constant innovation. An educated workforce well-grounded in STEM skills was seen as critical to maintaining a competitive edge in the global economy and prosperity (National Academy of Sciences, 2005). Since then, the STEM-focused curriculum has been spreading around the world with increased expectations of its contribution to national development and productivity (Freeman et al., 2019). Despite STEM

T. Ratnam (✉)
Independent Researcher and Teacher Educator, Mysore, Karnataka, India

K. Plakitsi, S. Barma (eds.), *Sociocultural Approaches to STEM Education*, Sociocultural Explorations of Science Education 21,
https://doi.org/10.1007/978-3-031-44377-0_2

education and industries' diversity outreach efforts, STEM is beset by socioeconomic, gender and racial inequity (McGee, 2020). Addressing these issues points to the need for going beyond the narrow instrumental and competitive economic goals that drive STEM to embed it in the more encompassing social and environmental goals of education. The latter aims at "sustainable development" (UNESCO, 2015) that is ecologically sound, economically viable and socially just. Taylor (2016) asserts that the real challenge to STEM education is not preparing students for STEM jobs, but promoting in them their higher-order abilities as "critical consumers, creative and ethically astute citizens" to address the global crises (inequity, loss of biocultural diversity, environmental disasters) impacting social and environmental wellbeing. This concern is articulated in the National Foundation on the Arts and Humanities Act of 1965 by the United States Congress: – "Democracy demands wisdom and vision in its citizens. It must therefore foster and support a form of education, and access to the arts and the humanities, designed to make people of all backgrounds and wherever located masters of their technology and not its unthinking servants." The promise for this vision seems to lie in broadening the base of STEM by embracing the arts-and-humanities-infused STEAM approach.

The STEM curriculum is supposed to be interdisciplinary, making learning adaptive and applicable to present and future workplace needs by promoting capacities such as collaboration, communication, problem-solving, critical thinking, creativity and innovation. These so-called "twenty-first-century skills" are evinced by the Arts (Anderson, 2016). Broadening the scope of STEM to include the Arts can provide a channel through which these capacities can be promoted, making children participate successfully in the global knowledge-based society. There are several studies with findings that support the value of arts education for innovation (e.g., Avvisati et al., 2013). Martin et al. (2013) found that students engaged in the Arts did better academically in their non-Arts subjects such as language, Mathematics and Science when compared to students who did not take part in the Arts. STEAM education was found to enhance the students' ability to develop conceptual understanding (Ozkan & Topsakal, 2020). The impact of STEAM education from research findings shows its positive effect on academic achievement, cognitive functioning, and social and emotional learning making STEM education more inclusive (Weyer & Dell'Erba, 2022).

Despite research evidence which shows how arts integration serves the goals of STEM education by making it more inclusive and effective, it has not been given its due place in STEM education. Increased funding for STEM disciplines has given them an esteemed status (Wilson et al., 2006) marginalizing the Arts (Catterall, 2017) and leading to student disengagement (Feldman, 2015), thus defeating STEM's raison d'être. A STEAM curriculum would enrich STEM education by better engaging students and promoting their skills of creativity and improvisation (Anderson, 2016). More importantly, the arts and humanities serve an epistemological function (Eisner, 2008) by providing students with tools to understand and interpret the world and enabling them to view their learning against the broader scheme of things. It would not only enable them to make STEM knowledge more adaptive and applicable to workplace needs, but also provide opportunities to harness their creative energy for addressing global challenges through "positive and productive" social action (Taylor, 2016) as this study demonstrates.

2.2 STEAM in the Indian Context

Science and technology policy in post-independent India has evolved from its earlier focus on the cultivation of science and scientific research in Higher Education Institutions for attaining technological competence (Technology Policy Statement 1983) (Government of India [GOI], 1983) to an increase in scientific and engineering research (Science and Technology Policy 2003) (GOI, 2003). The more recent policy focuses on strengthening a science-and-technology-led innovation ecosystem by universalising access to STEM learning to all students at all levels of education (Science Technology and Innovation Policy [STIP] 2013) (GOI, 2013). India's STIP 2013 is in alignment with the United Nations- Sustainable Development Goals (UN-SDGs). It links science and technology innovation to socioeconomic and environmental priorities and its broader STEAM orientation is also articulated in the National Education Policy (GOI, 2020), which asserts that integrating humanities and arts in education with STEM promotes "higher-order thinking capacities" among other twenty-first century futuristic skills (p. 34).

STIP 2013 took an actionable form through Atal Innovation Mission (AIM). AIM is India's flagship program to foster a culture of innovation and entrepreneurship in the country. The program has established Atal Tinkering Laboratories (ATLs) in schools across India since 2016. To date there are over 7000 ATLs with several more in the pipeline.

ATL is a makerspace meant for students from grades 6–12, providing them access to tools and technologies such as Internet of Things, 3D printing, rapid prototyping tools, robotics, miniaturized electronics, and do-it-yourself kits. These labs are envisaged as a hub for innovation, creating a pool of talent that is future-ready, fostering curiosity, creativity and imagination in young minds, by encouraging them to use these competencies – in tandem with new technologies – to address local and global problems (GOI, NITI Aayog, 2020).

Self-access materials provide well-designed and carefully curated content that teachers and students can use flexibly to guide their activities in the Tinkering Labs. These materials feature modules on digital literacy, ideation, design thinking, computational thinking, and physical computing. Besides, ATL schools organize human resources by identifying and developing partnerships with relevant stakeholders – mentors from among NGOs, industry and academic experts, and student volunteers from engineering colleges. While all the students in the school are introduced to the lab through an initial motivational orientation program, subsequent participation is voluntary, based on student interest. The ATL facilities are also open to nearby schools and communities to maximise outreach.

Thus far, the plans for implementation seem to align well with the ATL vision of providing universal access and motivating students to become solution-providers and engage in self-learning, that not only will lead to the development of higher cognitive skills, but eventually may spur discoveries, innovations and inventions. However, these well-intentioned plans and provisions have to be understood within the larger "political economy and sociology of iniquitous educational development" (Velaskar, 2013, p. 126) in which the schools are embedded.

2.2.1 Structural Inequality in the Indian Education System: Dual System of Education

Global neoliberalism found its way into India through the economic liberalisation of the early 1990s. The educational fall out of this neoliberal political and economic hegemony was a reversal of the policy of public provision of free education for all until the age of 14 (RTE, 2009) by allowing increased privatisation of education to serve neoliberal interests (Boucher, 2017). Since then, there has been a proliferation of private schools (Kingdon & Pal, 2014), which are expensive but prized by everyone for the quality education they purport to provide. If they have the means to do so, parents frequently choose private schools, beguiled by a sense of exclusivity (Ratnam, 2015). This pattern exacerbates social and educational inequalities by leaving the under-resourced, poorly-staffed and poorly-managed government schools as the only option for the majority of historically and socioeconomically disadvantaged students. State government elementary schools which embody the idea of mass education form 74% of Indian elementary schools catering to 200 million school going children (GOI, 2018). These children are denied access to ATL. The prerequisite of a 1500-square-feet built area in the school to house the ATL – along with other criteria (GOI, NITI Aayog, 2017) – disqualify State-run elementary schools from applying for ATL facility. These schools have no lab space and store such basic lab equipment as they have in a cupboard. It is the affluent private schools and schools affiliated to government controlled Central Board of Secondary Education (CBSE) that are the primary beneficiaries of ATL facilities.

2.2.2 Conflict Between Education and Culture

Besides structural inequality, there are issues of equity through pedagogy arising from patterns in teachers' thinking and practice (see Ratnam, 2013 for detailed analysis). The prevalent culture of schooling depends upon *knowledge transmission* as a means to fulfil the instrumental goal of teaching to the test. This imperative has continued from colonial times despite every academic position dwelling on the agency of the learner as significant in responding to the unmet need for multicultural sensitivity in our educational context. A lack of multicultural sensitivity makes learning alienating for culturally diverse and historically disadvantaged students (e.g., NCERT, 2005), contributing to poor school performance (Pratham, 2019) and increased dropout rates (Lewin, 2011). This prevailing tension between the goals and means of policy and practice also impacts the fulfilment of more recent STEAM-aligned goals espoused in policy statements (GOI, 2020). Enactment of these goals presupposes the availability of teachers to teach students *how to learn* in a process of *knowledge building* or co-construction of knowledge in teacher-learner and learner-learner interactions as they engage in joint exploratory activities. These interactions are as much or more crucial than material aspects in the day-to-day learning-teaching process.

2.3 Questions for the Study

This study juxtaposes two educational scenarios, one that is largely knowledge transmission-oriented, thus typifying the largely prevalent culture of formal school in India (Kumar, 2005) as elsewhere (Russell, 2021) – and another that encourages knowledge co-construction in joint inquiry-based activities, to which both teachers and learners contribute. The latter aligns with the principles underlying STEAM education by fostering creative thinking and problem solving. The study examines these practices within the larger sociopolitical and institutional culture to expose the terrain between espoused intentions and the social reality in which these intentions are enacted.

1. What constrains or facilitates the development of students' ability to think critically, to pose questions and to find out and thus to learn to learn meaningfully in the classroom?
2. What are some of the complexities involved in changing the long-standing traditional culture of pedagogy, making it more STEAM-friendly?

2.4 Theoretical Orientation

The focus of a well-rounded STEAM curriculum is on application of knowledge to solve complex interrelated problems in realistic situation that require multidisciplinary approach. STEAM makes STEM more visual and creative leading students to see connections and expand their thinking. The focus of STEAM education is future-oriented with its emphasis on critical innovative thinking "to work creatively with knowledge" (Scardamalia & Bereiter, 2003, p. 1370). STEAM education, which stimulates curiosity and engages students in inquiry and experimentation, entails a different image of learners and learning than the conventional school teaching, whereby learners are passive receivers who can be 'given' knowledge that they then reproduce in standard tests as an indication of their learning. The development of critical and creative thinking to engage innovatively in the production of knowledge for performing in the 'real world' puts ideas (Scardamalia &

Bereiter, 2006), interest (Davydov & Markova, 1982), subjectivity (Rey, 2007) and dialogic meaning making (Bakhtin, 1981) at the centre of teaching and learning. It is a process of finding an answer to a question, a problem or a challenge faced by learners and teachers alike (Ratnam, 2020).

A sociocultural perspective that includes cultural historical activity theory (CHAT) stemming from the work of Vygotsky, facilitates a conceptualization of teaching and learning whereby learners are not passive receivers of knowledge transmitted to them. In this view, learners are active agents contributing to the co-construction of knowledge in interaction with social others and cultural tools and practices available in the social settings while they participate in purposeful joint activities (Vygotsky, 1978). Joint explorative activity involves learners in active construction and reconstruction of knowledge as they think together both retrospectively and prospectively (Kozulin, 1998) to negotiate historically developing contradictions experienced while accomplishing goal-oriented actions (Sannino & Engeström, 2018). This perspective enables us to view teachers as engaging students in meaningful activities that rouse curiosity and motivate them to seek and create knowledge: to learn to use their voice based on their "living knowledge" (Vygotsky, 1987) – to theorize, justify their stance, find out what others think, evaluate their own understanding vis-à-vis others' points of view, and thus to push towards deeper understanding and transformative action. Getting students into a "knowledge producing trajectory" (Scardamalia & Bereiter, 2003) makes the joint activities students engage in personally relevant while also providing "an apprenticeship into valued activity systems that are of importance in the wider community beyond the school" (Wells & Mejía-Arauz, 2005, p. 2).

2.5 Method

2.5.1 Data Sources

This is part of a larger ongoing study (2020-present) using the CHAT framework of an informal educational outreach program for socioeconomically disadvantaged students by establishing new types of collaboration among diverse activity systems, corporate, government, NGOs, and individual and group volunteers to achieve common educational goals in an Indian context. For the purpose of this qualitative study, the data collected between January and June 2021 are used. The data used to compare the two orientations to teaching, viz. *knowledge transmission* and *knowledge co-construction,* are drawn from online observation of two consecutive math classes each from two different schools, and from in-depth interviews of teachers who took these classes including after class discussions, conversations with students (telephone and WhatsApp), interview of parents (two from each school) and principals of these schools. In addition, interviews with stakeholders of the ATAL initiative in ten schools covering four private schools, two government run state secondary schools and four CBSE schools have also fed into the study. These stakeholders include nodal officers who oversee the ATAL program in schools, principals of schools, teachers in charge of the ATAL program in schools, participating students, visiting ATAL mentors to schools, and scientists, engineers and art educators. The data also includes documents and literature related to the study and passages from my reflective diary.

All the data from class observation and interview have been recorded and transcribed. In my role as a participant observer, I participated in the classroom interactions to pose questions to the students. The parts where some students have used regional languages to respond have been translated into English with the help of the teachers involved. Respecting the wishes of participants to remain anonymous, the study does not name people or institutions. The recorded data from classes and the resulting analysis have been shared and discussed with teacher participants and two critical friends.

2.5.2 Data Analysis

According to Vygotsky (1978) development proceeds from the social plane to the individual plane mediated by culturally evolved artifacts and symbolic tools. Thus, social interaction in relation to practical activity is at the source of both human and cultural development. Vygotsky's basic model of tool-mediated action is represented by *subjects* engaged in activity, *objects* towards which they work and *tools* that mediate accomplishment of the activity (p. 40). This has been expanded by Engeström (1987/2015) to include *motivation* for the subject to engage in activity; *community of people* involved in the activity; *division of labour* indicating the role relationship within the community; *rules* that regulate within the activity system and *outcomes* that refer to the change resulting from the activity. All these features that structure human activity make up a holistic unit of analysis. It facilitates juxtaposing the elements of the two classroom activity contexts, AC1 and AC2, presented in this study to locate the opportunities and constraints there to promote the goals of STEAM education.

Since social interaction provides the means for coordinating thinking and acting together of teachers and students in the classroom community, the nature of social interaction in the classroom assumes an important focus of analysis in this study: how the nature of social interaction in the two activity contexts AC1 and AC2 opens and closes opportunities for achieving the STEAM-oriented goals envisioned in educational policies addressing question one of the study.

The educational significance of interaction can be understood in the relationship between language communication and ways of thinking postulated by Vygotsky (1987). There is growing research literature to support that open-ended meaning making interaction expedites deep learning through co-construction of knowledge promoting thinking, problem-solving and reasoning (Mercer et al., 2019; Matusov, 2020). This line of research also helps to expose the limitation of the use of the initiation-response-feedback (IRF) pattern of classroom interaction (Sinclair & Coulthard, 1975), which forms the mainstay of teaching and learning around the world (e.g., Gutiérrez, 2011). This IRF form, where the outcome is predetermined, is seen as restricting thinking by clipping the potential for extended thoughtful exchanges among teachers and students.

These two orientations to interaction, that shape thinking and development, are captured in Bakhtin (1981) by the distinction he makes between two types of discourses as defining the poles of a continuum: "authoritative discourse" and "internally persuasive discourse". Authoritative discourse is "monologic" (Lotman, 1988) and "demands that we acknowledge it, that we make it our own. . . . We encounter it with its authority already fused to it" (Bakhtin, 1981, p. 342), whereas internally persuasive discourse is "dialogic" (Lotman, 1988) and open to connect with other voices in dialogue democratically and thus open to growth and change (Bakhtin, 1981). These two discourses are associated with the two modes of pedagogy that Bakhtin alludes to: "When verbal disciplines are taught in school, two basic modes are recognized for the appropriation and transmission – simultaneously – of another's words (a text, a rule, a model): 'reciting by heart' and 'retelling in one's own words'" (Bakhtin, 1981, p. 341). 'Reciting by heart' corresponds to monologic 'authoritarian discourse' within a structure of hierarchical relationship, while 'retelling in one's own words' goes with dialogic internally persuasive discourse, where the words of others, playing a role in one's inner speech, gets reaccentuated based on one's own intent or authorial voice. In my analysis this distinction between 'monologic' and 'dialogic' discourses serves two important heuristic purposes: one, to analyze the two juxtaposed micro classroom episodes (responding to question one of the study) and two, to situate classroom micro processes in the macro "histories of cultural practices" (Gutiérrez, 2011, p. 30) to understand the implications of the constellation of voices in the larger cultural historical world presupposed in 'authoritative discourse' and 'internally persuasive discourse' (Bakhtin, 1981) on classroom interaction rendering it more or less effective for achieving STEAM oriented outcomes (addressing question two of the study).

The subthemes under which the findings are discussed emerged through a dialectical and iterative process of "visiting and revisiting the data and connecting them with emerging insights" (Srivastava & Hopwood, 2009, p. 77). The thematic analysis and interpretations went hand-in-hand with ongoing dialogue with teacher participants, critical friends and relevant literature. This social process has aided my theorizing, enhancing its validity and trustworthiness.

2.6 Findings

2.6.1 Juxtaposing the Elements of Two Activity Contexts, AC1 and AC2

AC1 is a typical state-run government school while AC2 is a private school. Both activity contexts are located within the organizational framework of formal schooling and the normative procedures set by it in the form of a fixed timetable, prescribed syllabus, materials and evaluation. The aspirational STEAM-aligned objective stated in the aforementioned national education policies applies alike to both schools.

The following findings use classroom excerpts from AC1 and AC2 to examine the nature of classroom interaction taking place there as seen on a continuum formed by monologic authoritative discourse and dialogic, internally persuasive discourse. The analysis helps locate the nature of social interaction that helps or hinders students' "active engagement as subjects" in the process of their learning transforming themselves (Davydov & Markova, 1982).

2.6.2 Problem-Solving Activity: Nature of Classroom Interaction in AC1

The following are classroom extracts from AC1 where the teacher was teaching algebraic equations to grade 7 students.

The dominant mode of interaction through which the teacher led students was IRF as the following example illustrates:

Teacher: What is 2x?
Students: (chorus) term.
Teacher: Yes term. Answer one by one.
Teacher: what is x? (calls out student1)
Student1: variable
Teacher: okay, what is 2?
Student2: Coefficient.
Teacher: Yes. Together variable x and coefficient 2 become 2x and 2x is a term.

In the next class, she illustrated, with examples, the algebraic representation of mathematical structures involving multiplication and division laying out the procedures to be followed:

Teacher: Suppose I want to buy 5 pens and each pen costs Rs20, how much money should I have to buy five pens? (explains both in English and regional language as she writes on the board)
Let the value be as X, if X = 20, then what is the value of 5x?
X = 20
5X =?
You have to cross multiply,
5 × 20 = 100.
X = 20

∴ 5X = 100. This is the answer.

Suppose you know the value of 5x = 100, and want to find the cost of one pen, then you put the equation like this:

$5x = 100$
$X = ?$
$X = 100 \times 1/5$

$\therefore X = 20.$

In this way we use algebraic expression to find the value of an unknown value.

After rounds of explanations in both English and regional language to help students "recognize and produce the pattern", the students had to "enact" it (Matusov, 2020) by using the pattern to solve a problem: If the cost of 10 mangoes is Rs 150, how much must I pay to buy one mango?

Following are some samples of students' responses:

Student 5: I will use variable for Rs 150. The value of mangoes is as x.
$1X = Y$

$\therefore 1 \times 10$, (10 mangoes)
$\therefore X = 10$
$\therefore 1X = 10$
$\therefore X = 150/10$

Students 6: 10 m = 150Rs
So, 1 m = X

$\therefore 10\ m = 150$
$\therefore 1\ m = X$
$\therefore 1 \times X$

Student 7: X is a variable
$10 = 150$

$\therefore 1 = X$

Cross multiply $10 \times X = 10X$

$\therefore 150 \times 1 = 150$
$\therefore 150/10X$
$\therefore X = 150/10$
$\therefore X = 15$

I asked student 7, why he had done cross-multiplication and he said, "We've been taught to do that." I also wanted to know how students 5 and 7 had arrived at the equation "X = 150/10" and both simply said that they had to divide in order to get the correct answer and could not provide a logical explanation as to why they used these operations. Such thought-provoking discussions were not part of the authoritative instructional discourse in the classroom. The teacher seemed to see no point in the questions I posed to students which made her ask me, "When the students have the right answer, is it necessary to ask so many questions?"

The students arrived at a dead end in their attempts to work out the steps by applying mechanically the pattern the teacher had showed them. Students who knew how to arrive at the answer intuitively skipped the algebraic steps leading logically to the answer and simply took a short cut to provide the "right" answer. This seemed to defeat the stated object of the activity, which was to promote the emergence of algebraic thinking by engaging students in framing the given problem and solving it through algebraic equations. Nonetheless, their confusion and the sheer incomprehension with which they seemed to be working threw up many potential teaching moments. It brought to view the nature of the challenge each met with while figuring out solutions and the nature of differentiated support they needed to accomplish the task in a process of meaning negotiation, based on their unique subjective locations. Student 5, for instance, had two variables, X and Y. She told me that X was the number of mangoes and Y was the unknown value of one mango that she had to find out. However, in working out the equation, she had used the two variables interchangeably (1X = Y) and got entangled. The impasse reached by several students in the class similar to student 5 and Student 6 ("∵1 × X") did not lead to any interaction to foster students' "thinking and reasoning of meaningful relations between objects and operations" (Schwarzkopf et al., 2018, p. 195). Instead, the teacher feedback that followed was reteaching of the algorithms to "correct" their "wrong" answers. It was a blanket response to the differentiated learning needs of the students.

2.6.3 Problem-Solving Activity: Nature of Classroom Interaction in AC2

In the class I observed, students of grade 6 were discussing how they had worked out the following task on ratio set for them by the teacher:

The table below shows the comparison of the amount of water to the amount of juice concentrate (JC) in grape juice made by three different people. Whose juice would taste strongest? Be sure to justify your answer.

Lavanya		Faraha		Manoj	
Water	JC	Water	JC	Water	JC
12	2	15	6	16	6
18	3	20	8	24	9
30	5	35	14	40	15
42	7	50	20	64	24

All the students had come up with the basic ratio of all the three juice concentrates on which the ratio table was built: Lavanya (1): 6:1; Faraha (2): 5:2; Manoj (3): 8:3. However, there were differences in the way they had rank ordered from strongest to weakest juice: (a) 3,1,2; (b) 2,1,3; (c) 3,2,1. The teacher solicited the students to explain the strategy they had used in arriving at these diverse conclusions and their rationale. The following are excerpts from the ensuing classroom interaction.

Excerpt 1

Student 1: There is already 6:1. 5:2 can be simplified to 2.5:1.

Teacher: You are simplifying it further?

Student 1: Yes, I did it for my understanding.

Student 2: (raises hand and jumps in to say something, not audible)

Teacher: I'll come to you. Let her [student 1] complete. (to student1) Go on.

Student 1: 8:3 can be simplified to 3:1. Since 8 is very close to 9, I rounded 8 to 9 and made it 9:3 and got 3:1. Now juice concentrate became same in all the three and I could decide easily based on quantity of water.

Teacher: (sums up Student1's reasoning then asks the class) Is there anyone who has a question or something to say about the way she has done it?

(Several hands go up with a chorus of "yes")

Student 3: Yes, can I?

Student 2 & 4: I have another way of solving

Student 5: Even I.

Student 6: I also have done it differently.

Teacher: Yes, yes, one by one. (to Student 2) Tell us what you have done.

Student 2: Instead of simplifying it further we can upscale and make the same amount of juice concentrate in all.

Teacher: All right, when you scaled, what did you get?

Student 2: Lavanya (1) is 36:6, Faraha (2),15:6 and Manoj (3), 16:6

Teacher: So, if you do it this way, what's your conclusion?

Student 2: Faraha, he has least cups of water, so he is first, Manoj, second and Lavanya, third.

Teacher: (to Student 7) You have given a different order. Did you listen to what she [student 2] said?

Student 7: Yes.

Teacher: Do you agree with what she has done and her way of ordering?

Student 7: No, because one ratio she has multiplied by one non zero number and another by another number.

Student 8: (raises hand). I want to reply to . . .[Student7].

Teacher: Okay, go ahead.

Student 8: If we want to get the same number for juice concentrate, then we have to multiply by different numbers 1×6, 3×2 and 2×3.

Student 2: I have multiplied both numbers of the ratio by the same number in each case, not by different numbers. I multiplied 1 by 6 and also 6 by 6 in the first ratio. I used 3 for both numbers in second ratio and 2 for third ratio. So, they are equivalent.

Student 7: Oh, got it, LCM [lowest common multiple]. I agree

Teacher: (to Student7) What were you thinking?

Student 7: I took all three ratios as one and thought you had to use the same numbers for all.

Student 9: I also agree with her [Student 2]. I did my calculation looking at only the juice concentrate in the simplified ratio 6:1, 5:2 and 8:3. So I put 8:3 as the strongest (laughs).

Student 3: You didn't look at the water quantity.
Student 9: Exactly. I should have looked at the cups of water also. They are not all the same like the first problem we did.
Tara: (raises her hand and takes over with the teacher's leave) I was thinking of what ... [Student 1] had done. Do you think it calls for more thinking?
Student 10: I think it's okay. She has scaled down instead of scaling up.
Teacher: I think she [Tara] is referring to the rounding off that ...[student1] did, from 8 to 9.
Tara: Yes. Does that lead to any problem?
Student 1: I think I was right in that situation.
Student 11: Yes, because even if you used 8 and got 2.66666, it is higher than 2.5 and rounding off to 9 didn't change the order.
Student 1: But I am not sure it will work every time. I think in single digits we should scale up. It is much easier. But in big numbers it [scaling down] is really useful. But about rounding off I'm not sure.
Tara: Can you think of some situations in your daily life where you use ratios and then see where rounding off might lead to complications or might not work?

Since the class time was up and the teacher had some housekeeping to do, she posted my question on the class forum where the discussion continued offline and spilled over to the next day's class.

I have provided a long excerpt from AC2 in order to capture at first hand for the reader some of the features of dialogic meaning making present in the connected flow of this classroom interaction as students and teacher together negotiated the contradictions in the joint activity emerging from their diverse perspectives and understandings. These points of view were not put forth as modular units of bricks to construct the edifice of knowledge (Eisner, 2002), but as responses to other points of view and, in turn, anticipating others' responses with the potential to influence each other to see things in a new light (Bakhtin, 1981). This was possible because everyone had the democratic right to participate as active and agentive subjects of teacher-learner and peer-peer intersubjectivity, posing questions, agreeing or disagreeing with what the others said (Bakhtin, 1981) from their unique cognizance. In this internally persuasive discourse, the teacher followed students' orientation to task creating opportunities for them to explore their subjective perceptions in interaction with others, reach new understanding and rethink their earlier stand in view of it, just as Student 7 and Student 9 did in the above excerpt. The teacher was not the sole authority in class. Students had equal rights to assume the role of more experienced peers – as Student 8 did – by offering assistance to one another to progress in the 'zone of proximal development' (ZPD) (Vygotsky, 1978).

The questions posed by the teacher or me were not "display" questions with known answers, but genuine open-ended questions that stimulated thinking and the curiosity to find out. Meaning was not finalized by the teacher. Student 1's equivocal remark, "But about rounding off I'm not sure," left everyone with something to puzzle out beyond the class. The teacher used this unfinalized meaning as a bridge to her next class around the topic of estimation and rounding-off in math. Although this was not a new concept per se, its situational discovery by student 1 was innovative. This and all the other rethinking that the knowledge co-construction in the class occasioned can be characterized as what Bereiter (1994) called "progressive". According to him, knowledge co-constructed is "progressive in the sense that understandings are being generated that are new to the local participants and that the participants recognize as superior to their previous knowledge" (p. 9).

The teacher's decision to teach estimation and rounding-off in the next class was deliberate. She told me that it was meant to create opportunities for building more complex understanding by "capitalizing on their interest and curiosity of the moment roused by your question" [about situations where rounding off could be problematic]. This question, which engaged students in digging into their everyday experience, made the teachers' choice topical and personally relevant by linking to their preoccupation. There was a continuity in meaning in the sequencing of topics, with each class forming "a link in the chain of meaning" (Bakhtin, 1986, p. 146) connecting the past to the present purposefully and providing an orientation to the future. This continuity in the "making", "creating", or "generating" meaning is the crux of STEAM education. Its prospective social goals hinge on the development of human potential for higher order thinking, innovation and everyday creativity by associating the present meaning with the past in imagining new possibilities for future action.

2.6.4 The Nature of Classroom Interaction in AC1 and AC2 Compared

The classes in both AC1 and AC2 were engaged in object-oriented problem-solving activities (tasks). However, the findings show that engaging in activities per se is not directly linked to the development of transformative theoretical thinking. This finding accords with the hypothesis formulated by Davydov and Markova (1982), who pointed out that development depends upon creating "conditions that will enable activity to acquire personal meaning, to become a source of the person's self-development and comprehensive development of his (sic) personality, and a condition for his entry into social practice" (p. 57). The difference in the mediational means between the dominant authoritarian discourse of AC1 and the more internally persuasive discourse of AC2 leading to different developmental outcomes seems to elucidate this.

In both contexts, students were dealing with a structured math problem and the prompts teachers used to find out what they had done had features of IRF. However, the way they organized the communicative process of teaching and learning were different. The teacher in AC1 used questions to elicit responses at the factual level rather than find out what students were thinking and why as the teacher in AC2 did. The follow up move in the IRF structure in AC1 brought a closure. This end point stamped with teacher authority blocked the scope to think further and engage students in joint meaning making whereby the activity would gain personal relevance for students promoting learning and development. Within this discourse structure, students' responses were always addressed to the teacher and peer interaction was conspicuously absent (Wells & Mejía-Arauz, 2005), whereas in AC2, the teacher's follow-up opened the potential for a new "thirdness" (Sannino & Engeström, 2018) in the dialogic negotiation of meaning among conflicting voices of students set off by it. For instance, the critical knowledge building dialogue of difference around the contribution student 1 made from her subjective location led students to "perceive reality differently" (Matusov, 2017, p. 106) making them aware of the contextual nature of knowledge implied in the concept of rounding off and epitomized in the uncertainty expressed by student 1: – "I'm not sure it will work all the time". This 'progressive' (Bereiter, 1994) 'thirdness' opened students to question their assumption that science is universally objective. It excited their curiosity to find out about how scientific theories that once had the status of "facts" have been displaced by new discoveries. The importance of developing a historical perspective on science emphasized by Rohit Prasanna in the opening vignette of this chapter is relevant here. It sensitizes students to the developmental potential of the so called "scientific facts" and the need to hold common sense beliefs, including beliefs about science, to greater scrutiny.

2.6.5 Weaving STEAM into School Curriculum: Putting Learning into the Hands of the Students

In AC2, children's school learning is not confined to the classroom. It takes place in many places in many ways both inside and outside the classroom and school: in the school theatre, street, tinker lab, and art galleries; in nature, the local community, and care homes; and in visits between neighboring schools including a school for the visually impaired. Students engage subjectively with their emotions, curiosity, questions, interests and goals in joint STEAM related inquiry activities. A teacher pointed out, "In a class of 40, however hard we try, it is difficult to reach out to everyone. There are students from different backgrounds and not all respond. We tend to think they are kind of slow. But STEAM projects bring out their real ability. They take part so actively." Another teacher added, "Surprising [is], the difference we see in them, how well they can learn and support each other."

STEAM projects are not subject based but thematic facilitating organic interdisciplinary connections. Engaging in practical activity provides students a context to "ascend from the abstract to the concrete" (Engeström, 2020) by linking their "theoretical thinking" (Davydov & Markova, 1982) dialectically to practical activity as they spiral through several rounds of trial and error. While students harness all the learning in art, science and humanities for design conceptualization, this culturally and historically afforded knowledge is reconstructed in their practical activity to achieve their ideational intent, which, in turn, influences their thinking:

> We learnt about seed dispersal in biology class, but when we prepared the model [seed model] it included physics, math, art, thinking and many more things... The concepts helped us to design, but there was more learning as we refined the model, by trial and error. In spite of all the ideas and thinking we had used, the first model fell down without drifting in the air. Our teachers didn't tell us what went wrong, they wanted us to find out. We even studied how parachute was made, how it stays for so long in the air. Our main finding was to have maximum surface area to create air resistance and keep the weight of the model as low as possible. We redesigned the model increasing the surface area. We examined the seeds we had collected on our nature walk and made our playdough seed more flat and thin... (a student participant)

The STEAM project is not just about accomplishing a task. It doesn't end with designing and developing a model. Students review their project critically, the challenges faced and the learning outcome. They also pursue the new questions it raises for them: "How does seed dispersal happen in rain forest where trees and plants grow close to each other? I am curious because there is rain forest close to my hometown."

"What problems plants and trees face in dispersing seeds, because we are cutting down so many trees for power projects, buildings. Forests are becoming less." These questions and concerns are linked to follow up actions undertaken with co-participation of the community. Thus, the STEAM activities, in which students engage, will contribute to achieving the broader social and sustainable goals of education by providing "apprenticeships" into activity systems of value to the community (Wells & Mejía-Arauz, 2005).

The following are a few more excerpts from student interviews about interfacing art, science and technology:

> In designing we need scientific knowledge, engineering knowledge and creativity. There is art in actual making it [design implementation] ... Art and science are mixed in all the stages [of the learning activity]. (a student)

> Art puts things in a nice way. We like it and appreciate and understand easily. We can use it to make people understand environmental problems like global warming in a better way than science experts. In our STEAM project, we did a street play on biodiversity using posters about how we can protect our nature. People watched with interest. After that we discussed how our way of life affects nature and how we can make changes. They [community participants] also cooperate with us in environmental conservation projects, like garbage sorting and recycling, harvesting greywater in school and homes. (a student)

One of the teachers pointed out,

> Art provides children many ways to express themselves freely and imaginatively. All their experience and knowledge are in it. The project helps them use it for planning and implementing, plus a lot of new things, all the science knowledge they need for solving the problems they face in the project, is connected to what they have done. This makes learning science interesting, because they want to find out solutions. It connects what they know and what they want to know. As I said earlier, we see all the children take part with interest.

Empathy motivating action

> All of us love going to the blind school... we sing songs, dance, play hop scotch, read stories, listen to them, what they want to say. They are very capable even if they can't see. I

> couldn't believe when one of them told me she cooks at home. They teach us how to read using braille. We learn how they see the world and how we can communicate with them. (a student)

> They [the visually challenged students] are so interested to know about many things. I want to help them to fulfill all their ambitions. (a student)

Some of the students from AC2 are participants in an informal online creative computing forum of which I am also a part. They are passionately engaged in a project that involves developing "search, listen and learn" applications exploring various technology and hardware options to help the visually impaired children access various knowledge resources. Posts of their innovations keep popping up on the online forum: "Please check this voice assistant I created for the visually impaired[1]" (This from the student who said he wants to help the visually impaired).

Integrated STEAM Education in the hands of the learner is action oriented. Students become self evolving learners as they contribute collectively to object-oriented activities transforming the self as well as the social practice they engage in.

2.7 Significance of Art in STEM Education

Art, as Eisner (2008) argues cogently, is not just an "ornamental" aspect of human production that simply appeals to the emotions. It is deeply connected to epistemological issues in the significant role it plays in "enlarging human understanding" (p. 3). Its aesthetic appeal also makes it an effective medium of communication to help people make sense of scientific wisdom in a personally and culturally relevant manner, inspiring them to improve practice as seen in the account of the student who spoke of using street play and posters for awareness raising and transformative social action.

Learning as a generative process which transforms the self to see the world in new ways is an agentive, creative, artistic process (Eisner, 2002; Shugurova, 2019). Art and humanities provide tools to interpret the world, sharpen students' sensitivity, and to place the siloed curriculum within the broad spectrum of life, to illumine its different angles and connect it meaningfully to their personal life and social practice.

Art, according to teachers' experience in AC2, is inclusive as it engages students "ontologically" (Matusov, 2011) in giving expression to their "selves," and their values as they create and respond to artistic forms of expression. This authorial creative act is grounded in students' experience, "their personal ways of knowing and situated forms of knowledge" (Shugurova, 2019, p. 2). This opens opportunities for historically underserved students from diverse backgrounds to partake in school learning (including science learning) by linking it to their own meaning making process (Ratnam, 2018). The means provided by art serves to transform students'

[1] These innovative apps students have created using Scratch, code.org and Python are being taken further with support from mentors in this informal forum to make them fully self running apps.

worldviews and helps them develop the capacity to live and work sustainably (Taylor, 2020).

STEAM teaching and learning works across the curriculum fostering creative thinking and problem solving required for effective STEM teaching and learning. The beneficial effect of participating in STEAM project design on STEM teaching and learning can be seen authentically in the interactions among students and teachers (e.g., the earlier classroom excerpt from AC2) and the accounts of teachers and students in AC2.

So far in this chapter, I have tried to establish two interrelated points by using data from the study and supporting literature: (1) Activities mediated by the dialogic meaning making internally persuasive discourse that acknowledges students' agency as subjects is conducive to transformative learning and development, and (2) Art and humanities have a significant epistemological role in education (Eisner, 2008), including STEM education. However, the paradox remains as to why the internally persuasive discourse has not become part of the common cultural practice in schools, despite the robust possibility it holds for achieving the desirable STEAM goals in education. This paradox, which addresses the second question of the study, will be discussed in the following section.

2.8 The Problem of Change

Enhancing the meaningfulness and effectiveness of the learning experiences of students drawing on internally persuasive discourse is essentially a matter of pedagogy. This is seen largely as the responsibility of the individual teachers – their ability – and they become easy targets of criticism and of a deficit view when school reforms fail (Ratnam, 2021). However, a sociocultural perspective helps us see that teachers' practice is not the isolated product of their volition (Smagorinsky, 2020), but part of a complex object-oriented activity system. All the elements of this activity system, such as the object of activity and the outcome expectations, the people and artifacts are historically contingent as they are embedded in multiple and diverse strands of history (Engeström, 1999). However, the historical antecedents of teachers and the dilemmas experienced by them in the day-to-day institutional practice – including classroom practice – are prone to "genesis amnesia" (Bourdieu & Passeron, 1990, p. 9) and therefore not obvious. They become visible through historical analysis both at the immediate institutional and more distant societal level, providing a deeper understanding of the dynamics by which change in practice is inhibited or facilitated with implications for teacher education and school reform.

Engeström (1999) argues that the object of an activity is its main motivating force in determining "the horizon of actions" (p. 381). The institutionally authorized object guiding teachers' actions are different in AC1 and AC2, and so are the actions of teachers. While AC1, which typifies the mainstream conventional schools, emphasizes the goal of promoting students by teaching to the test, AC2 works with the broader STEAM purpose of promoting learning, that enables students to become well-integrated human beings who are emotionally mature, reflective and capable of making innovatively meaningful contributions to society. The instrumental goals of the formal curriculum are subsumed under its broader vision in AC2. These are goals set by the institution and not by teachers. However, since teachers are employed by the institutions, they become complicit with the institutional norms and values that tacitly involve the larger sociopolitical and economic interests governing them (See Ratnam, 2021 for the sociopolitical agenda and its historical links governing education in India). This is evident in the neoliberal hold on education in India – also true more globally – which hijacks teachers' autonomy and reduces them to transmitters of externally mandated common curricula through scripted instruction, as seen in AC1. On one hand, teachers are exhorted to use their agency to bend the curriculum to address the needs of diversity, while on the other hand, they are expected to erase differences and get all students to the same finish line – the standard tests. There is a contradiction between the espoused object and outcome expectations. Teachers' sense making of this contradiction is influenced by the "audit culture" (Apple, 2007) to which they are subjected to in order to improve their performance. They come to see uniformity, rigidity and compliance as the valued norms of the institution. These institutional norms and values are driven by the claims that equality of opportunity is ensured by standardization and uniformization thus making it acceptable to teachers. The extent of the damage this manipulation does to teachers' capacity to think and use their "personal practical knowledge" (Clandinin, 1985) is exemplified in the question posed by the teacher in AC1 to me, "When the students have the right answer, is it necessary to ask so many questions?" Working with the narrow goals of teaching to the test over time has made teachers accept getting the "right answers" as the goal of learning (Wells, 2015). These beliefs are reinforced in teacher support programs that focus largely on facilitating pattern recognition and its production instead of helping teachers to question the "given", negotiate the contradictions in their institutional settings and develop their professional autonomy to work towards the transformative goals of education.

The dynamic works differently in AC2, where there is closer affinity between the object and outcome. The leeway that AC2 enjoys being a private institution acts as a buffer – an in between space – for teachers to exercise their professional autonomy and explore new ways of engaging students in achieving the broader STEAM goals:

> We were also once ordinary traditional teachers. But we have changed a lot after coming here. When we joined, we didn't go straight to the classroom. We observed senior teachers for about two weeks and got to know the culture of the school. The management has very high expectations, we have to innovate, come up with new ideas for classroom and other activities. We can't say all the ideas are ours. We search, read and use these ideas and build on them. We have teacher professional development session every week where we share our ideas and discuss plans for STEAM projects together. So, our learning is ongoing with student learning. (a teacher).

From what this teacher says it is plain that teachers, regardless of the institutional pressure, have the potential and disposition to be innovative and want to use it for students' benefit. However, in a conservative school environment such as AC1 they lack the space to capitalize on it whereas, the encouragement they receive in alternative progressive environment such as AC2, pushes them to realize their potential capacity to invent. This is evident in what another teacher said:

> The management is strict, but supports us. We have freedom to experiment. They also get us whatever we want for the lab- all the latest technologies available. I should say it's the management expectations and support that's the driving force for us. In the previous school, where I worked, the HM [Head Master] called me to the office and asked me to follow "regular teaching" when I tried to bring in some change.

Neither the teachers nor the students to whom I spoke seemed obsessed with board examination as in AC1. One of the teachers explained, "Parents do expect good results. But our students are well prepared for it [exams], because they learn much more than the syllabus through all the activities they do."

Teachers in both activity contexts have tensions to negotiate as they strive to attain the contradictory demands placed on them in their diverse work settings. These contradictions in the settings have their impact on teachers in the gaps that develop between their intentions and professional values on one hand and their action on the other hand without their conscious awareness. The teacher in AC1 said that she wanted to "help all children learn" and she followed the institutionally scripted uniform teaching to achieve this personally and professionally valued goal, without questioning its effectiveness for teaching to diversity. The institutionally valued instrumental goals driven by the belief that uniform input begets uniform output provided no motivation for such questioning. Expectations of "progressive" (Bereiter, 1994) outcomes in AC2 foregrounds a different imperative. Here, the development of teachers' agency is closely related to their acknowledgement of the need to help students become independent – self evolving – learners through inquiry and the concomitant need for teachers themselves to be active self-evolving learners and researchers. It is the discovery of this need that motivated teachers in AC2 to collectively design and transform the activities historically and culturally constructed in educational settings.

The educational aspiration of achieving the futuristic STEAM goals universally through the introduction of ATLs in schools must be understood within the broader realm of sociopolitical dominance and the economic interests governing education as well as the culture – including teachers' cultural beliefs and practice – that prevails in most schools. For teachers enculturated in text and tradition it is difficult to understand what it means to be a learner through inquiry and much less throw themselves into such learning:

> Many things are totally new for us; we have to learn a lot. We can learn, but for that they have to arrange a course, give us all the knowledge.... But with all the regular work we do, where is the time for this extra learning? (an ATL in charge science teacher)

> We do it [ATL work] because we are told. We don't really feel qualified for it. (a science teacher)

The teachers in charge of ATLs I interviewed, as well as the principals, were unanimous about the need for a full-time teacher with engineering and technology background to run the ATL successfully. They tended to see the ATL program as an add on to the curriculum load and not as something to be integrated into it: "It's not one of the subjects in the examination." (a principal).

As Cole (in Davydov & Markova, 1982, p. 51) has emphasized alluding to Marx's dialectical materialist philosophy, "freedom is the recognition of necessity." The foregoing analysis shows this to be the most formidable challenge for teacher education – to help teachers recognize the potential value of ATL and the necessity of change in their practice to leverage it by becoming consciously aware of the ways in which their current practice, which they believe benefits students, is in fact contributing to the stability of the neoliberal control in education eroding their own professional values. This again is circumscribed by the surrounding institutional culture and the extent to which other stakeholders in the institutional milieu are engaged collectively and critically with teachers to raise questions about, "Who is benefiting most or least from the way learning is mediated in the classroom?" (Peter Taylor, personal communication, March 12, 2021).

2.9 Conclusion

In addressing the questions of the study about adopting the STEAM approach in mainstream formal school education I have used a sociocultural lens and, in particular, the CHAT framework, based on the premise of the social nature of learning in joint activities where human action is mediated by cultural tools. I have highlighted the significant epistemological function that art plays in achieving the social and economic goals of STEM education. I have illustrated empirically the centrality of knowledge building internally persuasive social interaction in the classroom for promoting students' ability to learn and develop the higher cognitive skills of creative thinking, reasoning and problem solving among others. Finally, I have linked the micro classroom interactions to the larger institutional and societal dynamics in which they are embedded, as a way to locate the contradictions that inhibit change in practice. A nuanced pedagogical understanding of art-integrated STEAM education and the constraints posed to its adoption by historically developed cultural practices enhances the prospect of improving these practices.

References

Anderson, M. (2016, October 14). Creativity as the innovation literacy. In *Keynote speech at Australian Curriculum Studies Association conference, STEM, STEAM or HASS? Interrogating models of curriculum integration*, SMC Conference and Function Centre, Sydney, Australia.

Apple, M. (2007). Education, markets, and an audit culture. *International Journal of Education Policies, 1*(1), 4–19.

Avvisati, F., Jacotin, G., & Vincent-Lancrin, S. (2013). Educating higher education students for innovative economies: What international data tell us. *Tuning Journal for Higher Education, 1*(1), 223–240.

Bakhtin, M. (1981). *The dialogic imagination: Four essays by M.M. Bakhtin* (M. Holquist, Ed., C. Emerson & M. Holquist, Trans.). University of Texas Press.

Bakhtin, M.M. (1986). *Speech genres and other late essays.* (C. Emerson & M. Holquist, Eds., V. W. McGee, Trans.). University of Texas Press.

Bereiter, C. (1994). Implications of postodernism for science, or, science as progressive discourse. *Educational Psychologist, 29*(1), 3–12.

Boucher, E. (2017). Colonial liberalism, global neoliberalism, and education in India. *The Global Studies Journal, 10*(2), 51–66.

Bourdieu, P. & Passeron, J. 1990. *Reproduction in education, society and culture* (R. Nice, Trans.) Sage.

Catterall, L. G. (2017). A brief history of STEM and STEAM from an inadvertent insider. *The STEAM Journal, 3*(1), 5. https://doi.org/10.5642/steam.20170301.05

Clandinin, D. J. (1985). Personal practical knowledge: A study of teachers' classroom images. *Curriculum Inquiry, 15*(4), 361–385. https://doi.org/10.1080/03626784.1985.11075976

Davydov, V. V., & Markova, A. K. (1982). A concept of educational activity for schoolchildren. *Soviet Psychology, 21*, 50–76. https://doi.org/10.2753/RPO1061-0405210250

Eisner, E. (2002). What can education learn from the arts about the practice of education? *Journal of Curriculum and Supervision, 18*(1), 4–16.

Eisner, E. (2008). Art and knowledge. In J. G. Knowles & A. L. Cole (Eds.), *Handbook of the arts in qualitative research* (pp. 3–12). Sage.

Engeström, Y. (1987/2015). *Learning by expanding: An activity-theoretical approach to developmental research* (2nd ed.). Cambridge University Press.

Engeström, Y. (1999). Innovative learning in work teams: Analysing cycles of knowledge creation in practice. In Y. Engeström, R. Miettinen, & R. L. Punamaki (Eds.), *Perspectives on activity theory* (pp. 377–404). Cambridge University Press. https://doi.org/10.1017/CBO9780511812774.003

Engeström, Y. (2020). Ascending from the abstract to the concrete as a principle of expansive learning. *Psychological Science and Education, 25*(5), 31–43. https://doi.org/10.17759/pse.2020250503

Feldman, A. (2015). *STEAM rising: Why we need to put the arts into STEM education.* http://www.slate.com/articles/technology/future_tense/2015/06/steam_vs_stem_why_we_need_to_put_the_arts_into_stem_education.html

Freeman, B., Marginson, S., & Tytler, R. (2019). An international view of stem education. In A. Sahim & M. J. Mohr-Schroeder (Eds.), *STEM education: Myths and truths – What has K-12 STEM education research taught us?* (2nd ed., pp. 350–363). Brill Sense. https://doi.org/10.1163/9789004405400_019

Government of India (GOI). (1983). *Technology policy statement.* Department of Science and Technology (DST), Ministry of Science and Technology. https://dst.gov.in/st-system-india/science-and-technology-policy-2013

Government of India (GOI). (2003). *Science and technology policy.* Department of Science and Technology (DST), Ministry of Science and Technology https://www.india.gov.in/science-and-technology-policy-2003-department-science-and-technology

Government of India (GOI). (2013). *Science Technology and Innovation Policy (STIP).* Department of Science and Technology (DST), Ministry of Science and Technology https://dst.gov.in/st-system-india/science-and-technology-policy-2013

Government of India (GOI). (2018). *Educational statistics at a glance.* MHRD Department of Education.

Government of India (GOI). (2020). *National policy on education.* Ministry of Human Resource Development.

Government of India (GOI), NITI Aayog (2017). Guidelines for setting up of Tinkering Laboratories under Atal Innovation Mission – 'Atal Tinkering Laboratories'. ATL – Final Guidelines – 2017-18.pdf (niti.gov.in).

Government of India (GOI), NITI Aayog. (2020). Atal Innovation Mission (AIM). https://www.aim.gov.in/overview.php

Gutiérrez, K. D. (2011). Teaching toward possibility: Building cultural supports for robust learning. *Journal of Educational Justice, 3*(1), 22–38.

Hardiman, M. M., & JohnBull, R. M. (2019). From STEM to STEAM: How can educators meet the challenge? In A. J. Stewart, M. P. Mueller, & D. J. Tippins (Eds.), *Converting STEM into STEAM programs: Methods and examples from and for education* (pp. 1–10). Springer.

Kingdon, G., & Pal, S. (2014). Can private school growth foster 'education for all': Tracing the aggregate effects at the district level. *SSRN*. https://www.researchgate.net/publication/262911606_Can_Private_School_Growth_Foster_Education_for_All_Tracing_the_Aggregate_Effects_at_the_District-level

Kozulin, A. (1998). *Psychological tools: A sociocultural approach to education*. Harvard University Press.

Kumar, K. (2005). *Political agenda of education* (2nd ed.). Sage.

Lewin, K. M. (2011). *Beyond universal access to elementary education in India: Is it achievable at affordable costs?* University of Sussex.

Lotman, Y. M. (1988). Text within a text. *Soviet Psychology, 26*(3), 32–51.

Martin, A. J., Mansour, M., Anderson, M., Gibson, R., & Leim, G. A. D. (2013). The role of arts participation in studnets' academic and non-academic outcomes: A longitudinal study of school, home and community factors. *Journal of Educational Psychology., 105*(3), 709–727.

Matusov, E. (2011). Authorial teaching and learning. In E. J. White & M. Peters (Eds.), *Bakhtinian pedagogy: Opportunities and challenges for research, policy and practice in education across the globe* (pp. 21–46). Peter Lang.

Matusov, E. (2017). Examining how and why to engage practitioners from across the learning landscape in research enterprise: Proposal for phronêtic research on education. *Integrative Psychological and Behavioral Science, 51*(1), 94–119.

Matusov, E. (2020). Pattern-recognition, intersubjectivity, and dialogic meaning-making in education. *Dialogic Pedagogy: An International Online Journal* I*, 8*, E1–E24. https://doi.org/10.5195/dpj.2020.314

McGee, E. O. (2020). Interrogating structural racism in STEM higher education. *Educational Researcher, 49*(9), 633–644.

Mercer, N., Hennessy, S., & Warwick, P. (2019). Dialogue, thinking together and digital technology in the classroom: Some educational implications of a continuing line of inquiry. *International Journal of Educational Research., 97*, 187–199. https://doi.org/10.1016/j.ijer.2017.08.007

National Academy of Sciences. (2005). *Rising above the gathering storm: Energizing and employing america for a brighter economic future*. available at http://nap.edu/11463

National Foundation on the Arts and the Humanities Act of 1965 (P.L. 89-209) (1965). https://www.neh.gov/about/history/national-foundation-arts-and-humanities-act-1965-pl-89-209

NCERT (National Council of Educational Research and Training). (2005). *National Curriculum Framework*. Secretary, Publication Department, NCERT.

Ozkan, G., & Topsakal, U. U. (2020). Investigating the effectiveness of STEAM education on students' conceptual understanding of force and energy topics. *Research in Science and Technological Education, 39*(4), 441–460.

Pratham. (2019). *The annual status of education report (ASER) for 2018*. Pratham.

Ratnam, T. (2013). Engaging India's social history to understand and promote teacher change. In C. Craig, P. Meijer, & J. Broeckmans (Eds.), *From teacher thinking to teachers and teaching: The evolution of a research community, advances in research on teaching* (Vol. 19, pp. 527–554). Emerald Publishing.

Ratnam, T. (2015). Pedagogies of social justice: An Indian case. In L. Orland-Barak & C. Craig (Eds.), *International teacher education: Promising pedagogies (part B) advances in research on teaching* (Vol. 22, pp. 255–282). Emerald Group Publishing Limited.

Ratnam, T. (2018). Why do we lose students in India? The importance of tuning into students' ZPD. *English Teachers Association Journal (ETAS J), 35*(3), 20–22.
Ratnam, T. (2020). Provocation to dialog in a third space: Helping teachers walk toward equity pedagogy. *Frontiers in Education, 5*, 569018.
Ratnam, T. (2021). The interaction of culture and context in the construction of teachers' putative entitled attitude in the midst of change. In T. Ratnam & C. J. Craig (Eds.), *Understanding excessive teacher and faculty entitlement: Digging at the roots, advances in research on teaching* (Vol. 38, pp. 77–101). Emerald Publishing Limited. https://doi.org/10.1108/S1479-368720210000038006
Rey, G. (2007). The topic of subjectivity in psychology: Contradictions, paths and new alternatives. *Theory of Social Behaviour, 47*(4), 502–521.
RTE (Right to Education). (2009). Retrieved from http://mhrd.gov.in/rte
Russell, T. (2021). Exploring teacher entitlement: Perspectives from personal experience. In T. Ratnam & C. J. Craig (Eds.), *Understanding excessive teacher and faculty entitlement: Digging at the roots, advances in research on teaching* (Vol. 38, pp. 35–46). Emerald Publishing Limited.
Sannino, A., & Engeström, Y. (2018). Cultural-historical activity theory: Founding insights and new challenge. *Cultural-Historical Psychology., 14*(3), 43–56.
Scardamalia, M., & Bereiter, C. (2003). Knowledge building. In *Encyclopedia of education* (pp. 1370–1373). Macmillan Reference.
Scardamalia, M., & Bereiter, C. (2006). Knowledge building: Theory, pedagogy, and technology. In K. Sawyer (Ed.), *Cambridge handbook of the learning sciences* (pp. 97–118). Cambridge University Press.
Schwarzkopf, R., Nührenbörger, M., & Mayer, C. (2018). Algebraic understanding of equalities in primary classes. In C. Kieran (Ed.), *Teaching and learning algebraic thinking with 5- to 12-year-olds* (pp. 195–212). ICME-13 Monographs. Springer. https://doi.org/10.1007/978-3-319-68351-5_8
Shugurova, O. (2019). Teaching teacher candidates about social transformations through arts and place: "Wait, but what does it have to do with me as a teacher?". *Inquiry in Education, 11*(1), 6.
Sinclair, J., & Coulthard, M. (1975). *Towards an analysis of discourse: The English used by teachers and pupils*. Oxford University Press.
Smagorinsky, P. (2020). *Learning to teach English and language arts: A Vygotskian perspective on beginning teachers' pedagogical concept development*. Bloomsbury.
Srivastava, P., & Hopwood, N. (2009). A practical iterative framework for qualitative data analysis. *International Journal of Qualitative Methods, 8*(1), 76–84.
Taylor, P.C. (2016). Transformative STEAM education for the 21st century. In *Proceedings of the Australian conference on science and mathematics education (formerly uniserve science conference)*. https://openjournals.library.sydney.edu.au/index.php/IISME/article/view/10343
Taylor, P.C. (2020). Transformative STE(A)M education for a sustainable world. In *Proceedings of the international joint conference on arts and humanities*. Atlantis. https://doi.org/10.2991/assehr.k.201201.001
UNESCO (2015). *Sustainable goals*. https://unescoghana.org/about-us/sustainable-goals/
Velaskar, P. (2013). Sociology of educational inequality in India: A critique and a new research agenda. In G. R. Nambissan & S. S. Rao (Eds.), *Sociology of education in India: Changing contours and emerging concerns* (pp. 103–135). Oxford University Press.
Vygotsky, L. (1978). *Mind in society. The development of higher psychological processes*. Harvard University Press.
Vygotsky, L. S. (1987). *The collected works of L. S. Vygotsky. Vol. 1. Thinking and Speech.* (Eds.), R. W. Rieber & A. S. Carton (Trans.), N. Minick. Plenum Press.
Wells, G. (2015). Dialogic learning: Talking our way into understanding. In T. Dragonas, K. J. Gergen, S. McNamee, & E. Tseliou (Eds.), *Education as social construction contributions to theory, research and practice* (pp. 62–90). A Taos Institute Publication.
Wells, G., & Mejía-Arauz, R. (2005). Towards a dialogue in the classroom.Learning and teaching through inquiry. *Working Papers on Culture, Education and Human Development, 1*(4), 1–45.

Weyer, M., & Dell'Erba, M. (2022). *Research and policy implications of STEAM Education for young students. Education Commission of the States*. Available at: https://www.ecs.org/wp-content/uploads/Research-and-Policy-Implications-of-STEAM-Education-for-Young-Students.pdf

Wilson, K., Lambright, K., & Smeeding, T. M. (2006). School finance, equivalent educational expenditure, and the income distribution: Equal dollars or equal chances for success. *Education Finance and Policy, 1*(4), 396–424.

Dr Tara Ratnam is an independent teacher, educator, and researcher based in India. She pursues classroom research with communities of teachers to help negotiate the tension-laden path created by the sociocultural and situated dynamics of their institutional milieu in order to provide socially sensitive learning support to the culturally diverse student populations in an inclusive learning environment. Her career-long quest to understand the puzzle of why teachers hold on to their well-worn practice in a world that demands change led her to the idea of "excessive teacher entitlement." This groundbreaking concept in Teacher Education is the theme of the book: https://books.emeraldinsight.com/page/detail/Understanding-Excessive-Teacher-and-Faculty-Entitlement/?K=9781800439412

Chapter 3
Developing Science Education Through Developmental Teaching: Theoretical Thinking, Personality Development, and Radical-Local Teaching and Learning

Seth Chaiklin

In the late 1920s and early 1930s Lev Vygotsky elaborated a research tradition that focused on explaining human development. This tradition, which came to be called cultural-historical, provided conceptual foundations for an approach to developmental teaching, elaborated initially by D.B. El'konin and V.V. Davydov in the late 1950s. There are still many conceptual ideas found within these traditions which have not been received in contemporary research, but which deserve further examination and consideration. One aim of this article is to introduce some of these concepts, including the idea of personality development, and the role of theoretical thinking for school children's development, and discuss their relevance for science education. A second aim is to introduce the idea of radical-local teaching and learning, which both continues this line of thinking, but elaborates a more situated curricular perspective. These general ideas are concretised and illustrated in relation to an example of teaching about electromagnetic phenomena, including a new approach to working with theoretical thinking in relation to subject-matter concepts. The article concludes with a brief response to some common concerns about working with theoretical models in teaching.

S. Chaiklin (✉)
Frederiksberg, Denmark
e-mail: seth@c.dk

K. Plakitsi, S. Barma (eds.), *Sociocultural Approaches to STEM Education*, Sociocultural Explorations of Science Education 21,
https://doi.org/10.1007/978-3-031-44377-0_3

3.1 Introduction

Some curricular objectives in science education are widely acknowledged among science education researchers and educators—without these objectives being dependent on a particular theoretical perspective about teaching and learning processes. For example, a persistent and central problem in the practice of science education is how to help pupils and students learn to think with disciplinary knowledge in relation to meaningful and relevant problems and situations in their lifeworld. This central problem can be recognized in actual examples of curricular objectives (see Table 3.1), where it is notable that all the examples in this somewhat arbitrary selection include a focus on the utility of scientific knowledge in relation to a pupil's life situation. These examples stretch over 100 years, and are geographically and societally diverse, coming respectively from the United States of America, Finland, and Australia. The first selection comes from a physics professor at the University of Chicago, who was the first person in the United States of America to write about physics education at the secondary level. The quoted example comes from his book about the role of physics education in general education (Mann, 1912). He had previously produced a high school physics textbook (Mann & Twiss, 1905), which suggests that he needed to reflect about the objectives of physics education. The second example is an extract from the 2004 curriculum document from the Finnish National Board of Education for the aim of physics for grades 7–9, while the last selection is the full statement of aims of science education from kindergarten to year 10 in Australia. A quick skim of these three examples should help to give a deeper understanding of the complex set of issues that might be involved in helping pupils to learn to think with disciplinary knowledge in relation to their lifeworld.

When attention shifts from formulating general curricular objectives, such as in Table 3.1, to addressing them in actual practice, then a theoretical perspective can be useful, perhaps even necessary. The present chapter is motivated by the assumption that the theory and practice of developmental teaching provides productive ways for designing educational activities that engage with curricular objectives in general, including the central problem about using disciplinary knowledge in relation to pupils' lifeworld, mentioned in the previous paragraph. The main aims of the chapter are to introduce some key concepts from the theory of developmental teaching that are relevant for addressing the central problem, and to illustrate how these concepts might be used in relation to that problem.

The chapter starts with a brief historical introduction about the idea of developmental teaching and learning, primarily to highlight its key assumption that education should be conceptualized as a developmental process. Then the concepts of theoretical thinking and personality development are introduced and discussed. In addition to being relevant to the central problem, these concepts were also selected for attention because (a) they provide general orienting points and relevant objectives that can be used for all subjects in primary and secondary education, and (b) they deserve more attention among researchers who use cultural-historical theory in relation to educational research. Finally, the chapter introduces the idea of

Table 3.1 Examples of science curricular aims

Mann (1912)	Finnish National Board of Education (2004)	Australian Curriculum, Version 8.4 (2018)
The purpose of teaching physics is to assist the pupils in acquiring the benefits of physics to the fullest possible degree. The benefits of physics are of two kinds: they consist in the acquisition of: 1. Useful knowledge of physical phenomena. 2. Discipline in the methods of acquiring this useful knowledge. (p. 213)	**Grades 7–9** The core task of physics instruction in the seventh through ninth grades is to broaden the pupils' knowledge of physics and their conception of the nature of physics, and to strengthen skills in experimental acquisition of information. ... The study of physics supports the pupil in developing his or her personality, and for forming a modern world view. It also provides capabilities for making everyday choices, especially in matters related to environmental protection and the use of energy resources. (p. 188)	The Australian Curriculum: Science aims to ensure that students develop: • An interest in science as a means of expanding their curiosity and willingness to explore, ask questions about and speculate on the changing world in which they live • An understanding of the vision that science provides of the nature of living things, of Earth and its place in the cosmos, and of the physical and chemical processes that explain the behaviour of all material things • An understanding of the nature of scientific inquiry and the ability to use a range of scientific inquiry methods, including questioning; planning and conducting experiments and investigations based on ethical principles; collecting and analysing data; evaluating results; and drawing critical, evidence-based conclusions • An ability to communicate scientific understanding and findings to a range of audiences, to justify ideas on the basis of evidence, and to evaluate and debate scientific arguments and claims • An ability to solve problems and make informed, evidence-based decisions about current and future applications of science while taking into account ethical and social implications of decisions. • An understanding of historical and cultural contributions to science as well as contemporary science issues and activities and an understanding of the diversity of careers related to science.

(continued)

Table 3.1 (continued)

Mann (1912)	Finnish National Board of Education (2004)	Australian Curriculum, Version 8.4 (2018)
		• A solid foundation of knowledge of the biological, chemical, physical, earth and space sciences, including being able to select and integrate the scientific knowledge and methods needed to explain and predict phenomena, to apply that understanding to new situations and events, and to appreciate the dynamic nature of science knowledge.

radical-local teaching and learning as a way to work with these concepts in relation to the central problem in science education about thinking with disciplinary concepts, and finishes with some concrete illustrations of these ideas for the topic of electromagnetism.

3.2 Historical Origins of Developmental Education

In the late 1920s and early 1930s Lev Vygotsky initiated and elaborated a research tradition that focused primarily on characterizing and explaining human development. An important feature of Vygotsky's theoretical framework is the idea that psychological capabilities arise primarily from interactions between a person and the socially-structured material environment in which a person acts (e.g., Vygotsky, 1931/1998b, p. 198). This idea may seem obvious or trivial, when considered from the point of view of the early twenty-first century, but at the time this idea reflected a radical shift from then-existing psychological conceptions of human development. And this idea of development through social interaction may still be a radically different perspective relative to contemporary conceptions that view development as largely determined or limited by genetic, neurological, or cultural characteristics of individuals. An important implication of this perspective is that responsibility is placed on teachers and researchers to consider the conditions and opportunities provided for school children to develop, including the role of content learning in that development.

Vygotsky's research tradition, which came to be called *cultural-historical*, provides essential conceptual foundations for a theoretical conception of education. A key idea in this foundation is Vygotsky's idea (or hypothesis) that some kinds of learning "go in front of" (or "lead") development. Vygotsky (1934/1987) uses a stimulating analogy to suggest the generative potential that results from well-chosen learning: "we have given the child a penny's worth of instruction and the consequence has been a dollar's worth of development" (p. 198). Not all learning is developmental. But sometimes, through educational activities, persons acquire particular intellectual functions (e.g., thinking with theoretical concepts) or conceptual models (e.g., learning about ecosystems), which engender qualitative changes in a person's relation to their social environment, and would be considered development in this theoretical perspective. It is unlikely that these developmentally-significant learning achievements would have happened spontaneously, without some kind of educational intervention. This distinction – between learning that is developmental and that which is not – serves as a bearing element in the idea of developmental education, where the main aim or focus is to support learning that will have developmental consequences. This general or global idea of "learning as developmental" motivates research efforts to elaborate a theory of developmental teaching and learning, and to develop educational practices that realize these theoretical visions.

Starting in the late 1950s, different research groups in the Soviet Union started to elaborate such theories in connection with experimental efforts to operationalize these ideas in concrete teaching practices. The same Russian word, *обучения* (obuchenie), is used to refer to the object of this theory. This word, in English translation, has been rendered in a variety of different ways including *teaching, teaching/learning, instruction,* and *education,* with corresponding labels including *developmental education*, *theory of developmental teaching/learning,* and *developmental teaching.* The present chapter uses the term *developmental teaching* to refer to theoretical perspectives that address the key idea of learning that leads development. No theoretical significance is intended with this choice, which was made primarily because of length and simplicity. The meaning of the theoretical perspective lies in its conceptual ideas, which cannot be conveyed adequately by its label, no matter which one is chosen. In all the translation variations of this term, the key idea of learning leading development is the basic problem that animates the development of theoretical analysis and educational practice.

This brief historical chronicle is meant to highlight that "developmental teaching" is a general conceptual idea that refers to the general idea of the relation between education and development. Both that conceptual idea and the practical actions needed to operationalize that idea for particular disciplinary content are challenges to be addressed. Different theoretical perspectives and partial attempts to address these challenges have been pursued since at least the 1930s (see Davydov, 1998, pp. 17–18 for an enumeration). In choosing to work with developmental teaching, one must also consider the particular theoretical perspective that is being pursued.

This chapter focuses specifically on the tradition and theory of developmental teaching initiated by D. B. El'konin and V. V. Davydov in the late 1950s, and further elaborated theoretically and practically for several decades by Davydov together with many colleagues and collaborators (e.g., Repkin, 2003). Although relevant texts that introduce and explain this tradition have been available in various languages (including English) for at least 30 years (e.g., Davydov, 1986/1988a, b, c), many important conceptual ideas found within this tradition still do not enjoy sufficient attention and discussion among contemporary educational researchers who work in the cultural-historical tradition. Two of these conceptual ideas—theoretical thinking and personality development—are highlighted in this chapter for further consideration.

3.2.1 A Comment on System

Over the past decade, it has become more common to see phrases like "the system of developmental instruction" (e.g., Moxhay, 2008, p. 5) or "the Elkonin-Davydov system of developmental education" (e.g., https://en.mgpu.ru/developmental-teaching-history-and-perspectives/; Puentes et al., 2018; Gordeeva, 2020). The word *system* is polysemous in English, where one of its meanings is oriented to the idea of a procedure or method for achieving some objective. For example, the Barton Reading System or the Wilson Reading System provide highly structured instructional activities for persons who have difficulties in learning to read. This idea of system—as a definite, often highly structured, procedure—is fundamentally incompatible with the spirit of Davydov's theoretical perspective on developmental teaching.

Another meaning of system sometimes refers to a complex of parts in an institutionalized structure. For example, especially in the 1990s, Davydov and his colleagues developed teacher training programs, textbook systems (i.e., for the first five school years), and inservice support to schools who were working with these ideas. It is often convenient to refer to this complex as a "system", where again the label is an indefinite reference to this complex of elements that embody an approach to developmental education and not the internal nature of the educational activity.

Davydov was working and writing at a time when other research groups were trying to elaborate the general idea of developmental education, and he used the Russian word *система* [sistema] to refer to his particular approach to developmental education (i.e., where the word *system* is referring to the implementation of the theoretical ideas in an institutionalized school practice, and not to a particular method or technique within the approach itself). This phrase provides a way to differentiate his approach from those developed by Zankov and colleagues or Gal'perin, Talyzina and colleagues (see Davydov, 1998, p. 18 for a clear illustration).

3.2.2 A Comment on Present Status

One of the challenges in communicating about and learning to work with the developmental teaching tradition is its wholistic quality. The theoretical and practical ideas as elaborated by Davydov and colleagues have been intensively developed for several decades. This theory and practice of developmental teaching is a comprehensive perspective for thinking about any educational activity. It has a self-sufficiency for conceptualizing and forming educational goals and objectives, in that the theory can be used to formulate (a) objectives for educational activities in general and in relation to specific disciplinary and topical areas, and (b) concrete, practical ideas for how to realize those objectives in teaching practices.

In other words, the perspective can be used for any kind of disciplinary content, and its considerations about learning and development must be addressed in any kind of teaching unit. This comprehensive quality is particularly desirable in relation to STEM education, because it can be adapted readily to different disciplinary perspectives that may be included in a STEM teaching unit, but it has implications or consequences for curricular goals, which must also be confronted, whether working in a STEM perspective or with a particular natural science. Some examples are discussed later.

Some of the ideas in the theoretical perspective are not radically different or novel in relation to other educational traditions (e.g., viewing education as a process of assimilating the historical accomplishments of material and spiritual culture), while other perspectives (e.g., a focus on modeling and theoretical thinking) are less common or unfamiliar.

Although this theoretical perspective is self-sufficient, it usually can be used adaptively in ways that preserve its objectives while addressing or elaborating formal or institutionalized curricular demands from other sources (e.g., state policy). In other words, the theoretical perspective can be viable in a wide range of practical institutional contexts.

A practical challenge for researchers. A practical challenge in pursuing research from a developmental teaching perspective arises because of a tension between central theoretical objectives of developmental teaching and typical organizational structures and working conditions for researchers in relation to school practice. Consider, for example, the idea that development of learning activity is the main educational objective to be realized for school children (e.g., Davydov et al., 1992/2003). Learning activity is a general capability, not tied to any particular subject-matter content, but which can be expressed in any particular subject matter content. In pursuing that developmental teaching ideal, one might prefer to transform the entire educational program for school children in order to realize the ideas of developmental teaching, on the assumption that a generalized learning activity will need to be used consistently across the entire educational program for a learner, not simply for a single subject, content area, or a topic. It is rare however for researchers to have both the institutional authority and the practical resources for making such comprehensive changes. More common is for researchers to select a particular topic within a particular discipline, with greater attention directed to pupils' learning of that content. This more common approach is the perspective adopted in this chapter, which may be considered a pragmatic solution in relation to the goals of developmental education, responding to limits on possible actions within existing institutional conditions, but better reflecting the possibilities for practical action in science education. Even with this limited focus, it is still possible and meaningful to engage with the general issues of developmental education, because each particular instance of educational activities, whether in science, history, or language, must still be oriented to the question of the relation between learning and development, such that particular cases of subject-matter teaching must still engage with the general goals of developmental teaching. In this way, teaching approaches focused on particular contents are useful for testing and advancing aspects of the theoretical perspective, while the theoretical perspective may serve to challenge how one conceptualizes particular teaching content and approaches.

3.3 Theoretical Thinking as an Ideal for School Science Teaching

Two issues should be clarified by the end of this section: (a) the meaning of theoretical thinking, and (b) reasons for considering this meaning as an ideal aim for science education. As will become apparent, theoretical thinking is understood in a way that is highly compatible with the central aim of helping pupils to think with disciplinary knowledge (i.e., the problem introduced at the beginning of the chapter). Supporting the development of theoretical thinking among school pupils serves, in effect, to developing thinking with disciplinary knowledge. This is the main reason for including a focus on theoretical thinking.

Before discussing the idea of theoretical thinking in particular, it is useful to (a) discuss what is meant by thinking more generally and (b) introduce theoretical considerations about the nature of knowledge. These clarifications will make it possible to then explain the meaning of theoretical thinking.

3.3.1 What Is Thinking?

The meaning or nature of thinking has been an object of philosophical and empirical study for thousands of years (e.g., Plato, Hegel, Dewey). The aim of the present discussion is to have a concept of thinking that is concrete enough to support practical and scientific work aimed at developing this capability among learners in relation to natural science. This concept may not satisfy those who are seeking a comprehensive or universal definition, according to some philosophical or theoretical criteria, but this is not a problem as long as the concept is sufficiently precise to support communication and motivate applications and investigations in relation to teaching and learning. In this perspective, the idea of thinking here might be considered a semi-scientific concept, drawing in part from everyday understandings and interests while moving toward a systematic, scientific concept.

The basic idea of thinking, of any kind, theoretical or not, "is nothing other than the ability to deal with each object intelligently—that is, in accordance with its own nature and not in conformity with one's fantasies about it. Thinking is really functioning knowledge" (Ilyenkov, 2007, p. 76). Ilyenkov's characterization of thinking provides an excellent way to concretize the meaning of "thinking with disciplinary knowledge in relation to meaningful and relevant problems and situations" (i.e., the central aim named at the beginning of the chapter). What would it mean to "deal with an object according to its own nature"? While the general idea seems clear enough, the practical problem (for the educator) is to be able to operationalize that idea in relation to the wide range of phenomena and situations typically found in science education.

Additional problems are embedded in this aim to help pupils to know objects according to their nature. Evald. V. Ilyenkov (2007) highlights these problems in his observation that: "it is impossible to "know" in general; it is possible only to know something in particular, this or that object, and truly knowing an object means being able to handle and understand it *independently*" (p. 76). In other words, the conception of thinking depends on knowing, and the conception of knowing depends on being able to act appropriately in relation to an object. In the context of science education, this appropriate action may take the form of explanations, predictions, empirical measurements, creative use or modification of natural processes, and so forth. These objectives should be readily familiar to science educators. They reflect the kinds of objectives found in the aims described in Table 3.1, and in the central problem addressed here about using disciplinary knowledge in relation to pupils' lifeworld. The reason to highlight this understanding of thinking and knowledge is that teaching approaches need a way to conceptualize what kind of knowledge is needed to enable thinking as described here.

Ilyenkov's definition or equation between knowledge and thinking does not acknowledge the existence of different kinds of knowledge. It was a tremendous accomplishment by Davydov (1972/1990) to outline the difference between empirical thinking and theoretical thinking. The basic or general idea of theoretical thinking is that all disciplinary areas in school subjects have a core of underlying relationships that, if mastered, can be used to reason critically about many situations, beyond those that are presented in teaching situations. The capability to think theoretically in relation to particular content areas reflects the central aim named at the beginning of the chapter and the ideal of thinking described by Ilyenkov. In contrast, empirical thinking (which unfortunately still characterizes much of primary and secondary school teaching around the world) is largely descriptive, based on appearances and classification of different properties of an object. For some purposes, this is a valuable and necessary form of knowledge, but empirical knowledge and empirical thinking is not adequate for realizing appropriate, independent action (including thinking) with phenomena and objects. The concept of theoretical thinking provides an ideal of science education, where the achievement of theoretical thinking by pupils will require that they learn appropriate action in relation to an object (i.e., think independently).

3.3.2 Substantive Generalizations and Theoretical Thinking

The hypothesis that Davydov (1972/1990, p. 339) pursued is that it is possible to develop instructional materials and teaching approaches that would enable pupils to develop substantive generalizations and concepts about the essential or central ideas in a problem area, where these relationships could be used to "ascend to the concrete" (i.e., to exemplify general conceptual relations through the use of particular concrete examples). The use of substantive generalizations in ascending to the concrete, provides the foundation for thinking (appropriate action with objects) and relating disciplinary knowledge to the lifeworld.

> Theoretical consciousness and thought is based on a substantive generalization. A person analyzes some evolving system of objects to then discover in it a principle that is generally prior, essential, or universal (a relationship). The identification and recording of this principle are a substantive generalization of the particular system. (Davydov, 1998, p. 25)

A brief clarification of the meaning of *substantive* is needed in this context, because there is no commonly-established word to express the intended meaning. There is always a substantive content when we reason about phenomena. For example, predicting the appearance of rain depends on an analysis of material relations between temperature, air pressure, amount of water vapor contained by air, and a solid on which water vapor can condense, where rain would be explained by interactions among these relations. The "logic" of the interactions in this case is a matter of the substance or content of these relations (e.g., water vapor needs a solid to condense on). One can speak about the substantive logic in this situation. There is not a formal logic. To reason about most phenomena, especially when encountering situations that one has not studied before, it is necessary to understand the underlying relations (substantive logic) in the problem area. A substantive generalization describes the content-based "logic" in a problem area, where these basic relations can be used to analyze the infinite number of concrete situations that reflect these underlying relations. In short, *substantive* refers to the concrete content in a problem area, while *generalization* is possible with the identified essential or central relations that characterize the logic of the problem area. Perhaps it is possible now to see the connection between the use of substantive generalization and the ideal of thinking—appropriate action with the true nature of an object—described by Ilyenkov. This use of substantive generalization is also a manifestation of theoretical thinking.

Consider the following (simple) example of a substantive generalization, which can be used to ascend to the concrete. An electromagnetic interaction occurs between charged bodies. This generalization can be used as part of an explanation for why some seeds "hang" inside a plastic bag, a charged balloon can "hang" on a wall, or a charged plastic rod can attract small bits of paper. The expression "ascending from the abstract to the concrete" is meant to convey the image that the "abstract" (i.e., the substantive generalization) can be used in many different concrete situations, where it is necessary to concretize the general relation in the abstraction to the particular situation being analyzed. In the present example, the abstraction is the idea of electromagnetic interaction, where ascending to the concrete would focus on explaining particular situations (e.g., rubbing a glass rod with silk will strip some electrons from the glass, leaving it with a positive charge, which can polarize small bits of paper – where the strength of the interaction is sufficient to move these small bits toward the rod).

Two aspects in this example must be highlighted. First, the idea is that pupils should develop a generalized capability to construct comparable explanations without intervention from the teacher for many other kinds of electrostatic situations (e.g., recognizing that the concrete case of getting a shock when touching a metal doorknob can be understood using the same general principle that explains the polarization of the rubbed glass rod), and not just to explain (or memorize) a particular case (e.g., why a charged glass rod attracts paper). This kind of generalized capability reflects "thinking" (as defined here). Second, this particular example – of a generalization and its application – is illustrative or representative of the more general hypothesis that appropriate substantive generalizations can be formulated and acquired for (more or less) all topical areas included in science curricula.

These two aspects—substantive generalizations and their use for all topics—are central features of the theory of developmental teaching. At the same time, these aspects provide a concrete way to operationalize the vision of supporting pupils to develop the capability to think independently, critically and creatively in relation to problems in everyday life. That is, in being able to move back and forth between substantive generalizations and concrete situations, pupils have intellectual tools that can used to interpret, explain, and analyze concrete situations. Similarly, by using generalizations, which in effect are models, it is also possible to generate hypotheses or predictions about observable phenomena. In short, the idea of theoretical thinking can be understood as a theoretical elaboration of the more general curricular goal that is expressed in everyday terms as "scientific thinking".

Models and theoretical thinking. Central to the issue of theoretical thinking is the idea of a theoretical model. A theoretical model is an abstraction and idealization of some real phenomenon, with a particular focus on identifying underlying universal (i.e., always present) relations that are involved in the observable appearance of a particular instance of the phenomenon. Theoretical thinking always involves some form of modeling because theoretical concepts are general relationships related to particular instances. Theoretical thinking is manifest both in the formulation of a model in relation to a particular situation (i.e., identification of central, underlying relations), and the modification of that model by various means (e.g., generating new hypotheses from implications of relationships in the model, interpreting new situations by rising to the concrete using relationships in the model).

A terminological clarification. In the Davydov quote about substantive generalization, the word *substantive* is used to translate the Russian word *содержательный* [soderzhatel'nii] that Davydov used. Sometimes this word is also translated as *content-oriented* (e.g., Davydov, 1972/1990, p. 282), though *content-oriented* was also used once for *существенный* [sushchestvennii] (p. 160), which is usually translated with words like *essential, significant*, or *substantial.* The important conceptual point here is that all these terms—essential, substantive, significant—can be understood as trying to convey the basic idea of underlying, essential relations that are always present in phenomena (i.e., as universal or general), where these general relations are used to understand or interpret particular individual examples of a general phenomenon. Each term, in its own way, is trying to convey the idea that understanding scientific phenomena involves learning the relationships that underlie the phenomenon (i.e., its nature), where the logic depends on the substance of the object, not formal relations.

That there are variations in how these words are translated only reflects, in my view, the lack of an established, standard theoretical term to refer to the idea of substantive generalization, even though the idea itself is clear enough. Ultimately, the term used to refer to this idea is probably less important than having a clear understanding of the issue being addressed here, namely, the ontological and epistemological assumption that all phenomena (i.e., individual, particular appearances) can be understood as the result of a process or development from underlying (essential) relationships. This is the key idea that motivates the orientation to theoretical thinking as an educational ideal. In effect, theoretical thinking involves both learning essential relationships and how to use these relations (ascending to the concrete – which can proceed in both directions, in that a particular instance could be interpreted in terms of essential relations or essential relations could be used to generate a particular instance) in working with concrete problem situations. This idea of underlying relations can be seen in the following statement by Davydov (1972/1990), where the word *essence* refers to the same thing as underlying relations: "essence is an internal connection, which, as a single source, as a genetic base, determines all of the other particular features of a whole" (p. 288). In effect, Davydov is explaining this ontological assumption that "the particular features of a whole" (i.e., the appearances of an individual case) are determined (i.e., can be understood as necessary consequences) of the underlying relations (i.e., the essence as an internal connection). In the same paragraph, Davydov goes on to note explicitly that a substantive abstraction is the essence that is used to interpret an object: "a genetically original, informal[1] abstraction expresses the essence of its concrete object" (p. 288).

Generality and availability of theoretical models. In principle theoretical models can be made for any kind of natural, social, or cultural phenomenon (e.g., Vygotsky's, 1971, discussion of art in terms of the uncertainty of outcomes in response to a conflict between opposing interests is an example of a theoretical model, where he abstracted from individual novels, stories, plays to describe a central theoretical relation). This general idea of a theoretical model can be applied to the natural sciences as well as the spheres of issues that might arise under a STEM perspective. (A proper discussion of the issues involved in assessing the validity and universality of this general idea goes beyond the scope of the present chapter.) While generating theoretical models can be a challenging and time-consuming task, it is assumed for now that it is possible, in general, to develop plausible theoretical models that can be used productively in educational settings. My experience of

[1] *Author's note*: poor translation choice in this source, should be *substantive*.

generating several models in the natural sciences—especially in physics and mathematics, but also in relation to socioscientific issues—leads me to believe that this assumption is realistic. (Therefore, it seems more productive to focus on the critical issue of generating useful models for specific topical areas, rather than waiting to resolve largely philosophical questions about the meaning of theoretical models. The value of the models will be confirmed in practice.)

Ideally the discussion in this section has given sufficient insight into the idea of theoretical thinking to support the idea that the aim of helping pupils to think critically and independently in relation to meaningful situations will depend critically on the availability of theoretical models in relation to topical areas of interest. The important contribution of the idea of theoretical thinking is that it gives a concrete idea for what educational activities should be aiming towards in order to realize the kinds of aims outlined in Table 3.1.

At present, it is usually necessary for educators to generate their own theoretical models for use in instructional situations, because this way of understanding the nature of knowledge is not commonly found in the natural sciences, or only appears implicitly on occasion, so it is unlikely that one can find adequate theoretical analyses within existing disciplinary or pedagogical sources. An important research task is to analyze disciplinary content in the natural sciences in a way that gives a platform for developing concrete teaching approaches that support the development of theoretical thinking. These analyses are needed to provide a way to use ideas from the developmental teaching tradition to engage with the kinds of curricular objectives noted in the beginning of this chapter.

3.4 Personality Development and Science Education

An important reason for paying attention to the concept of personality development in developmental teaching is the assumption of the integrated nature of intellectual development and personality development. Given that developmental education is motivated by the assumption that learning can be developmental, it is necessary to have some conception of development and the role of learning in relation to that development. In cultural-historical theory, development is understood in relation to personality development (Bozhovich, 1968/2009; Vygotsky, 1931/1997, pp. 242–243), which depends in part on the development of intellectual capabilities (i.e., reflecting the integrated nature of learning and personality development). In short, development is not equivalent to the idea that all accumulated factual and procedural knowledge is developmental. Rather it is necessary to focus on the development of intellectual capabilities that have consequences for personality development. The idea of personality development can be understood as ideal toward which all instruction should be developed.

The discussion that follows is concerned to illustrate how one can think about personality development in relation to science education. The intent for now is to convey a general feeling for the spirit of the idea of personality and personality development, without trying to give formal definitions or a systematic development of the underpinnings for the arguments about science education. The main points to be highlighted are:

1. Acquiring the ability to think theoretically in relation to natural scientific topics can and should contribute simultaneously to personality development.
2. Planning of teaching units should be designed and evaluated with attention to personality development.
3. Further critical discussion and elaboration of the meaning of personality development in relation to science education should be pursued.

3.4.1 Meaning of **Personality Development** *in Relation to Science Education*

A standard meaning of personality, as a theoretical concept in cultural-historical theory, is not well-established. A variety of different views have been proposed about its meaning (cf. Asmolov, 1984; Davydov, 1986/2008, pp. 61–65). Despite the variations among the cultural-historical proposals, all diverge significantly from popular and predominant scientific meanings of personality (which tend to focus on traits, temperaments, and often lean toward biologically-based explanations such as heredity, without giving much or any significance to socialization or enculturation).

Although the general meaning of the personality concept may still be under development within the cultural-historical tradition, it is still possible and meaningful to work with this concept in relation to science education. There are sufficiently well-developed ideas within this theoretical focus that are worth considering, and some pragmatic considerations should make it possible to navigate through or around some of the open questions in a general theory of personality development.

The general idea of personality and personality development in cultural-historical theory presented here, draws primarily on ideas from A. N. Leontiev (1975/1978) and Elkonin (1971/1999), which are interpreted in Seth Chaiklin (2001). Personality is a quality that an individual acquires through participating in social activity, including the demands and conditions of institutionalized life situations. More precisely, an individual's *personality* is reflected in their motives and personal sense and associated actions that manifest and further elaborate those motives. Motives and personal sense develop through participation in societal practices and, more precisely, through the development and control of psychological functions (e.g., Vygotsky, 1931/1998a, p. 171).

The discussion of motives cannot be expressed in a compact way because it involves an ontological shift, where motive is a relation between person and world, not a property of a person (see Chaiklin, 2012a, p. 212). For now, three important features of motives are highlighted in an informal way (i.e., using the theoretical concepts with minimal explanation of their meaning, while indicating the practical implications for working with science education).

1. The development of motives, which in turn are reflected as personality development, involves the development of new or elaboration of existing relationships between a person and societal practices. As a consequence, instructional activities should be considered in terms of what relationships to societal practices are being developed.
2. Motives are realized and expressed through action, which depends on the ability to act, which often includes an ability to analyze and generate (i.e., think) in concrete situations. In other words, the development of motives is dependent on intellectual development (e.g., ability to work with specific problems and natural scientific content in relation to particular problems and topics), which is why it is possible to consider their relationship.
3. Motives always have a personal sense (i.e., an individual's relation to their motive), which is relevant to personality development. A practical implication is the need to provide opportunities for pupils to formulate or express their personal sense in relation to the content with which they are working.

A fuller discussion of the personality concept would require a consideration of the concept of age period (cf. Chaiklin, 2019), and the meaning of development for school age children in relation to age periods. For now, it may be sufficient to work with an approximate or general understanding of personality development, which can be used for interpreting the significance of the intellectual content of teaching activities. For the sake of the argument, let us consider that science education in school is usually focused on persons from about 6 to 18 years-old, so it is sufficient to consider a cultural-historical personality concept in relation to these age periods, without having to resolve the wider range of questions and disputes that appear in relation to a general personality concept. Further, let us assume that for school-age children, the main developmental task is to learn to think theoretically in many different subject-matter areas. The development of these intellectual capabilities has broad consequences for a person's relationship to the world. This interactive relation where learning is understood in relation to personality development is a key characteristic for a developmental teaching approach. For practical and theoretical purposes, it is important to have a functionally useful conception of personality development, which provides a way to think about how to approach this interactive relation.

The primary aims of science education may still be focused on intellectual development, but the idea of personality development will influence the ways in which these intellectual goals are pursued (e.g., drawing on considerations about the significance of the learning in relation to a person's relationship to their lifeworld). Educators and researchers should consider possible consequences of this dynamic relationship for the design of teaching situations. Obviously, this pragmatic concept can be further refined and developed through attempts to use it in the conduct of educational experiments.

3.4.2 Role of Theoretical Thinking in Personality Development

Theoretical thinking involves the ability to creatively analyze problems and generate ideas for evaluation in relation to concrete situations. Learning to think theoretically gives the possibility for persons to engage critically and independently in relation to different problem situations in the lifeworld, which in turn usually involves changes in a person's relationship to the world, which is taken as evidence for personality development. This capability can be understood as an important foundation for the development of motives. The actions involved in thinking theoretically are also needed to realize many of the kinds of aims listed, for example, under the Australian Curriculum in Table 3.1. The main point here is that a focus on theoretical thinking in relation to personality development, which is an important issue for developmental teaching, is also directly relevant to addressing or realizing general curricular objectives in science education.

3.4.3 Should Schools Be Engaged in Personality Development?

Different countries have different traditions and policies about the ideological functions and objectives of schooling. For example, in Denmark, the idea of personal development has long been expressed explicitly in the first paragraph of the national school law, which states the "purpose" of the school (see https://eng.uvm.dk/primary-and-lower-secondary-education/the-folkeskole/the-aims-of-the-folkeskole for an English translation). Other countries may be more reluctant to give such responsibilities or objectives to schools. For the purposes of this chapter, and for researchers more generally, it does not matter whether a state schooling system has an objective for personality development, or not. Rather, the focus here is to understand more concretely what such an ideal might look like, and how one might approach it in relation to an actual classroom teaching practice. The elaboration of concrete ideas reflecting this ideal make it easier for various actors in educational practices to evaluate the appropriateness or possibilities to work with or adapt these ideas in relation to specific schooling systems. From examining or working with school curricula in many different countries (and continents), it is my impression that it is possible and culturally appropriate to work with the idea of personality development, even if it is not formulated in these terms in formal school policy. Of course, such evaluations must be made situationally, so this task must be left to individual practitioners and researchers to evaluate in relation to their concrete conditions.

3.4.4 *Need for Further Discussion*

While the call for "more research" is always relevant for all research areas, it seems particularly relevant in relation to this topic, in part because the topic is barely recognized among cultural-historical researchers and in part because it is difficult to elaborate a theoretical concept of personality development that is comprehensible and operational. In spite of these difficulties, it seems important to be able to confront the issue of personality development, because it is always present in teaching situations – whether conceptualized or not. In this connection, it is interesting to examine the aims listed in Table 3.1. When I read the list of aims for the Australian Curriculum, I get the impression of a long list of capabilities that pupils are supposed to acquire, but no sense of the pupil as a whole person. Why should pupils (from their point of view) have "an interest in science" or "an ability to solve problems and make informed, evidence-based decisions about current and future applications of science while taking into account ethical and social implications of decisions"?—to take just two of the aims that are listed. No doubt many interesting answers could be proposed, but one point being raised here is that the curricular document does not seem to give any indication of the need for pupils to find their own relationship to any of the long list of aims being proposed. A second point (or hypothesis) is that it will be necessary for pupils to have an opportunity to find their relationship to these aims (i.e., personal sense), which is part of developing motives that make these actions meaningful. This brief analytic discussion is meant to illustrate concretely how and why it can be useful to think about personality development in relation to intellectual development, where those considerations are likely to have immediate practical consequences for how educational activities are formulated and conducted.

3.5 Radical-Local Teaching and Learning

A main goal for radical-local teaching and learning is to support the development of creative, productive, knowledge-based thinking in relation to pupils' lifeworld (Hedegaard & Chaiklin, 2005). This goal is identical to the central problem for science education named at the beginning of this chapter—to help pupils and students learn to think with disciplinary knowledge in relation to meaningful and relevant problems and situations in their lifeworld.

The radical-local teaching and learning approach builds centrally on theoretical and practical ideas from Davydov (1986/2008), Mariane Hedegaard (1988), and Lada Aidarova (1982).The main extension in the radical-local approach is to consider the relation or interaction between the content of teaching and the learner's lifeworld explicitly and centrally.

The phrase *radical-local teaching and learning* refers to the idea that schooling should help pupils develop their theoretical thinking (in the sense outlined by Davydov). This is the meaning of "radical" in the label. The meaning of "local" refers to the idea that pupils are working with questions and problems that are meaningful to their lifeworld. Note that local does not necessarily mean that the content is always immediately visible or physically local. Often it requires the instructional planner to have a deep knowledge of the societal conditions and practices in which children are living, where the theoretical challenge is to understand what content is significant in the lifeworld of pupils. Note that significant here does not mean that the content is always immediately meaningful to the pupils from their initial personal perspective. Rather, the point is to draw on content that is significant to children's lifeworlds, and their development as persons in a particular historical context. Some "local" topics may catch pupils' attention immediately, but sometimes it may be necessary to help pupils to discover the "local" significance of the disciplinary content.

The designation *radical-local* is meant to emphasize the special feature or concern to interrelate disciplinary content (e.g., general intellectual concepts) with meaningful content for the learner, where both aspects are dominant. That is, rather than viewing teaching as a conflict between satisfying the needs or demands of disciplinary concepts on the one hand and satisfying the interests or needs of the pupil on the other, the idea is to work with subject-matter content in a way that is meaningful to learners, which at the same time involves working on tasks that are likely to result in learners mastering relevant disciplinary concepts.

The main aim of the radical-local approach to teaching and learning is to support personality development. This includes, for example, helping children acquire intellectual tools that allow them to engage with conditions and practices in their life situation. Theoretical thinking is particularly relevant here, because it gives a way to work with disciplinary concepts, where the particular cases for ascending to the concrete involve problems and situations that are meaningful to the learner. At the same time, this aim can be understood as reflecting an attempt to confront ideas of full human development (in terms of developing motive orientation, knowledge and skill) that support freedom[2] (as a capability to act competently in chosen situations that one needs to be in).

3.6 Electromagnetism in Lower-Secondary School

To this point, the chapter has discussed some general concepts within the theory of developmental teaching, highlighting their significance for designing concrete teaching approaches. Now, attention turns to discussing an example of subject-matter

[2] See Chaiklin (2012b, pp. 33–37) for a discussion about the significance of freedom in the cultural-historical tradition, particularly in the sense of self-realization of one's human potential. As Vygotsky noted: 'The cent[ral] problem of all psychology is freedom' (quoted in Zavershneva, 2010, p. 66). The point for the present discussion is that developing capabilities of thinking theoretically with disciplinary content can be understood as an important aspect in developing freedom.

analysis that is part of a radical-local approach to science teaching. The main objective is to present some concrete illustrations of general ideas presented so far—primarily for the purpose to show how these ideas might be operationalized. The examples are an early report from a work-in-progress about how to teach electromagnetism at the lower-secondary school level, and therefore should be considered primarily as hypotheses for further investigation. At the same time, if the presentation is sufficient to communicate how one might work with theoretical thinking and/or personality development, or inspires interest to explore these ideas further, then the presentation has served its primary purpose.

3.6.1 Personality Development

Why teach electromagnetic concepts in lower-secondary school? It is interesting to consider this question from the point of view of personality development. One of the significant features – in the historical development of scientific knowledge about electromagnetism – is the introduction of the field concept. As a quick indication of its profound significance, consider the following assessment:

> A new concept appears in physics, the most important invention since Newton's time: the field. It needed great scientific imagination to realize that it is not the charges nor the particles but the field in the space between the charges and the particles which is essential for the description of physical phenomena. (Einstein & Infeld, 1938, p. 259)
>
> This change in the conception of Reality is the most profound and the most fruitful that physics has experienced since the time of Newton. (Einstein, 1931, p. 71)

The electromagnetic field was the first example of a field concept in the history of physics. If part of personality development involves developing a deep appreciation of the physical world, then it would be meaningful, even essential, as part of learning about electromagnetism to learn about the electromagnetic field. Additionally, it would be relevant, at some point in the lower-secondary curriculum, to generalize the idea of the field concept in physics. (A Russian textbook has provided one example of relating the field concept between electromagnetic field and gravitational field, Lvovsky & Gruk, 2008.)

From the point of view of personality development, developing a generalized understanding of field as a concept may also have profound significance. A generalized concept of field introduces the idea of analyzing situations in terms of interaction, and the need to focus on the interaction rather than the individual components in the interaction. (Note that field concepts are used in psychology, e.g., Stivers & Wheelan, 1986, and sociology, e.g., Fligstein & McAdam, 2012). While there is no expectation that pupils can automatically or effortlessly generalize from an electromagnetic field to a generalized field concept, it is expected that working concretely with relational thinking in relation to electromagnetic interactions provides experience and a concrete example that can be used as an analogy or scaffolding for learning to think relationally in other situations, both physical and non-physical. If pupils acquire a generalized idea of field (e.g., becoming aware of and practicing a relational way of thinking as expressed in a field concept), which they are able to use far beyond electromagnetism, and not only within physics (e.g., as a model for thinking about social and societal interactions), then it may be appropriate to consider this capability and orientation as a genuine example of personality development, because of the transformations in their way to relate to their lifeworld.

3.6.2 Theoretical Thinking and Radical-Local Teaching and Learning

The discussion of theoretical thinking and radical-local teaching and learning are considered simultaneously because the analysis of theoretical thinking about electromagnetism should be done in a way that also facilitates the possibility for lower-secondary pupils to think about meaningful situations in their lifeworld. As noted before, it is necessary to develop to an analysis of the conceptual structure of electromagnetism, to provide a theoretical model of the underlying relations that are central for understanding this disciplinary content, and which provide abstract relations that can be concretized in ascending to the concrete.

The concept of field was mentioned in the previous section as a key concept to be introduced, not only because of its importance for understanding electromagnetism, but also because of its potential significance for personality development. An important technique for developing theoretical models is to make historical analysis of the development of disciplinary concepts. In the case of electromagnetism, the concept of field does not have a distinctive event that marks its appearance (for historical details, see, e.g., Berkson, 1974). The main point for now is that while historical analysis remains important, it may not always be sufficient for developing theoretical models. Other approaches may be needed. In the present case, the idea of electromagnetic interaction (between charged matter and the electromagnetic field) is proposed as the primary essential relation (i.e., germ cell) of electromagnetism. The focus on electromagnetic interaction provides a way to bring the field concept into the approach for thinking about electromagnetic situations. This idea can be readily accepted by physicists, who often point out that Maxwell's equations are sufficient for describing all electromagnetic fields and their interactions—at least in principle, while actual calculations in specific cases often prove to be too difficult for various reasons (Fano et al., 1960, p. 105; Jackson, 1999, p. 13). This is an interesting (and challenging) situation from the point of view of analyzing the content of electromagnetism for developing theoretical models. On the one hand, it may be possible to elaborate a system of relations that describe electromagnetic interactions, but it may be difficult or functionally impossible to use those analyses to reason about meaningful situations in one's lifeworld.

As a way to address this problem, I have formulated five intermediate categories that serve as a "bridge" between everyday life situations and physical concepts.

These categories are called cultural-historical, because they reflect actions and practices that are meaningful in human life. Some of the categories were formulated initially as a result of historical analysis (e.g., looking at the kinds of problems or questions that were being investigated by researchers particularly in the eighteenth and nineteenth centuries, when knowledge about electromagnetic processes was being developed). Subsequently, some of the categories were added or differentiated by surveying current applications that involve the use of electromagnetic processes.

The categories are transmission, energy, storage, information/signal, and detection. *Transmission* concerns the process and nature of movement of electromagnetic elements. *Energy* concerns the transformation of energy from non-electrical forms (e.g., mechanical, heat) into electrical forms and back to non-electrical forms. *Storage* concerns the storing of potential energy using electromagnetic properties. *Information/signal* concerns ways of using electromagnetic properties to encode information (e.g., in a telegraph, a hard disk, a television picture), while *detection* is concerned with detecting and interpreting the significance of electromagnetic patterns (in particular contexts).

These five categories can be understood as perspectives for conceptualizing electromagnetic systems, where more than one category might be used or needed in relation to any particular physical system. For example, some computer harddisks involve a rotating disc, where magnetized patterns on the disc are used to store information. From an energy perspective, one can describe how the disc is rotated using electrical energy; from an information perspective, one can understand how particular characters are stored on the disc, while from a detection perspective, one can consider how these discs can be read.

The categories are meant to be more cultural and functional than logical, reflecting the informal ways that humans think about phenomena, often with a mixture of everyday and scientific concepts. In reasoning about a particular situation, a category is chosen that reflects the main aspect of interest, while other necessary aspects are assumed implicitly or simply not considered. For example, an information perspective may consider how it is possible to use electromagnetic properties to represent the Morse code, while ignoring the technical aspects that might be of interest from a transmission perspective. The value of a category comes from providing a simple way to direct interest or attention in a general way to a particular significant aspect of a physical system, that one wants to explain or understand, without requiring the category to be formed from systematic physical concepts and principles.

Interesting and important features of the cultural-historical categories are that there are only five categories to learn, while they appear to be comprehensive, in the sense that more or less all practical situations that involve electromagnetism can be interpreted using these categories. Finally, it is possible to understand the physical basis for each of these categories by learning a few basic principles. This situation has profound implications because (a) it becomes possible to develop a conceptual understanding of electromagnetism that can be applied to any kind of situation in the lifeworld, but (b) where a fairly small number of physical principles are involved, which (c) should be achievable within the typical number of hours available for instruction in institutionalized educational systems.

By learning these five categories and understanding them in terms of underlying disciplinary concepts about electromagnetic interaction, a person has a way to approach thinking about any kind of situation that involves electromagnetism – reflecting the kinds of goals described in Table 3.1.

In terms of theoretical thinking, the introduction of cultural-historical categories as a bridge between the lifeworld and disciplinary concepts is an innovation. It does not reject or deny the original idea of conceptualizing knowledge in terms of essence (i.e., substantive abstractions) and appearance (i.e., substantive concretizations of these abstractions), but it opens up for the possibility that more complicated relationships may be necessary sometimes for relating disciplinary concepts to everyday situations. Unlike the original idea, where abstractions in the disciplinary concepts are related directly to appearance (i.e., ascending from the abstract to the concrete), the idea of abstraction is extended to the cultural-historical categories, which mediate between the underlying physical relationships (essences) and practical situations (appearances).

The cultural-historical categories may have another important function, in that they can support the ability to "see the whole before the parts", which Davydov (1992) highlights as an important feature of being able to work creatively in relation to practical situations. The ability to think creatively and productively often depends on being able to grasp situations in their totality before being able to differentiate the parts. The cultural-historical categories give a way to start this structuring process.

In relation to radical-local teaching and learning, using cultural-historical categories in educational activities makes it possible and easy to engage in any kind of situation that is meaningful in the pupils' lifeworld. Concrete, meaningful problems can be investigated by highlighting relevant cultural-historical categories, and their underlying disciplinary concepts, which in turn helps pupils to develop a general strategy for approaching all problems involving electromagnetism.

Consider, for example, a teaching unit that focuses on learning how electricity is available in the electrical socket where one lives (meaningful to lifeworld). Lower-secondary pupils are likely to be able to formulate central and relevant questions (e.g., what is the role of the power plant, how does electricity come from the power plant to the house), reflecting the cultural-historical categories of transmission and energy. In investigating these questions, there is an opportunity to introduce disciplinary concepts that concretize these cultural-historical categories with disciplinary concepts. This concrete investigation is focused on energy and transmission from a powerplant to an electrical socket, where substantive generalizations of energy and transmission could involve batteries and motors and other situations in which transmission and/or energy is involved.

This quick sketch is meant to show how a focus on theoretical thinking and the pupil's lifeworld makes it possible to develop an educational approach that can help pupils and students to learn to think with disciplinary knowledge in relation to meaningful and relevant problems and situations in their lifeworld. Of course, there are likely to be a variety of practical and technical challenges to realize this vision in actual teaching situations, but the main point for now is to highlight the value of this theoretical perspective for enabling researchers to develop a conceptual framework that has realistic possibilities to address a central problem in science education practice.

3.7 Concluding Comments

The primary intention of this chapter has been to draw attention to some important concepts in the developmental teaching tradition—theoretical thinking and personality development—that deserve more attention from researchers, and to point out the relevance of using a radical-local teaching and learning approach for engaging with these concepts. This intention is motivated by the belief that these ideas will be relevant for addressing central aims of science education, such as the development of critical thinking in relation to relevant problems, as well as other aims, such as listed in Table 3.1. This concluding discussion addresses some of the concerns or objections that have been raised against this approach to theoretical thinking.

3.7.1 Is It Important to Work with Germ Cells or Core Relations?

There are often ancillary questions that lie behind or motivate this question, including concerns that it will take too long or be too difficult to work with core relations. If it is going to be too difficult to find germ cells, then, wouldn't it be sufficient to simply work with different important topics, so that pupils learn content that is relevant? The critical issue that must be confronted in answering these questions is: What is your understanding of how pupils can develop a generative understanding of disciplinary concepts, so that they can reason independently and creatively with them. Will those capabilities develop from a selection of different topics—especially if their relations and implications are not transparent, as would be expected with core relations.

Notice that the object here is theoretical thinking, not simply learning theoretical concepts. The difference is that theoretical thinking involves the use of theoretical (or abstract) concepts, which means, among other things, rising to the concrete, by interpreting concrete situations in relation to the theoretical concept, considering the consequences of variations in the situation, and drawing implications based on the general relations.

A traditional[3] view about theoretical concepts is that if persons learn basic concepts in a content area, then they will be able to use or apply these concepts in thinking critically and productively in a problem area. Both systematic research and personal experience do not support this idea. Similarly, in relation to theoretical thinking, it is not sufficient to simply teach the theoretical abstractions and then expect that learners can apply these abstraction (e.g., rise to the concrete) once the abstractions are learned. It is necessary to learn theoretical abstractions in a way that also enables theoretical thinking. Often, in teaching experiments, these two aspects—theoretical concepts and theoretical thinking—are being developed at the same time.

3.7.2 Is it Possible to Find Germ Cells or Core Relations in All Content Areas?

In principle it should be possible to find core relations in all content areas. The reason for this expectation is based on the assumption that all observable phenomena reflect some underlying relationships, where variations in these relations may result in different surface appearances, or different surface appearances can be understood in relation to these underlying relations. In other words, content areas are not random or arbitrary lists of facts and descriptions, but reflect systematic relations.

The previous paragraph started with an "in principle" because there is no guarantee that one is able or capable of finding these relations, but in my experience, it is usually possible within a few hours to start to get useful ideas that can be explored and developed further. The reasonable likelihood that analyses can be made for any content area is important because it removes a critical barrier or objection to working along these lines.

References

Aidarova, L. (1982). *Child development and education* (L. Lezhneva, Trans.). Progress.

Asmolov, A. G. (1984). The subject matter of the psychology of personality. *Soviet Psychology, 22*(4), 23–43. https://doi.org/10.2753/rpo1061-0405220423

Australian Curriculum. (2018). *F-10 Curriculum. science. Aims.* https://australiancurriculum.edu.au/f-10-curriculum/science/aims/

Berkson, W. (1974). *Fields of force: The development of a world view from Faraday to Einstein.* Routledge and Kegan. https://doi.org/10.4324/9781315779386

[3] The use of *traditional* in this context is meant to refer to common or widespread beliefs about how critical thinking is developed. It is likely to be referring to unexamined, taken-for-granted beliefs or expectations, rather than a matter of referring to a specific theoretical perspective or educational philosophy.

Bozhovich, L. I. (2009). The struggle for concrete psychology and the integrated study of personality. *Journal of Russian & East European Psychology, 47*(4), 28–58. (Original work published 1968). https://doi.org/10.2753/RPO1061-0405470402

Chaiklin, S. (2001). The category of 'personality' in cultural-historical psychology. In S. Chaiklin (Ed.), *The theory and practice of cultural-historical psychology* (pp. 238–259). Aarhus University Press.

Chaiklin, S. (2012a). A conceptual perspective for investigating motive in cultural-historical theory. In M. Hedegaard, A. Edwards, & M. Fleer (Eds.), *Motives in children's development: Cultural-historical approaches* (pp. 209–224). Cambridge University Press. https://doi.org/10.1017/CBO9781139049474.016

Chaiklin, S. (2012b). Dialectics, politics and contemporary cultural-historical research, exemplified through Marx and Vygotsky. In H. Daniels (Ed.), *Vygotsky and sociology* (pp. 24–43). Routledge. https://doi.org/10.4324/9780203112991-8

Chaiklin, S. (2019). Age as a historical materialist concept in cultural-historical theory of human development. *Obutchénie, 3*(3), 10.14393/obv3n3.a2019-51707.

Davydov, V. V. (1988a). Problems of developmental teaching. *Soviet Education, 30*(8), 6–97. (Original work published 1986). https://doi.org/10.2753/RES1060-939330086

Davydov, V. V. (1988b). Problems of developmental teaching. *Soviet Education, 30*(9), 3–83. (Original work published 1986). https://doi.org/10.2753/RES1060-939330093

Davydov, V. V. (1988c). Problems of developmental teaching. *Soviet Education, 30*(10), 3–41. (Original work published 1986). https://doi.org/10.2753/RES1060-939330103

Davydov, V. V. (1990). *Types of generalization in instruction: Logical and psychological problems in the structuring of school curricula* (Soviet studies in mathematics education, Vol. 2) (J. Kilpatrick, Ed.; J. Teller, Trans.). National Council of Teachers of Mathematics. (Original work published 1972).

Davydov, V. V. (1992). Genezis i razvitiye lichnosti v detskom vozraste [Genesis and development of personality in childhood]. *Voprosy Psikhologii, (1–2)*, 22–33.

Davydov, V. V. (1998). The concept of developmental teaching. *Journal of Russian & East European Psychology, 36*(4), 11–36. https://doi.org/10.2753/rpo1061-0405360411

Davydov, V. V. (2008). *Problems of developmental instruction: A theoretical and experimental psychological study* (P. Moxhay, Trans.). Nova Science. (Original work published 1986).

Davydov, V. V., Slobodchikov, V. I., & Tsukerman, G. A. (2003). The elementary school student as an agent of learning activity. *Journal of Russian & East European Psychology, 41*(5), 63–76. (Original work published 1992). https://doi.org/10.2753/RPO1061-0405410563

Einstein, A. (1931). Maxwell's influence on the development of the conception of physical reality. In *James Clerk Maxwell: A commemoration volume, 1831–1931* (pp. 66–73). Cambridge University Press.

Einstein, A., & Infeld, L. (1938). *The evolution of physics: The growth of ideas from the early concepts to relativity and quanta*. Cambridge University Press.

El'konin, D. B. (1999). Toward the problem of stages in the mental development of children. *Journal of Russian & East European Psychology, 37*(6), 11–30. (Original work published 1971). https://doi.org/10.2753/RPO1061-0405370611

Fano, R. M., Chu, L. J., & Adler, R. B. (1960). *Electromagnetic fields, energy and forces*. MIT Press.

Finnish National Board of Education. (2004). *National core curriculum for basic education 2004*. Author.

Fligstein, N., & McAdam, D. (2012). *A theory of fields*. Oxford University Press. https://doi.org/10.1093/acprof:oso/9780199859948.001.0001

Gordeeva, T. O. (2020). Когнитивные и образовательные эффекты системы развивающего обучения Д.Б. Эльконина—В.В. Давыдова: возможности и ограничения Портал психологических изданий [Cognitive and educational effects of the Elkonin—Davydov System of developmental education: Opportunities and limitations]. *Kul'turno-istoricheskaia psikhologiia, 16*(4), 14–25. https://doi.org/10.17759/chp.2020160402

Hedegaard, M. (1988). *Skolebørns personlighedsudvikling set gennem orienteringsfagene* [The development of schoolchildren's personality viewed through the social science subjects]. Aarhus Universitetsforlag.
Hedegaard, M., & Chaiklin, S. (2005). *Radical-local teaching and learning: A cultural-historical approach*. Aarhus University Press.
Ilyenkov, E. V. (2007). Knowledge and thinking. *Journal of Russian & East European Psychology, 45*(4), 75–80. (Original work published 2002). https://doi.org/10.2753/rpo1061-0405450407
Jackson, J. D. (1999). *Classical electrodynamics* (3rd ed.). Wiley.
Leontiev, A. N. (1978). *Activity, consciousness, and personality* (M. J. Hall, Trans.). Prentice-Hall (Original work published 1975).
Lvovsky V. A., & Gruk V. Y. (2008). *Fizika v sisteme D.B. El'konina – V.V. Davydova. 8 kl* [Physics in the system of D.B. Elkonin – V.V. Davydov Grade 8]. 1C.
Mann, C. R. (1912). *The teaching of physics for purposes of general education*. Macmillan.
Mann, C. R., & Twiss, R. G. (1905). *Physics*. Scott, Foresman.
Moxhay, P. (2008). Translator's introduction. In V. V. Davydov, *Problems of developmental instruction: A theoretical and experimental psychological study* (pp. 5–9). Nova Science.
Puentes, R. V., Coelho Cardoso, C. G., Prudente Amorim, P. A., & Musiychuk, M. V. (2018). Elkonin-Davidov system: Historical aspects (1958-2015). *Гуманитарно-педагогические исследования, 2*(2), 6–13.
Repkin, V. V. (2003). From the history of research into the problems of developmental teaching in Kharkov. *Journal of Russian and East European Psychology, 41*(5), 77–96. https://doi.org/10.2753/rpo1061-0405410577
Stivers, E., & Wheelan, S. (Eds.). (1986). *The Lewin legacy: Field theory in current practice*. Springer. https://doi.org/10.1007/978-1-4615-8030-0
Vygotsky, L. S. (1971). *The psychology of art*. MIT Press.
Vygotsky, L. S. (1987). Thinking and speech (N. Minick, Trans.). In R. W. Rieber & A. S. Carton (Eds.), *The collected works of L. S. Vygotsky: Vol. 1: Problems of general psychology* (pp. 43–285). Plenum. (Original work published 1934). https://doi.org/10.1007/978-1-4613-1655-8_4
Vygotsky, L. S. (1997). Conclusion; further research; development of personality and world view in the child (M. J. Hall, Trans.). In R. W. Rieber (Ed.), *The collected works of L. S. Vygotsky: Vol. 4: The history of the development of higher mental functions* (pp. 241–251). Plenum (Original work written 1931). https://doi.org/10.1007/978-1-4615-5939-9_15
Vygotsky, L. S. (1998a). Dynamics and structure of the adolescent's personality (M. J. Hall, Trans.). In R. W. Rieber (Ed.), *The collected works of L. S. Vygotsky: Vol. 5. Child psychology* (pp. 167–184). Plenum. (Original work published 1931). https://doi.org/10.1007/978-1-4615-5401-1_5
Vygotsky, L. S. (1998b). The problem of age (M. J. Hall, Trans.). In R. W. Rieber (Ed.), *The collected works of L. S. Vygotsky: Vol. 5. Child psychology* (pp. 187–205). Plenum (Original work published ca. 1933). https://doi.org/10.1007/978-1-4615-5401-1_6
Zavershneva, E. I. (2010). The way to freedom. *Journal of Russian and East European Psychology, 48*(1), 61–90. https://doi.org/10.2753/RPO1061-0405480103

Seth Chaiklin Research interests are dialectical approaches to subject-matter teaching and learning, conceptual and methodological issues in cultural-historical psychology, theoretical approaches to analyzing practice, and interventions for developing practice. Most recent publication is: Can there be multicultural science education policy in a country that does not recognize multicultural science education? The case of Denmark and the *folkeskole*. In *International Handbook of Research on Multicultural Science Education* (2022, Springer).

Part II
Early Years Science Education from a Cultural Historical Perspective

Chapter 4
A Cultural-Historical Study of Teacher Development: How Early Childhood Teachers Meet the Demands of a Theoretical Problem in STEM for Practice Change

Marilyn Fleer

4.1 Introduction

This chapter contributes to answering the fundamental question of what is developing for the early childhood teacher when engaged in professional practice change in support of STEM teaching. It builds on previous research (Fleer, 2021) by focusing on how the method of engagement in research – an educational experiment – creates developmental conditions for teachers. In drawing upon cultural-historical theory, this chapter also examines previous studies and theoretical papers on teacher development in order to better understand what might be the suite of concepts needed for explaining observed early childhood teacher development during the process of participating in an educational experiment of introducing STEM teaching into play-based settings.

Because Vygotsky's system of concepts was oriented to child development, new rules for relationally bringing existing concepts together are needed. But also new concepts are needed to fill gaps. Therefore, this chapter begins with what is known about a cultural-historical conception of teacher development, followed by a study of early childhood teacher development over 2 years, where a new conceptual framework is introduced to give directions for explaining what might be the source and content of early childhood teacher development when engaged in STEM teaching.

M. Fleer (✉)
Conceptual PlayLab, School of Educational Psychology and Counselling, Faculty of Education, Monash University, Melbourne, VIC, Australia
e-mail: Marilyn.fleer@monash.edu

K. Plakitsi, S. Barma (eds.), *Sociocultural Approaches to STEM Education*, Sociocultural Explorations of Science Education 21,
https://doi.org/10.1007/978-3-031-44377-0_4

4.2 A Cultural-Historical Conceptualisation of Teacher Development

In order to achieve the goal of this chapter, consideration is given to bringing forward what is known about teacher development from a cultural-historical perspective. Five core ideas are discussed.

First, Barohny Eun (2008) has insightfully argued that because Vygotsky's theory was on child development, researchers studying professional development (PD) need to proceed with caution. For instance, Eun stated that, "the applications of his theories to professional development, which lies in the realm of adult learning, may seem to be stretching the scope of theoretical implications" (Eun, 2008, p. 151). It is suggested by Eun (2008) that when an analysis of the core concepts of child development is considered alongside of the foundational principle of studying the process rather than the product of development, "the process inherent in learning and development is essentially the same for adults and children alike" (Eun, 2008, p. 151). But herein lies a theoretical contradiction. Vygotsky himself wrote, "pedology is the science of the child in development and not that of the person's development to the end of life (reproduced xv, Vygotsky, 2019; David Kellogg and Nikolai Veresov). In stating this he drew attention to the point that one cannot simply place the child and the adult on the same psychological trajectory. This key theoretical point is worthy of elaboration, and is reproduced here. Vygotsky said,

> I think that those who wish to extend pedology from the cradle to the grave, those who wish to put on the same plane of development that of the child and that development which adults go through, are, without realizing it, doing the same thing as the authors of antiquity who affirmed that the child is only a midget adult, that is to say, they deny the qualitative uniqueness of the processes of development of the child is comparison to the processes and changes that are produced in a relatively stable situation (reproduced xv, Vygotsky, 2019; Kellogg and Veresov).

Vygotsky's point is that the child is not a miniature adult, and therefore an adult's development cannot be conceptualised as though it is the full and complete form of what it means to be human simply because the person is no longer a child. If Vygotsky himself identified this as problematic, the challenge that arises for researchers is how can we study and theorise teacher development from a cultural-historical perspective when engaged in STEM teaching?

Second, researchers need a way of conceptualising development of a teacher over time. In Lev Semenovich Vygotsky's (1998) conception of periodisation he discussed how there are periods of development where crises give new conditions for development, which is evidenced when children's motive orientation changes over their life course. But Vygotsky did not go beyond children in his categorisation of motives to discuss adult development. In using the metaphor of a passport, Vygotsky (2019) said that it is "not the passport age of the child, but his [sic] pedalogical age" (Vygotsky, 2019, p. 6) that we should be considering in understanding children's development. This is because in Vygotsky's writing, he framed development in terms of 'cultural age periods' and not 'biological age' periods.

Could the idea of 'pedalogical age' as part of conceptualising adult age periods, foreground the significance of the institutional practices in which a person is located? Further, could 'cultural age period' be associated with the institution, such as the university, the apprenticeships or professional practice or work settings? Thereby in extending this idea to the development of adults, the cultural age could be tied to different institutional pathways. This opens up a conceptual point on what might be the institutional contexts that act as a source of development for adults, and additionally, what might be the motives that develop for teachers engaged in professional practice in STEM? Vygotsky offers some guidance when he said,

> I do not think that adults do not develop, but I believe that they develop according to other rules and that for this development there are other characteristics lines than those of the child; (reproduced xv, Vygotsky, 2019; Kellogg and Veresov).

But what might be these rules and characteristic lines of development for teachers? Vygotsky did not offer any suggestions on this but did introduce the concept of transformation (see Grimmett, 2014; Vygotsky, 2019). Taken together, we could ask what then are the motives that emerge and change during adult transformation and what might be the institutional contexts that act as a source of teacher development?

Third, there are studies that have grappled with the problem of using cultural-historical concepts when studying teacher development. Of significance is the research of Anne Edwards et al. (2019) who said, "we have refashioned the concept of the social situation of development to help us with the task of understanding teachers professional learning" (p. 212). Like Eun (2008), Edwards et al. (2019) are cautious in their use of Vygotskian concepts when studying teacher development. What they point to is the gap in concepts. In Table 4.1 below is summarised those studies/theoretical works where cultural-historical concepts (column 2) and how they have been used to theorise teacher development (column 3) are presented as a useful beginning point for the study of early childhood teacher development. The selection of papers for Table 4.1 was based on articles that broadly used cultural-historical/activity theory/sociocultural concepts and were oriented to teacher PD (in-service primarily) and are indicative of what is in the literature.

The studies presented in Table 4.1 draw on cultural-historical theory to either theorise or research teacher development, such as Edwards et al. (2019), Murphy et al. (2015) and Grimmett (2014). But like Vygotsky, most struggle to illuminate what might be the concepts and the rules needed for explaining the lines of teacher development. For instance, Eun (2008) sought to give an "overview of Vygotsky's theories of development, emphasizing the aspects of the theory that have direct relevance for professional development" (p. 135). He sought to ground PD "models within the sociocultural developmental theories of Vygotsky, in order to understand the mechanism underlying the process of teacher development" (p. 135) and in his later work he gave "a firm theoretical foundation in which to ground professional development" (Eun, 2011, p. 320). But he did not do empirical work to take this forward. However, Grimmett (2014) did, and she makes a compelling case for Vygotsky's system of concepts, but goes one step further by bringing the concepts together as a model of PD that theorises the outcomes of her model of WITHIN

Table 4.1 Cultural-historical conceptions of teacher development generally

Researchers	Cultural-historical concept/terms as named and deployed by the researcher	Claims theorised in relation to cultural-historical concepts
Tasker et al. (2010)	Vygotsky's dialogic process of external artifacts into internal representations	Inquiry-based
Shabani (2016)	Vygotsky's social origin of mental functions, unity of behaviour and consciousness, mediation, and psychological systems; learning precedes development; zone of proximal development (ZPD); social situation of development and everyday and scientific concepts	Teacher development includes *cognitive*, *affective*, *social* and *contextual* dimensions
Eun (2008)	Vygotsky's social origin of mental functions; unity of behaviour and consciousness; mediation; psychological systems, serves as evidence of development	Psychological systems that focus on changing attitudes and instructional practices
Eun (2011)	Vygotsky's ZPD; mediation; inter and intra-psychological functioning	Cultural tools in mediating instructional interactions; development; theory-into-practice and practice-into-theory dynamic
Shi (2017)	Vygotsky's ZPD; mediation; activity theory	Micro-structure of the person's context and the macro-structure of the sociocultural model
Ebadi and Gheisari (2016)	Vygotsky's internal mediation, everyday and scientific teaching concepts and Lave and Wenger's peripheral participation of teachers	Classroom discourse and moment-to-moment complexities of teaching
Grimmett (2014)	Vygotsky's general genetic law of cultural development, including, crisis, ZPD, the social situation of development, intersubjectivity, perezhivanie and obshchenie, agency, imagination, motives, and concept development Mariane Hedegaard's concept of the double move	Motive 1: To merely attend the PD Motive 2: To change practice Motive 3: To develop as a professional
Brown and Mowry (2017)	Primarily Hedegaard's concept of double move	Double move – personal experience and theoretical sense-making; teacher inquiry in groups for future visions and historical imagination of practices
Murphy et al. (2015)	Vygotsky's dramatic collision, zone of proximal development, ideal and real form, imitation, unity of affect and intellect, regression/recursion	Coplanning, coreflection, copractice and coteaching
Edwards et al. (2019)	Vygotsky's social situation and social situation of development; motives; agency; dialectics between person and practice; Hedegaard's societal, institutional and personal perspectives, where demands and motives within activity settings are part of the practices of institutions.	SSD was realised through how teachers positioned themselves; motive orientation, demands of practice
Ellis (2007)	Lave and Wenger's conception of communities of practice, Shulman's typology and personal constructs	Social situation of teachers' subject knowledge; collaborative professional enquiry to understand and to transform subject knowledge

practice PD. She studied teacher transformation in one school setting and deployed most of Vygotsky's concepts on child development, in order to explain teacher motive development for practice change.

What can be observed across those theoretical works shown in Table 4.1 is a broad set of concepts: mediation, inter- and intra-psychological functioning, social situation of development (SSD), ideal and real form of development, imitation, everyday and scientific concepts, and the zone of proximal development (ZPD). These concepts are directly related to Vygotsky's system of concepts for theorising child development.

Fourth, if we drill down into those papers summarised in Table 4.1 we can identify some studies that advance new concepts to locate teacher development within the institutional practice traditions of schools. For instance, in empirical research Edwards et al. (2019) examined both pre- and in-service teacher development. In drawing upon the concepts of motives and demands (Marianne Hedegaard, 2012) and the institutional practices of secondary schools, they brought into focus the Vygotskian concept of the social situation of development (Vygotsky, 1998) to theorise teacher agency and teacher motive orientation when in challenging situations as in-service teachers or during professional placement of pre-service teachers. In examining the recurrent demands in practice, Edwards et al. (2019) reveal the dialectical formation of teachers as professionals, and in so doing, bring out some important differences between professional learning and development. They argue that when considering teacher SSD there is a tension that is held constant between the dialectical relation of agency and demands. This is showcased through how the teachers meet the demands of the profession (over controlling) or where no demands are made on them (lack of interest). They suggest that the teachers' agency is in response to the recurrent demands, which in turn creates a developmental niche located within the institutional practices of the schools. Edwards et al. (2019) nicely show how cultural-historical concept of SSD and Hedegaardian conception of motives and demands within institutional practices, can reveal teacher development. The motive orientation of the teachers is aligned with the practice tradition, and the content of development appears to be expressed as teacher agency within the recurrent demands of professional practice. This sophisticated theorisation of teacher development goes beyond teacher learning of practice and introduces a suite of concepts for understanding the nature and content of teacher development. They take forward Vygotsky's conception of adult transformation and advance new rules that are located in the practice tradition of the school. However, secondary schools are different to preschools, because preschools have play-based programs that give different kinds of social situations and practice traditions. The institutional practices are different and therefore the concepts and practices that are held in tension within the SSD may also be different.

Related to Edwards et al. (2019) study is that of Ellis (2007) who also theorised teacher development. Ellis was interested in the problem of professional knowledge that is often formulated as recipes in a context of subject matter knowledge. He specifically shows how complexity and dynamic social situation of teachers' subject knowledge is accessed and developed through culture, practice and agents. It is more than a recipe of action. In collaborative professional enquiry between school subject departments, interns and a university-based teacher educator, they sought to understand and to transform subject knowledge. In line with Hedegaard's (Edwards et al., 2019) work, Ellis (2007) illuminates the institutional demands of practice found in secondary contexts, as well as identifying the motives of the teachers who enter into the practice tradition of secondary subject matter knowledge. But all these studies are located within different contexts to that of preschools where teachers' motives and the expected institutional practices will be different.

Unsurprisingly, the literature on cultural-historical conceptions of teacher development appears to be anchored in the practice traditions in which teacher development is taking place. But under what conditions do we see adult transformation in play-based settings? Are there key psychological functions developing for teacher in the different social situations of their practices? Can we study teachers' SSD as suggested by Edwards et al. (2019)? Can we conceptualise professional development and teachers' practice context as a relation between the ideal and real form of teacher development as suggested by Grimmett (2014)? Is teacher professional development located in the knowledges of the different discipline areas as proposed by Ellis (2007)? What might be the unique professional practice conditions that support teacher development of early childhood teachers?

Fifth, in agreement with Eun (2011), is that Vygotsky's system of concepts gives a very good starting point from which to theorise teacher development, and from which to draw concepts for the analysis of teacher transformation in context of PD. As stated so eloquently by Eun (2011), "Vygotsky's theory of development has provided a fertile ground to explore the mechanism of teacher development as this theory pays special attention to the role that culture and its tools play in human interactions" (p. 330). However, given the unique institutional practices of early childhood (social situation), what have cultural-historical studies of teacher development identified? In Table 4.2 are the studies (Column 1), the cultural-historical concepts used (Column 2) and how they have been drawn upon or theorised in relation to the early childhood PD (Column 3).

What we know from this summary of research papers shown in Table 4.2, is a recognition that "Amidst debates about the nature of professional development in early childhood education, the most neglected aspect seems to be what is meant by 'development' (Nuttall et al., 2015, p. 225). This is in keeping with Vygotsky's (2019) claim that teacher transformation needs different rules to that of studying children's development. Directly relevant to the focus of this paper, Nuttall et al. (2015) noted contradiction, individual-collective dynamic, and motive object, as core theoretical foundations for the study of early childhood teachers' development. The contradictions pointed to the object motive in relation to the content of the PD. As part of conceptualising new practices in the context of the motivate for being involved in PD, Nuttall et al. (2015) identified that when using a cultural tool to create a contradiction during PD that this supported consciousness-raising and problem solving for positive teacher practice change. This work gives renewed emphasis to motives but does so by identifying motivating conditions in relation to contradictions in the beliefs and practices of the early childhood profession.

Table 4.2 Cultural-historical conceptions of early childhood teacher development

Researchers	Cultural-historical concept/terms as named and deployed by the researcher	Claims theorised in relation to cultural-historical concepts
Nuttall (2013)	Engestrom's developmental work research (DWR) with the relational dimensions of contradiction; cultural tools; division of labour; rules; motives, etc	PD is collective, situated, historically accumulating, and multi-vocal; Contradiction between object of effective teaching and division of labour
Nuttall et al. (2015)	Engestrom's object motives of teachers	Motive objects Development: teacher development cannot be thought of as an individual phenomenon
Johnston et al. (2020)	Apprenticeship and guided participation as presented by Barbara Rogoff	Practitioner inquiry utilising Rogoff's planes of analysis, with concept of intersubjectivity
Caudle (2013)	Communities of practice as described by Jean Lave and Etienne Wenger	Community of inquiry

Therefore, teacher motives and contradictions appear to be important concepts for understanding early childhood teacher development.

Taken together, we believe that the literature presented in this section gives some direction for the theoretical problem of what is developing during PD of the early childhood teachers. But to answer this more completely, we need to identify what concepts and the new system of rules that are needed to explain early childhood teacher development. To contribute to filling this theoretical gap, we investigated *the process of teacher development* where like Edwards et al. (2019), the *social situation* of the unique *institutional practices* of early childhood education conditions were foregrounded, and where the concept of crisis and a change in motives, were used to point to the moments of possible *teacher development.*

4.3 Study Design: An *Educational Experiment*

An educational experiment is a collaboration between researchers and teachers on a theoretical problem, rather than just a problem of practice (Marianne Hedegaard, 2008). The researchers and teachers worked over four period on the theoretical problem of how to bring concepts into children's play.

The research questions that drove the study were: What is developing when early childhood teachers participate in practice change in a context of STEM teaching? What is the content of their development and under what conditions does development take place?

4.3.1 Participants

Rather than a problem of practice dealt with by teachers engaging in PD, this study sought to bring teachers and researchers together to work on a theoretical problem of how to bring into play-based settings authentic STEM concepts and problems that children solve through play. The two teachers (Ruth and Olivia) who participated in educational experiment each had a 4-year university degree qualification and had over 10 years of teaching experience. Two classrooms – preschool (4 years old) and the first year of school (5 years old) – came together twice a week for each of the playworlds that were implemented (a series of 5 playworlds, which later in research was theorised as a Conceptual PlayWorld).

The researchers supported the teachers each week in Period 1 (6 months), then the teachers worked independent of the researchers in Period 2 (6 months), and then the researchers collaborated with the teachers weekly for Period 3 (12 months). In Period 4 the teachers participated in leading PD programs on a Conceptual PlayWorld (see Fleer, 2021).

A Conceptual PlayWorld is a model of practice that begins with the reading of a story (teachers select a book which has contradiction and drama). The theme of the story is used to change an area in the preschool (inside or outside) into an imaginary space, such as, the fair ground in the story of Charlotte's Web or Sherwood Castle and Forest in the legend of Robin Hood. Children and teachers enter the Conceptual PlayWorld and imagine being characters from the story, reliving the adventures of the story, but also meeting problems that teachers introduce (a letter is sent from a character in the story) which need solutions by researching and using STEM concepts. The characters ask the children for help and the children solve the problem in the imaginary situation – such as, using google earth to prepare a map and escape route, meeting the castle engineer back in time to learn about pulleys and drawbridges, and in turn using what they learn to continue their play in the Conceptual PlayWorld.

The researchers supported the teachers weekly planning and evaluation by suggesting ideas and asking questions, provided PD on Gunilla Lindqvist's (1995) playworld model in Period 1, shared information about Vygotskian concepts, responded to email queries, and were given digital devices and resources to support program development throughout the 2 years. The focus of the collaboration was on the theoretical problem, and as mentioned previously, this was how to introduce into children's play conceptual learning that is authentic to children, and which helps them to keep their play going. That is, the theoretical problem is how concepts act in service of children's play. The conceptual problem to be learned is different to the theoretical problem. For example, the concept of design is shown through drawing a design for the escape route in a plan view perspective or exploring the concept of Force when testing the chain's strength in a drawbridge, or understanding pulley systems, gears, load, and a fulcrum when modelling how to bring down the drawbridge in their play.

4.3.2 *Data Collection*

The process of data collection involved both digital video recordings of teacher practices when team teaching (152 hours), digital recording of weekly reflections and planning (32 hours), copying planning documentation, emails between researchers and teachers, photographs of displays/practices, children's designs, drawings, and products, and an interviews over the 2 years (Period 2: 2 hours), at the end of the 2 years of the educational experiment (Periods 3: 1 hour), and again 2 years later (Period 4: 1 hour).

4.3.3 *Analysis*

The study looked at how the teachers entered into, were shaped by, and also shaped the activity settings as part of the educational experiment. In order to determine how the teachers met the new demands during points of crisis, the concepts of motives/motivation (Hedegaard, 2012), and social situation/social situation of development (Vygotsky, 1994) were used to support the analysis of the collective practices and participation of teachers in the educational experiment.

4.4 Study Findings

The study sought to understand what was developing for early childhood teachers when engaged with the central theoretical problem of how to bring conceptual learning of STEM into children's play.

4.4.1 *Working on a Theoretical Problem Created New Psychological Conditions for Teachers*

Whilst the literature showcases processes, such as, mentoring (Trevethan & Sandretto, 2017), inquiries (Johnston et al., 2020) and the use of an outside expert in relation to problems of practice (Hadley et al., 2015), it does not bring out educators' engagement in a theoretical problem. Predictably, the theoretical problem began in relation to teacher practice of implementing a playworld where STEM concepts could be intentionally taught to children as part of their play.

In Period 1, it was found that the demands placed on the teacher as they entered into the new STEM practices, brought forward a level of conceptual engagement with the children. In the second period this came out more explicitly as the teachers sought to build conceptual knowledge through supporting children to create a relational model of the concepts they were learning.

> . . .we introduced the idea of the diorama. It's not something I had introduced to children before, and they just loved it. . . . it was an opportunity to show how all of the concepts interrelate [in the playworld]. So, they understood. They made the forest, the floor, nocturnal, endangered animals, solar system. They put in the threats – wood cutter, fire. They wrote messages for their families about re-cycling. . .in terms of developing their own model of a particular eco-system (Ruth).

During Period 2, the teachers created new conditions for the children. The teachers used the structure of a diorama to represent in miniature form the relational concepts within an ecosystem – and through this they were able to show how changing one part of the system, as 'a threat', changed the other parts of the system. Interestingly. In Period 3 the teachers' theoretical problem of how to bring STEM concepts into children's play, meant that the teachers conceptualised and planned their teaching beyond isolated teaching of a single concept or setting up a science experience. In Period 4, however, they focused more on their own science knowledge and what was the STEM concept to be taught to children:

> Olivia: I think about the many possibilities of how science can be weaved in the everyday for children. In everyday experiences, and I am understanding the essential role we have as early childhood educators in developing those science concepts with children, and having the language.
>
> Ruth: We didn't have the knowledge to support the science concepts. So really without a few people helping us, I am not sure how we would have refined the knowledge enough to teach it to the children with such complex concepts. That information is very hard to access, often online it is inaccurate. The Learning Frameworks [curriculum documents] talk more about learning dispositions than science content.
>
> If you are not scientifically trained, it's very hard to distil it, into one sentence that young children are capable of understanding. Whereas, if you are a trained scientist, I find, they have that skill. Because we would just go too deep, and get off track [from the essence of the concept].

What emerged from the study was how the teachers researched to build their science knowledge in relation to the problem and what would be the essence of the big ideas in science (Period 4). They moved towards distilling meaningful concepts for children. That is, from the 'noise of the science knowledge and potentially misinformation available or incomplete', they sought out what might be the essence of a particular concept. But also, what would be credible to very young children and therefore meaningfully integrated into the playworld as a problem to be solved, as Olivia explains:

> The other thing was thinking about the problem and how credible it was to the children. Going back to the story, and even though children were empathising with the characters, we were always thinking about the problem, was it credible and was it motivating children. This is the ongoing discussions that Ruth and I have had, and reflecting on this after each session, and in our planning.

The essence of the conceptual content across periods is shown in Table 4.3. The teachers' brought to the theoretical problem in Period 1 a focus on STEM concepts in

Table 4.3 Solving the theoretical problem associated with STEM learning in children's play

Dialectical relations	Theoretical problem Period 1	Theoretical problem Period 2	Theoretical problem Period 3	Theoretical problem Period 4
Teacher motives	What STEM concepts could be used with the narrative of the chosen book?	What was the STEM knowledge needed?	What was the essence of the STEM concept?	Was the STEM problem credible to the children?
Children's motives as motivating conditions for teachers	Moving from the book to generating the narrative that brings forward STEM play and problems to be solved.	Dialectical relations between everyday knowledge and scientific knowledge	Modelling: Children generate a model to capture the essence of STEM concept	Motivating conditions for children to want to solve the problem.

relation to the book. The same theoretical problem was then interpreted in Period 2 in relation to the general STEM knowledge they needed. However, in Period 3 more attention was directed to the essence of the concepts they were bringing into the playworld where the children needed a particular STEM concept for enriching the play or for maturing the play. However, it was during Period 4 that the teachers were able to identify what mattered for solving the theoretical problem of how to bring concepts into children's play.

In summary, it was found that working on a theoretical problem across 2 years created a different kind of context for the teachers planning and teaching. What dominated their weekly planning and evaluation of the children's learning and their pedagogical practices, was not a problem of practice. But rather, the theoretical problem of how to meaningfully bring STEM concepts into children's play at the same time as distilling the big ideas in STEM as credible concepts for children's learning, and the development of their play. While this was an ongoing point of crisis for the teachers, it was found, like Edwards et al. (2019), that the drama of the problem gave motivating conditions for their development. That is, the teachers viewed the same practice situation differently at different points over the 2 years. The teachers entered into the theoretical problem differently, as their own social situation of development was changing.

We now turn to the second theme where a different theoretical orientation to the concept of play was needed for professional practice change.

4.4.2 The Educational Experiment Brought Changes in the Dominating Motives of the Teachers

Studies following a cultural-historical tradition have identified that early childhood teachers object motive is towards their practice (Nuttall et al., 2015). One distinguishing feature of early childhood teachers' practices is planning educational programs for play-based settings. In this study it was found that although teachers created new motivating conditions towards learning concepts in children's play, they also met another theoretical demand in relation to their role in children's play. During Period 1 the teachers expressed a crisis that they had to resolve if they wanted to solve the theoretical problem of how to bring concepts into children's play. For instance, Ruth identified her anxieties about how to be a character that was convincing to children:

> [initially] I felt nervous that I had to convince that child that I was that character. But the children know we are pretending. They have different levels depending on their age and their ability. So, some will say, "You're not really the Mad Hatter", whilst another will call you [putting up her hand], "Mr Mad Hatter, Mr Mad Hatter". So they will relate to you at different levels, and all of that is fine. It's all part of the play. I didn't have to convince them to play with them now, and I understand that now (nodding head). (Ruth)

This is consistent with the practice tradition of many centres in Australia. Teachers believe that play is the domain of children and teachers do not play with children, as noted by Olivia in Period 2.

> ...obviously there is always a debate about how much to interfere with children's play, but in terms of this approach, it gave us a lot of confidence too, in when they get stuck, and then how more complex the play can be (Olivia).

Both teachers identified a shift in their role in a playworld:

> I feel much more confident now. So if the children are playing and they get stuck, I will jump in and say, "We need a ...." or "No we forgot the... rain. Who is going to be the rain?" When I can see that something is missing, I just take the above position. I feel really comfortable to do that. Because I know that the play will continue to develop. Otherwise it gets stuck. If you just ask open ended questions or you watch children's free play, it reaches a stage where it just gets stuck. With the teachers feeling confident to jump in, when its needed, then you can continue the play (Ruth).

The expressed nervousness about being a play partner is in keeping with Pentti Hakkarainen et al. (2013), who identified that adults had to learn how to become play partners with children. But different to Hakkarainen et al. (2013), was the theoretical thinking of the teachers when planning to enter teachers and children playworld.

In Period 3, the teachers explicitly brought this forward. In building on Hakkarainen et al. (2013), this study found that what was specifically unique to entering and exiting of a playworld was the mind shift from individual or pair play to whole group play. This needed play planning, as Olivia identified:

> It's important to plan the play with the children. With the children we share the same goals. When you are going out to play, if we are all playing a different game it wouldn't work. To have that common motive, like we are saving the dragon, and sometimes some children will go off, like dig a bone, and that's fine. It is still connected. Circle time is also, where we gather the knowledge to go deeper [in play]. We have a whole shared [play] history (Olivia)

In this illustrative example of this finding, a shared motive for collective imaginary play was brought forward by Olivia. This did not mean that children did not have agency to bring new problems and plots into their play. Rather, it meant that the children showed behaviours suggestive of being in the same imaginary situation.

And this was afforded when entry and exist as a whole group was planned. Olivia explained in Period 3, the need for the planned motivating condition of a shared imaginary play situation, where a common motive to solve a problem in play, or to relive the story in some way, was key. This established a shared play history for everyone.

> I didn't really understand the intention at the beginning. And understanding how important it was for building the collective in the imaginary space. It took a while. I don't know why professionally? But that is something that has really stood out, in terms of reflecting now. Maybe I didn't understand or appreciate the intention of it, that we all go together into the imaginary space. And even though we were going through the tunnel like in Alice in wonderland (rabbit hole), or going through the trees (The secret Garden) using a key, and we were thinking more and more about the scientific concept and the knowledge, and our position. I have realised how crucial and important that [planning for whole group entering and exiting] is (Olivia).

During Period 4 of the educational experiment a crucial conceptual link between the shared imaginary play with a problem to be solved, and the maturing of play in ways that brought out the imagining of the STEM concepts were identified. The teachers suggested that the link between imagination in play and therefore imagination of STEM concepts was the foundation for the abstract dimensions of the intentional teaching of concepts. As Olivia notes:

> In imagination, the possibilities open up, helps with understanding. That imagination is the higher order thinking to develop those conceptual understandings. And think in the abstract, and encouraging that so well. And how we work with positioning, to support children's conceptual thinking. Are we in the above or below or with? I really loved how that was made so clear in this [playworld] (Olivia, Period 4)

The four periods of solving the theoretical problem of how teachers enter into the play are brought together in Table 4.4.

In summary, when the four period are considered together, it is possible to see how the theoretical problem changed over 2 years. Like Lindqvist (1995), we also identified in the educational experiment over the first year that it was hard for teachers to be with the children inside of the playworld acting as a play partner (Period 1 and 2). But the study found a marked shift from Periods 1 and 2, to Periods 3 and 4 (Table 4.4). In Period 3 the teachers conceptualised play theoretically. They identified that they needed a new conception of play. Play had to be seen as a collective imaginary situation. Different to Lindqvist (1995) who was interested in the aesthetics of a common playworld, in Periods 3 and 4 the theoretical problem became more focused on how to bring concepts into collective play. It was during Periods 3 and 4 that new insights into this problem were articulated by the teachers. They theorised collective imaginary play as foundational for motivating and developing abstract thinking for understanding/learning the STEM concepts associated with the problem in the story/play.

Table 4.4 Solving the theoretical problem associated with how teachers enter children's play

Dialectical relations	Theoretical problem Period 1	Theoretical problem Period 2	Theoretical problem Period 3	Theoretical problem Period 4
Teacher motives	How to change positions from the teacher role to a play partner? How to be a believable play partner from the children's perspective	Professional perspective of not interfering in children's play	How to motivate all children into collective play?	Links between imaginary play and abstract thinking in play
Children's motives as motivating conditions for teachers	Children know teachers are pretending	Interacting with teachers as play partners, rather than as teachers in charge	Finding the roles/ characters to bring individual children into the playworld	Motivating conditions where the concept/research solves the problem and keeps the play going

4.5 Discussion

A Vygotskian reading of development generally suggests that the formation of new psychological functions or consciousness is linked to the participation of the individual in specific forms of social practice or activity. In this study the educational experiment brought changes in the dominating motives of the teachers as they worked with a theoretical problem. The study found there was a doubleness in the educational experiment associated with the theoretical problem – for STEM concepts (Table 4.3) and for a new conception of play (Table 4.4), and when taken together, this appeared to create motivating condition for teacher development. In line with Nuttall (2013), the study repositioned teachers "...not as individuals whose practice needs to be 'fixed' through modelling or coaching" (p. 208) but as collaborators with researchers seeking to solve a theoretical problem. A genuine problem brought about by a societal expectation in Australia for preschools to deliver greater cognitive outcomes associated with the intentional teaching of discipline concepts in play-based programs. This societal expectation brought new demands upon the teachers. By bringing out the essence of the expectation as a theoretical problem, and using this as the foundation for an educational experiment, this recognised that teachers were, "participants in complex systems of collective activity" (Nuttall, 2013, p. 208) who go beyond problems of practice and engage in theoretical thinking.

What is different to Nuttall (2013), is that it was not just contradictions which act as the "springboard for developing practice" (p. 208), but it was the theoretical problem. Focusing on a theoretical problem meant that teachers were oriented differently to their practice. Over 2 years the teachers and the researchers could work towards theorising a new model of practice – one that solved the problem of how to bring concepts into play.

Moreover, if we examine the second rows in Tables 4.3 and 4.4, it becomes possible to see how the new motivating conditions that were introduced to children affected the teachers own motives. That is, teachers paid close attention to the children's motive development in relation to the motivating conditions they designed. The dialectical relations between motives of the children and the motivating conditions within the theoretical problem acted as the source of development for the teachers.

But to understand what is developing for teachers, we had to examine how the theoretical problem changed over time (Columns in Tables 4.3 and 4.4). In this study it was the unique content of the educational experiment with it's theoretical problem. The theoretical problem of how to meaningfully bring STEM concepts into children's play at the same time as distilling the big ideas in STEM as credible concepts for children's learning and development of their play emerged. The teachers' brought to the theoretical problem STEM concepts (Table 4.3) in relation to the book (Period 1), STEM knowledge (Period 2), what is the essence of the STEM concept (Period 3) and how credible is the problem (Period 4) when solving the theoretical problem of how to bring concepts into children's play.

The different periods capture how the theoretical problem developed over time. Theoretical thinking acted as an important psychological condition for the teachers for solving the theoretical problem (Fig. 4.1 below), where the content of that thinking was initially oriented to the book, then to the STEM knowledge needed, and then to examining the essence of the STEM concept, and back to designing a credible problem for children to solve as part of the story and play.

Conversely, the teachers also brought into the theoretical problem of the educational experiment how to be a play partner (Period 1), how to retheorise their

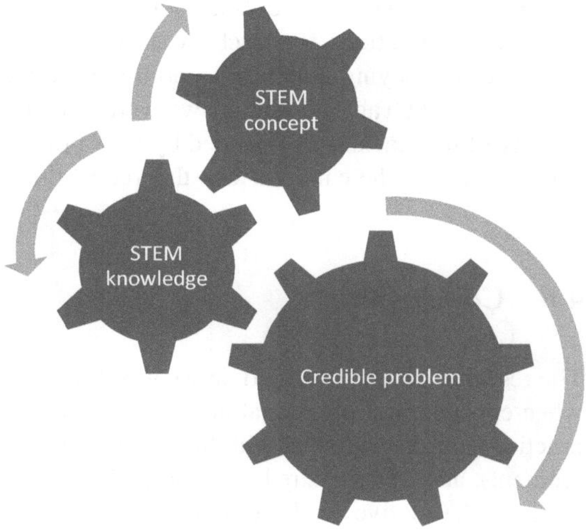

Fig. 4.1 Teacher theoretical thinking: relational model of how to bring STEM concepts into children's play

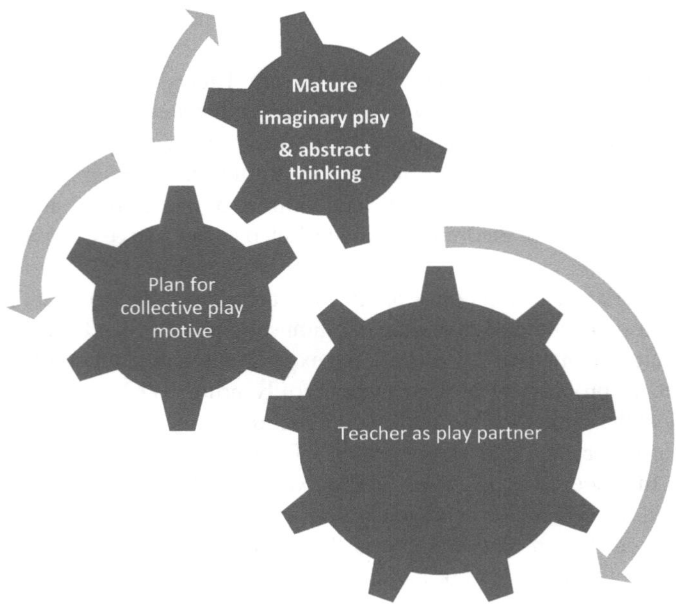

Fig. 4.2 Teacher theoretical thinking: relational model of imagination in play and imagination in STEM

conception of play (Period 2), how to create collective play conditions (Period 3), and how imaginary play brings abstract thinking for learning the STEM concepts (Period 4). We determined that theoretical thinking was developing for the teachers as they began to develop a new theoretical model of play that included STEM learning during the process of solving the theoretical problem. See Fig. 4.2 above.

We were mindful that the object of the activity for teachers had to be children and the pedagogy they created (see Nuttall et al., 2015), but children's motives as motivating conditions for teachers is not something that has emerged in the PD literature. By paying attention to both the motives and the motivating conditions (Row 2 in both Tables 3 and 4), we were able to begin to theorise that the leading activity of the teachers as bounded by their profession, and always in relation to the everyday life of the children, was theoretical thinking.

4.6 Conclusion

The research question centred on what is developing for the early childhood teacher when engaged in professional practice change in STEM. The recurrent demands of practice (source) were met with theoretical thinking about play and STEM concepts (content), and this appeared to be propelled forward through the dialectical tension between the motives and the motivating conditions of the educational experiment over the four periods of the study (development). When viewed holistically, the educational experiment appeared to create new developmental conditions for teachers, because the theoretical problem brought changes in the dominating motives of the teachers from practice to theoretical thinking. The source of development for the teachers in this study was found to be the demands made upon their practices through engaging with a theoretical problem. Therefore, teacher transformation must be conceptualised not as a simple problem of practice change, but as was shown in this study, as a maturing of theoretical thinking by early childhood teachers through solving the problem of play and STEM concept formation. These findings contribute to filling the gap in understanding about what are the conditions and content that act as a source of teacher development when engaged in STEM PD.

Acknowledgments Special acknowledgment is made of teachers who participated in the study and research assistance by Dr Sue March. This work was supported by the Australian Research Council for data collection [ARC DP180101030] and analysis through the Laureate Fellowship Scheme [FL180100161].

References

Brown, C. P., & Mowry, B. (2017). "I wanted to know how they perceived jail": Studying how one early educator brought her students' worlds into her standardized teaching context. *Early Childhood Education Journal, 45*, 163–173. https://doi.org/10.1007/s10643-016-0776-z

Caudle, L. A. (2013). Using a sociocultural perspective to establish teaching and social presences within a hybrid community of mentor teachers. *Adult Learning, 24*(4), 112–120. https://doi.org/10.1177/1045159513489112

Ebadi, S., & Gheisari, N. (2016). The role of consciousness-raising through critical reflection in teachers' professional development: A sociocultural perspective. *Cogent. Education, 3*(1). https://doi.org/10.1080/2331186X.2016.1147990

Edwards, A., Chan, J., & Tan, D. (2019). Motive orientation and the exercise of agency: Responding to recurrent demands in practices. In A. Edwards, M. Fleer, & L. Bottcher (Eds.), *Cultural-historical approaches to studying learning and development. Societal, institutional and personal perspectives* (pp. 201–214). Springer.

Ellis, V. (2007). Taking subject knowledge seriously: From professional knowledge recipes to complex conceptualizations of teacher development. *The Curriculum Journal, 18*(4), 447–462. https://doi.org/10.1080/09585170701687902

Eun, B. (2008). Making connections: Grounding professional development in the developmental theories of Vygotsky. *The Teacher Educator, 43*(2), 134–155. https://doi.org/10.1080/08878730701838934

Eun, B. (2011). A Vygotskian theory-based professional development: Implications for culturally diverse classrooms. *Professional Development in Education, 37*(3), 319–333. https://doi.org/10.1080/19415257.2010.527761

Fleer, M. (2021). An educational experiment into how to bring discipline concepts into play: How a theoretical problem acts as a source of teacher development, *Moscow University Psychology Bulletin 4,* [Vestnik Moskovskogo Universiteta. Seriya 14. Psikhologiya] pp. 72–103. https://doi.org/10.11621/vsp.2021.04.03

Grimmett, H. (2014). *The practice of teachers' professional development: A cultural-historical approach.* Sense Publishers.

Hadley, F., Waniganayake, M., & Shepherd, W. (2015). Contemporary practice in professional learning and development of early childhood educators in Australia: Reflection on what works and why. *Professional Development in Education, 41*(2), 187–202. https://doi.org/10.1080/19415257.2014.986818

Hakkarainen, P., Brėdikytė, M., Jakkula, K., & Munter, H. (2013). Adult play guidance and children's play development in a narrative play-world. *European Early Childhood Education Research Journal, 21*(2), 213–225. https://doi.org/10.1080/1350293X.2013.789189

Hedegaard, M. (2008). The educational experiment. In M. Hedegaard, M. Fleer, J. Bang, & P. Hvid (Eds.), *Studying children: A cultural-historical approach* (pp. 181–201). Open University Press.

Hedegaard, M. (2012). Analyzing children's learning and development in everyday settings from a cultural-historical wholeness approach. *Mind Culture and Activity, 19*, 127–138.

Johnston, K., Hadley, F., & Waniganayake, M. (2020). Practitioner inquiry as a professional learning strategy to support technology integration in early learning centres: Building understanding through Rogoff's planes of analysis. *Professional Development in Education, 46*(1), 49–64. https://doi.org/10.1080/19415257.2019.1647871

Lindqvist, G. (1995). *The aesthetics of play: a didactic study of play and culture in preschools.* (Doctoral dissertation). Uppsala University, Uppsala, Sweden.

Murphy, C., Scantlebury, K., & Milne, C. (2015). Using Vygotsky's zone of proximal development to propose and test an explanatory model for conceptualising coteaching in pre-service science teacher education. *Asia-Pacific Journal of Teacher Education.* https://doi.org/10.1080/1359866X.2015.1060291

Nuttall, J. (2013). The potential of developmental work research as a professional learning methodology in early childhood education. *Contemporary Issues in Early Childhood, 14*(3), 201–211. https://doi.org/10.2304/ciec.2013.14.3.201

Nuttall, J., Edwards, S., Mantilla, A., Grieshaber, S., & Wood, E. (2015). The role of motive objects in early childhood teacher development concerning children's digital play and play-based learning in early childhood curricula. *Professional Development in Education, 41*(2), 222–235. https://doi.org/10.1080/19415257.2014.990579

Shabani, K. (2016). Applications of Vygotsky's sociocultural approach for teachers' professional development. *Cogent Education, 3*(1). https://doi.org/10.1080/2331186X.2016.1252177

Shi, H. (2017). The theoretical interpretation of EFL teacher's professional development from the perspective of sociocultural theory. *Theory and Practice in Language Studies, 7*(11), 1059–1064. https://doi.org/10.17507/tpls.0711.14

Tasker, T., Johnson, K. E., & Davis, T. S. (2010). A sociocultural analysis of teacher talk in inquiry-based professional development. *Language Teaching Research, 14*(2), 129–140. https://doi.org/10.1177/1362168809353871

Trevethan, H., & Sandretto, S. (2017). Repositioning mentoring as educative: Examining missed opportunities for professional learning. *Teaching and Teacher Education, 68*, 127–133. https://doi.org/10.1016/j.tate.2017.08.012

Vygotsky, L. S. (1994). The problem of the environment. In J. Valsiner & R. van der Veer (Eds.), *The Vygotsky reader* (pp. 347–348). Blackwell.

Vygotsky, L. S. (1998). *The collected works of L. S. Vygotsky: Child psychology* (M. J. Hall, Trans. R. W. Rieber Ed. Vol. 5). Plenum Press.

Vygotsky, L. S. (2019). *L. S. Vygotsky's Pedological Works volume 1. Foundations of pedology translated with notes and lecture outlines by David Kellogg and Nikolai Veresov* (1st ed.). Springer.

Marilyn Fleer is a Sir John Monash Distinguished Professor at Monash University, Australia. She was awarded the 2018 Kathleen Fitzpatrick Laureate Fellowship by the Australian Research Council and was a former President of the International Society of Cultural-historical Activity Research (ISCAR). Additionally, she holds the position of an honorary research fellow in the Department of Education, University of Oxford, and a second professor position in the KINDKNOW Centre, Western Norway University of Applied Sciences (2018–2023), and has been bestowed the title of honorary professor at the Danish School of Education, Aarhus University, Denmark. She was presented with the 2019 Ashley Goldsworthy Award for outstanding leadership in university-business collaboration, elected as a fellow of the Australian Academy of Social Sciences, and inducted into the Honour Roll of Women in Victoria as a change agent.

Chapter 5
How Does Science Learning Happen During Scientific Play? A Case Example of the Dissolution Phenomenon

Eirini-Lida Remountaki, Glykeria Fragkiadaki, and Konstantinos Ravanis

5.1 Résumé

Plusieurs programmes d'enseignement pour les jeunes enfants à l'échelle internationale soulignent la nécessité de promouvoir l'apprentissage et le développement par le jeu. Cependant, nous ne savons pas grand-chose sur la manière dont l'enseignement des sciences par le jeu se présente dans la pratique, ni sur la manière dont l'apprentissage des sciences devient évident dans les environnements ludiques. Cette étude vise à explorer comment le jeu scientifique peut créer les conditions permettant aux enfants d'âge préscolaire de former le concept de dissolution au cours de la réalité éducative quotidienne. Elle présente un exemple indicatif d'une expérience d'apprentissage entre cinq enfants d'âge préscolaire, âgés de 5 à 6 ans, et deux enseignants de la petite enfance en Grèce. Des données empiriques qualitatives ont été recueillies par le biais d'enregistrements de dialogues et de dessins d'enfants, de notes de terrain et de photographies. Les résultats ont révélé que pendant le jeu scientifique: (a) les enfants ont commencé à penser au phénomène de la dissolution de manière abstraite, (b) leur réflexion sur le phénomène est passée d'un niveau macroscopique à un niveau microscopique, (c) ils ont réussi à distinguer les substances en deux catégories principales: les substances qui peuvent être dissoutes et celles qui ne peuvent pas l'être; ils ont également élaboré un récit autour de la conservation des substances, et (d) ils ont développé l'aptitude à tester si une

E.-L. Remountaki · K. Ravanis
Department of Educational Science and Early Childhood Education, University of Patras, Patras, Greece
e-mail: ravanis@upatras.gr

G. Fragkiadaki (✉)
School of Early Childhood Education, Faculty of Education, Aristotle University of Thessaloniki, Thessaloniki, Greece
e-mail: fragkiadaki@nured.auth.gr

K. Plakitsi, S. Barma (eds.), *Sociocultural Approaches to STEM Education*, Sociocultural Explorations of Science Education 21,
https://doi.org/10.1007/978-3-031-44377-0_5

substance peut être dissoute ou non. L'étude informe la pratique éducative quotidienne en fournissant un cadre pédagogique basé sur la cohérence, la consistance et l'équilibre entre les objectifs d'enseignement-apprentissage en sciences et les pratiques ludiques.

In many contemporary societies, play-based pedagogies are the baseline for the development of early childhood education curricula (Australian Government, 2010; Greek Ministry of National Education and Religious Affairs, 2002; Irish Government, 2009; UK Government, 2017). However, evidence has shown that early childhood teachers face difficulties when it comes to organizing play-based activities with specific learning outcomes (Tu, 2006; Miller & Almon, 2009) or do not conceptualize and categorize play in teaching methods for achieving specific learning outcomes (Lynch, 2015; Fesseha & Pyle, 2016). The conceptualization of play as a basic teaching approach becomes less evident when it comes to the field of learning and development in science in the early years (Fleer, 2006, 2009). This study aims to provide a better understanding of how play comes in line with concrete teaching goals and leads to learning outcomes in science. The study focuses on the dissolution phenomenon in young children's real-life experiences and seeks to explore how young children form the concept of dissolution through scientific play. The chapter begins by unpacking how scientific play is conceptualized in early childhood education. This is followed by a short overview of the empirical literature about how young children form the science concept of dissolution in early childhood educational settings. Details of the methodological framework of the study are presented and qualitative empirical data sets are analyzed. A discussion on how pre-schoolers approach the phenomenon of dissolution and develop their scientific thinking through scientific play is provided. The chapter concludes with insights into how science learning experiences for young learners can be improved through play during everyday educational reality in early childhood settings.

5.2 Conceptualizing Scientific Play in Early Childhood Education

From a cultural-historical standpoint, play has a fundamental and critical role in a young child's learning and development process (Vygotsky, 1966; Elkonin, 1999, 2005; Kravtsov & Kravtsova, 2010). In Vygotsky's writings (1966, 2004) play is conceptualized as the leading activity for a child's development and is dialectically interrelated to the higher mental function of imagination. During play, the child experiences a unique and usually complex imaginary situation that is created, led, and developed by him/her. Living through this imaginary situation the child interprets, stretches, transforms, and expands reality in multiple ways (Fragkiadaki et al., 2021b). Experiencing diverse imaginary situations, the child deepens his/her knowledge of the natural, technical and technological world as well as gains a better understanding of the social and cultural reality he/she lives in.

Importantly, imaginary play creates the conditions for the interrelation between everyday and scientific concepts. As Vygotsky (1987), there are two different types of concepts: everyday concepts and scientific concepts. Everyday concepts are formed through everyday experiences and routines as well as daily life understandings. Scientific concepts are formed through social interactions and usually, within formal settings such as schools and institutions. Everyday concepts are specific and related to objects, procedures, and situations. Scientific concepts are abstract and can be transformed into diverse settings and situations. For example, from the very early beginning of their lives, children realize that the day changes to night and night changes to day. They know that there is a day and night circle. This conceptualization is an everyday concept formed by the child based on the child's daily experiences. However, this does not mean that child knows the scientific explanation that lays behind this understanding, that is, the rotation of the earth on its axis and the sun's position. Daily experiences do not allow the child to reach this type of understanding and form the scientific concept. The formation of this scientific concept requires more abstract and academic knowledge. According to Vygotsky (1987), everyday concepts are the baseline for scientific concepts and scientific concepts are the means that deepen our understanding of everyday concepts. These two types of concepts are dialectically interrelated, and both are critical for the development of young children's thinking.

Empirical studies in the field of early childhood education (Fleer, 2009, 2011, 2017; Hakkarainen, 2008) have provided evidence about the dynamics of interrelating everyday and scientific concepts through imaginary play during everyday educational reality in early childhood settings. In these studies, the terms "conceptual play" or "scientific play" is introduced and used to describe the type of imaginary play that is oriented towards the development of conceptual thinking and the formation of scientific concepts by young children in play-based settings. Thus, within conceptual or scientific play learning and play are dialectically interrelated and develop as a unit. In the present study, the term "scientific play" is used to describe children's science-oriented imaginary play. Scientific play is conceptualized here as an experience where children's imagining, exploring, and learning about the natural and technical world are dialectically interrelated in a way that is meaningful and enjoyable for the child and creates the conditions for the formation of science concepts such as the concept of dissolution.

The empirical literature about learning and development in science during the early years is extensive (Andersson & Gullberg, 2014; O'Connor et al., 2021; Roth et al., 2013). These studies have shown that young learners in science can engage in science activities and science-related experiences, form science concepts, and develop an understanding of scientific processes such as problem-solving, experimentation, and observation. The social and cultural character of learning in science during the early years has been also highlighted (Fleer & Pramling, 2014). However, limited empirical studies (Akman & Özgül, 2015; Bulunuz, 2013; Fleer, 2009; Vartiainen & Kumpulainen, 2020) focus on how early childhood children form and develop their ideas about science concepts through scientific play. These studies have unpacked the critical role of play in teaching and learning science in a way that is tailored to young learners' interests, activities, and everydayness. Fleer (2017) underlines the importance of intentionally and systematically supporting young children with pedagogical practices that promote teaching science through play. Her research (2017, 2018) points out the interrelation between imagination in science education and imagination in play during the early years (see also Fleer et al., 2020a, b; Rai et al., 2021).

When considering scientific play in early childhood science education apart from the qualities of children's play such as the imaginative aspect of play as described above emphasis is also given to the content knowledge that is introduced and gained through play. The following section provides insight into young children's ideas, pre-conception, and cognitive challenges regarding the phenomenon of dissolution that the study focuses on.

5.3 Forming the Science Concept of Dissolution in the Early Years

In educational practice, it is required that teachers know the subjects they are going to teach and be able to make the subjects easy to comprehend by the children. The subject matter knowledge regarding the phenomenon of dissolution consists in conceptualizing that when two substances are mixed and the product of their mixture is a substance that its characteristics are not discernible by the eye or by using a plain microscope, this process is called dissolution and the substance-mixture is called solution (Adbo & Vidal Carulla, 2019).

Few studies focus on how young children form their ideas about the phenomenon of solid in liquid dissolution. Researching 3–5 years old children's thinking, Piaget and Inhelder (1974) found that children could not predict substances' conservation. Holding's (1987) research shows that children's ideas evolve with age so, a steadily growing percentage of children understand the conservation of substance. Rosen and Rosin (1993) found that young children understand that characteristics of the invisible dissolved substance will be transferred to the solution. Also, children understand, to a certain extent, the conservation of the matter as they acknowledge that the substance is split and still exists in the solution. Furthermore, it does not matter if the dissolved substance is familiar or not to the children because they acknowledge its presence anyways. Panagiotaki and Ravanis (2014) asked children to predict and draw sugar in different solutions. Children in their drawings seem to understand sugar's conservation whereas, in their predictions, many of them consider sugar not conserved.

From another point of view, there is a focus on mental representations of 4–7 years old children, when they are to predict solid into liquid dissolution. Christidou (2006) found that children use naturalistic explanations, namely, they rely on the natural properties of materials, more than they use teleological, intentional, metaphysical, or magical explanations. However, their naturalistic explanations are not compatible with the models used for dissolution in education, because they often focus on one substance to explain the phenomenon, instead of thinking about substances' interaction. Panagiotaki and Ravanis (2014) searched mental representations of 5–6 years old children about sugar's dissolution in water and oil. An important percentage of their sample does not see that there might be differences between the dissolution of sugar in water and oil.

Remountaki et al. (2017) explored the ideas of 5–6 years old children for the dissolution phenomenon of solid substances into liquids within play-based settings. By playing with a puppet, handled by the educator, children used their everyday knowledge and experience to form the concept of dissolution.

Despite the above empirical findings, more research has to be done to inform practice about how teaching science through play looks like in practice and how learning in science becomes evident in play-based settings. The present study seeks to make visible how young children approach the phenomenon of dissolution through scientific play as well as how scientific play can create the conditions for systematic engagement with science methodology such as the development of the trial skill by the children during everyday educational reality in kindergarten. To address this challenge the research question is posed as follows: How do preschoolers form the concept of dissolution through their scientific play?

5.4 Methodological Framework

5.4.1 Study Design

The study design was based on children's scientific play during everyday educational reality in early childhood settings. In line with the Greek curriculum (2011) objectives and statements about approaching natural phenomena children were anticipated to:

(a) express their ideas about the natural phenomenon and negotiate these ideas with others,
(b) form questions to investigate certain aspects of the phenomenon,
(c) form answers for the question they investigate,
(d) use the results of their research in new situations.

An improvised imaginary story inspired children's play. The story was about castles created by mud. Several problematic situations occurred throughout the process of creating the mud castles and children had to find out a solution to address the emerging problems. Children's scientific play was developed through four diverse play-based activities over 3 weeks. A detailed description of each play-based activity follows.

5.5 First Play-Based Activity: Mixing of Solid and Liquid Substances

In this phase, the challenge of creating an imaginary world in which children's toys, small dolls as characters of a plot, could live in, was introduced. The challenge was posed as follows: "*How can we build castles out of the mud?*". The teaching goals of the activity were to create the conditions and support children to:

(a) express their ideas about the phenomenon of dissolution,
(b) make hypotheses about mixing materials,
(c) explore the materials and the process of dissolving them,
(d) wonder about the analogies between the materials, and
(e) understand the need for efficient analogies of the materials to achieve a different quality in the mix.

Children handled, observed, experimented, and discussed the analogies of water and soil needed to build castles. After that, children brought their toys into the new space that they have created and organized their free play there.

5.6 Second Play-Based Activity: The Dissolution of Solid Substances in Liquid

As children and early childhood teachers were developing their scientific play the challenge of "*How can we make it as if it is snowing?*" was introduced within the imaginary situation. Two substances replaced "snow" (a) powdered sugar, and (b) glitter. The teaching goals of the activity were to create the conditions and support children to:

(a) express their ideas for the phenomenon of dissolution,
(b) wonder about the phenomenon,
(c) make hypotheses about mixing materials, and
(d) understand the conservation of a solid substance into a liquid.

Within their play, children experienced the dissolution of powdered sugar when in (a) soil, (b) wet soil, and (c) water. They also experienced the non-dissolution of glitter in the same situations and tried to explain why the two substances reacted differently.

5.7 Third Play-Based Activity: The Dissolution of Liquid Substances in Liquid

In this activity, children played with their toys. They were challenged to think "*What could happen if diverse materials got spilled in the lake?*". The teaching goals were to create the conditions and support children to:

(a) express their ideas for the phenomenon of dissolution,
(b) wonder about the phenomenon, and
(c) observe the conservation of a liquid substance into a liquid when the substance is visible and when it is not.

Within children's play, an imaginary wave turned a boat that was transferring materials upside down. Oil and vinegar got spilled and children experienced the non-dissolution and dissolution of the substances accordingly and try to reason their ideas about the phenomenon.

5.8 Forth Play-Based Activity: The Dissolution of Liquid Substances in Liquid

This activity followed the above activity. Children hypothesized what would happen with the red color if spilled in the lake too. The teaching goals were to create the conditions and support children to:

(a) express their ideas about the phenomenon,
(b) wonder about the phenomenon,
(c) use their previous observations in new situations to build a methodological tool, and
(d) develop trial skills.

Children explored the dissolution of red color and summarize what they had experienced in their overall play with the materials.

During the four play-based activities, the early childhood teachers created a learning environment with no restrictions on children's ideas and initiatives. Children had the opportunity to lead and transform the activities in a way tailored to their interests and needs. Early childhood teachers consistently posed questions stimulating children to craft a narrative around their understandings and better unpack their thinking about the phenomenon.

5.8.1 Participants and Data Collection

Empirical data of five children aged between 5 and 6 years old from one classroom in an urban area kindergarten in Greece are presented in this study. The data are part of a larger research project about understanding the way children conceptualize the phenomenon of dissolution in the early years (see for example Remountaki et al., 2017) All names mentioned in the text are pseudonyms. Parents' informed consent was given. Children had no previous engagement with the concept of dissolution within early childhood educational settings. Two early childhood teachers participated in the study. Both had previous teaching experience in play-based settings.

The set of play-based activities was developed in collaboration with the research team. Data were collected through (a) audio-recorded conversations (144 minutes of recorded conversations during conceptual play), (b) children's drawings (a total of 20 drawings), (c) field notes made by the early childhood teachers (a total of 4 pages of field notes), and (d) photographs of children's scientific play (a total of 80 photographs). Conversational analysis (Roy Pea, 1993; Harvey Sacks, 1995) of the transcripted dialogues and narratives was carried out supported by the documentation from the drawings, the field notes, and the photographs. The analysis aimed at capturing critical moments during the scientific play when children managed to start developing early forms of the science concept concerning the suggested teaching goals.

5.9 Findings

The findings of the study are reported first through an overview of how children's scientific play was developed in relation to specific learning outcomes through the set of play-based activities. This evidence is presented in a table (Table 5.1). This is followed by the analysis of how children gradually formed the concept of dissolution through the set of play-based activities. This evidence is presented through excerpts (Excerpts 5.1, 5.2, 5.3, and 5.4), drawings (Drawings 5.1, 5.2, and 5.3), and photos (Images 5.1, 5.2, and 5.3).

The following table (Table 5.1) illustrates how the emerged characteristics of children's scientific play and specific learning outcomes that were observed from the evidence presented were developed as a unit. In the left column, the characteristics of children's scientific play in each play-based activity are presented. The right column presents the learning outcomes of each play-based activity.

The above findings showcase how each play-based activity followed specific teaching goals that led to concrete learning outcomes through children's scientific play. A detailed description of how children's scientific play was elaborated through the diverse play-based activities follows.

5.9.1 First Play-Based Activity: Castles Made of Mud

The children and the early childhood teacher, Olivia, gathered at a corner of the classroom where soil and water were placed. Children conceptualized mud as a place where characters of their play, that is their pocket dolls, will live when the castles are built. They attempted to use different analogies of soil and water and they recalled summer experiences from their play with sand by the sea (Image 5.1).

While the children, as a team, were building the castle, Bruce started crafting a narrative around the mud castle. He created the following imaginary story: *"Once upon a time, too many castles, they were being built and people were coming, cars*

Table 5.1 Characteristics of children's play in relation to specific learning outcomes

Play-based activity	Characteristics of children's scientific play	Learning outcomes
	During their scientific play children:	Through their scientific play children:
First	Built castles out of the mud /create a place for their toys/plot-characters Tried different analogies of water and soil Observed what happens when soil is dry, wet, and very wet Hypothesized about why they could not build castles Discussed the possible reasons they cannot build castles under certain circumstances Recalled experiences Collectively created a plot story and play with their toys/plot characters	Formed, shared, and negotiated a set of arguments about the phenomenon Used drawings to elaborate on their ideas Used a wide range of everyday experiences to explain the phenomenon Made and tested hypotheses about the analogies of the materials Understood the need for efficient analogies to achieve a particular outcome for a mix
Second	Engaged emotionally with their story Imagined sugar as snow Observed sugar's dissolution Discussed and tried to justify sugar's conservation Observed the non-dissolution of glitter and tried to explain the difference between the substances Recalled experiences Changed their voice tone, used gestures	Formed, shared, and negotiated a set of arguments about the phenomenon Used drawings to elaborate on their ideas Used a wide range of everyday experiences to explain the phenomenon Understood the conservation of a solid substance into a liquid when the substance is not visible
Third	Engaged emotionally with their story Observed oil's non-dissolution and vinegar's (after hypotheses) dissolution Tried to make justifications about vinegar's conservation and the difference between the substances Collectively created a story of a battle between oil and vinegar Recalled experiences Changed their voice tone and used gestures	Formed, shared, and negotiated a set of arguments about the phenomenon Used drawings to elaborate on their ideas Used a wide range of everyday experiences to explain the phenomenon Made and tested hypotheses about the phenomenon Observed the reaction of a liquid substance into a liquid when the substance is visible Understand the conservation of liquid substance into a liquid when the substance is not visible
Fourth	Engaged emotionally with their story Hypothesized about what will happen if red color meets water Explained red color's conservation Recalled experiences Changed their voice tone and used gestures	Formed, shared, and negotiated a set of arguments about the phenomenon Used drawings to elaborate on their ideas Used a wide range of everyday experiences to explain the phenomenon Made and tested hypotheses about the phenomenon Developed the trial skill

Image 5.1 Using different analogies of soil and water to find the efficient analogy to build a mud castle

and then the queen. People were coming. Do you know that Ninja is coming from Japan?". Along with Lily, Bruce brought into their play some toys such as a "queen" and a "samurai". The following excerpt (Excerpt 5.1) indicates how children collaborated and interacted with each other and the early childhood teachers while trying to build their castles.

Excerpt 5.1 Finding a Solution to the Problem of the Efficient Analogy of Substances

Tom (T):	*Miss, it's leaking* (the mixture). *I'll add some soil to make it solid.*
Early childhood teacher 2 (E2):	*What do you mean by "make it solid"?*
T:	*Now, I need the soil. I need it. To make it solid, to make it strong.*
Early childhood teacher 1 (E1):	*Aaa, why isn't it solid now?*
T:	*It is watery.*
E2:	*What does it mean "strong" . . .?*
T:	*To be able to stick together.*
E2:	*To "stick together", namely. . .?*
Henry (H):	*It's leaking.*
E1:	*Do you want* (soil) *Lily, too? Tom, what is the reason, you think, this is happening?*
T:	*Yes, but it needs more soil to stick together.*
E1:	*"To stick together"? Why you said now "it is watery"?*

T:	*It looks like gray glue [. . .] it looks like water. Liquid.*
El:	*Liquid? So, you mean the castles were liquid before?*
T:	*Yes.*
Bruce (B):	*Yes.*
El:	*Aaaa! So, the castles were liquid?*
B:	*Yes.*
El:	*Now, aren't they?*
T:	*Now, it is still liquid, and it goes, look* (he adds water). *It's destroyed.*
El:	*Hmm. Why did that happen? Do you have any ideas?*
T:	*Because it wasn't as dehydrated as before.*

In the above excerpt, the early childhood teachers stimulated Tom to clarify his thinking and express argumentation about the phenomenon. Tom appeared to realize that castles can be built only by efficient analogies of water and soil. By associating his previous experience of dehydration with the mud, he was able to touch upon the problem of efficient analogies needed to dissolve the substances and create the mix. Through his scientific play after testing, observing, and recalling, Tom solved the problem and concluded that making castles out of the soil cannot be "dehydrated" nor "liquid". In the following drawing (Drawing 5.1), Henry's depiction of their play is illustrated. Henry shows that he added so much water to his soil that he could not build a castle. He also depicted all children playing and a castle he was planning to make.

In this play-based activity, the teaching goal was to promote children's understanding of the required analogies of the materials. Children experienced, observed, and discussed their ideas with each other and the early childhood teachers. They hypothesized possible solutions to the problem of analogies and solved the problem. At the end of the activity, children used drawings as a means to take notes of their scientific play. What is important here is that during their scientific play children used everyday experience and knowledge about hydration and dehydration to address the problem of finding an efficient analogy of substances. That allowed children to begin thinking about the phenomenon of dissolution more abstractly by focusing on the properties of the matter such as moisture and consistency.

5.9.2 Second Play-Based Activity: Snow Made of Powdered Sugar and Glitter

A week later, the children continued their scientific play. Olivia suggested it was winter and it could be snowing. She suggested powdered sugar could be imagined as

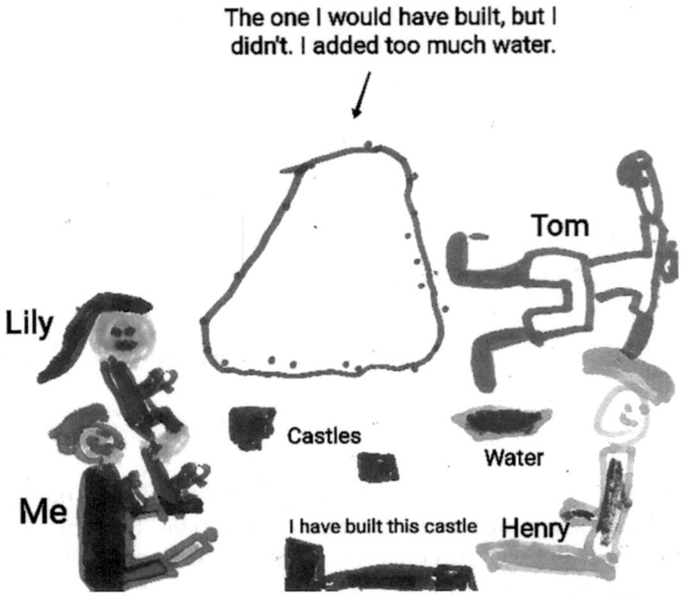

Drawing 5.1 Recording the process and the experience of children's scientific play

snow during play. Children incorporated powdered sugar in their play (Image 5.2). They observed sugar dissolution and expressed their ideas. They used glitter as an alternative substance for snow.

In the following excerpt (Excerpt 5.2), children's ideas expressed during their scientific play about sugar dissolution are presented.

Excerpt 5.2 Finding a Solution to the Problem of Sugar's Dissolution
(The childhood teacher is sprinkling some sugar to pretend it is snowing upon the castles.)

E1: *But where is it (means the sugar/snow? I sprinkled. Where is it?*

[...]

E1: *Tom, I don't know if you heard that. Lily told me that flour still exists in the soil. It's just, not visible. Do you agree with her?*
T: *With a naked eye.*
E1: *"With a naked eye"? What do you mean?*
T: *I mean, that it is not visible with the eye, with the eye being without.*
E1: *Without what?*
T: *Without something toooooo.*
E1: *So, we need something to better see if it is still into* (water)*?*
B: *I didn't know that for sure.*
T: *Yes.*

Image 5.2 The sugar dissolution during scientific play

In the above excerpt, during scientific play, Lily approached the concept of sugar conservation after dissolution. Bruce and Tom agreed with her even though they could not explain what happened to sugar. Tom made an argument about the eye's inability to see what exists in the microcosm. He used the everyday experience to understand that even though they could not see sugar, it was still there. He mentioned that an "intermediary tool" could help to see at a microscopic level. Bruce seemed content with Tom's idea. In the following drawing (Drawing 5.2), Steven depicted their play. He drew the children, the soil, the water in the lake, and one of their toys being sunk in it.

In this activity, the teaching goals were to create an encouraging environment for the children to wonder about the phenomenon of dissolution and through discussions to elaborate on their ideas and understand sugar's conservation into water. Through their scientific play children understood sugar was dissolved and conserved by trying to make their ideas clear to the rest of the children and early childhood teachers. Within the play, children expressed themselves by changing their voices and making a wide range of gestures. They depicted their play through drawings and crafted a narrative about their understandings while explaining to the early childhood teachers

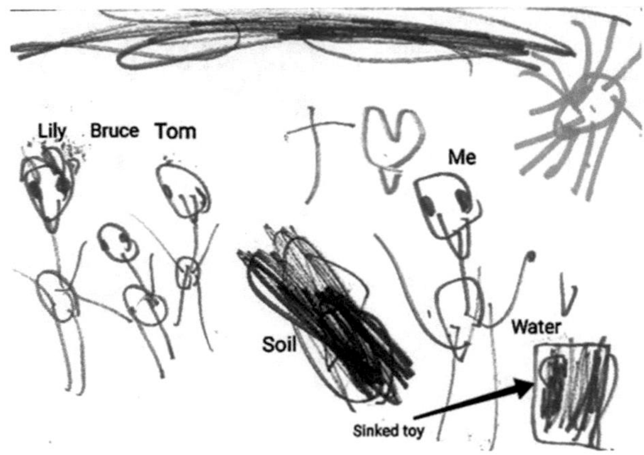

Drawing 5.2 Recording the experience of playing with each other and with toys within the imaginary plot

what they have drawn. In this vignette children appeared to approach a complex aspect of the dissolution phenomenon, that is conservation of the substances while playing with the materials. What is critical here is the transition of children's thinking about the phenomenon from a macroscopic to a microscopic level where substances still exist even though not visible to the naked eye. This understanding is a core idea in forming the concept of dissolution.

5.9.3 *Third Play-Based Activity: Oil and Vinegar Have a Battle*

A week later, children engaged in the concept of non-dissolution of oil into the water, observed the phenomenon and tried to explain it. They hypothesized about what will happen if vinegar meets water, too. When vinegar got spilled, they tried to explain oil's and vinegar's reactions. The following expert (Excerpt 5.3) documents children's conversations around that aspect of the phenomenon.

Excerpt 5.3 Finding a Solution to the Problem of the Non-dissolution of Olive Oil and the Dissolution of Vinegar

E1: *Is vinegar in the water now?*
Lilly (L): *It is, it is just transparent.*
E1: *Like flour?*
L: *Yes.*
E1: *Was flour in the water previously?*
L: *It was very transparent in the same way it has happened here with vinegar.*
E1: *Hmm, so vinegar and flour become transparent. Tom, do you agree?*

T: *Yes.*
El: *Yes? So, vinegar is in here?* (Pointing into the lake.)
T: *Vinegar.*
El: *Yes...*
L: *It is!*
T: *Um, um, um, I don't agree.*
El: *Oh, so tell me.*
T: *It is dissolved because of water and oil.*
El: *Yes...*
S: *Oil is more powerful.*
T: *No! It has more volume tooooo.*

[...]

T: *It is like oil and vinegar are fighting and oil wins.*

[..]

L: *Oil has arrived first.*
T: *Oil has a bigger volume than vinegar.*
L: *And for that reason, it means, it has more power.*

[...]

El: *But are both into...* (points to the lake)*? Is vinegar inside?*
T: *I... think so.*
El: *You think so. Oil* (is it present)*?*
T: *It is.*

In the above excerpt, Lily considered vinegar transparent. When Olivia asks her about vinegar's presence in the water, she says that it does exist. Tom disagreed with Lily when she claimed that vinegar is present in the water, but it seems that he tried to explain vinegar's dissolution. He also talked about a reaction, but the way he expresses his idea seemed as if he considered that oil is the substance responsible for the reaction. Steven considered oil "*more powerful*", without elaborating further on his idea. When children were trying to explain the difference between oil (non-dissolved) and vinegar (dissolved), Lily and Tom imagined a story of a battle between the two substances. Lily stated that "*oil wins*" because it has "*more power*" than vinegar. Tom stated that he is sure about oil's presence in the water and that he thinks vinegar is present, too. Lily tried to explicitly explain her idea and build on Tom's idea to explain why she considered oil and vinegar are present in the water. Lily concluded the whole science experience through her drawing (Drawing 5.3).

In Lily's drawing, the four phases of the scientific play are shown. During the activities, Lily stated that oil makes "*circles*" or "*balls*" and that vinegar is "*dissolved*". Initially, for Lily dissolved means "*it doesn't exist anymore*". Later, in her drawing, she mentions vinegar is presented everywhere because "*it is not visible, visible, but it is because we are very big*". She explained that if her peers

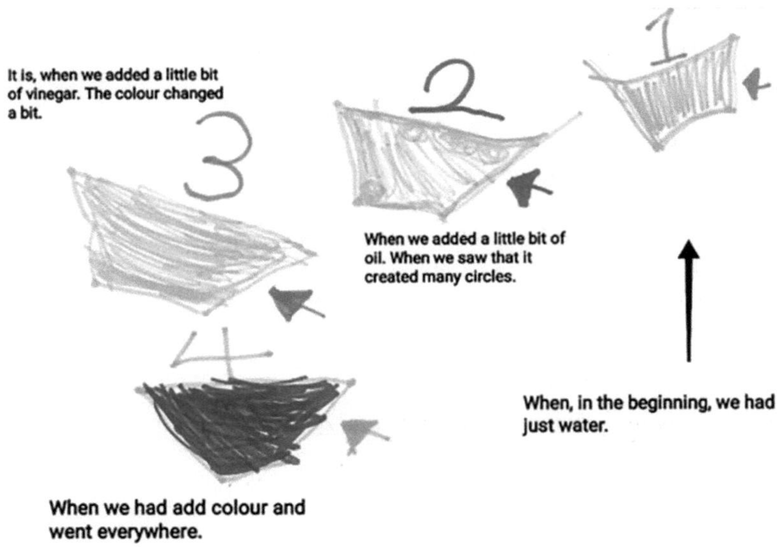

Drawing 5.3 Recording the experience of trying to dissolve oil and vinegar

were smaller and could emerge themselves into the water, perhaps they could see the vinegar. In her drawing Lily also explained vinegar's conservation.

In this play-based activity, the teaching goal was for children to understand the conservation of a liquid substance into a liquid when the substance is not visible. Most importantly the aim was for children to develop their ideas about the dissolution phenomenon by discussing, observing, and wondering about what a possible explanation of why the two substances react differently could be. The above evidence suggests that through their imaginary play children managed to distinguish the substances into two main categories: substances that can be dissolved and others that cannot be dissolved. During their scientific play children also crafted a narrative around the conservation of the substances in both cases. What is also important here is that children used their imagination to develop their understanding of the above aspects of the phenomenon. This is evident when for example Lily explained that if her peers were smaller and could emerge themselves into the water, perhaps they could see the vinegar.

5.9.4 Fourth Play-Based Activity: Would It Be Dissolved? We Should Do It First!

Having already seen the dissolution phenomenon in different situations, children hypothesize about what could happen if the red color would be spilled. Through a summary of what they have experienced in the previous activities, they build the skill of trial. Excerpt 5.4 illustrates how children reach this understanding.

Excerpt 5.4 Finding a Solution to the Problem of the Dissolution of Red Color and the Development of Trial Skill

El: *So now, that we are going to add the red color, will it be gone like vinegar?*
T: *No! It will stay, it will stay because it has a bigger color volume.*
El: *How do you know that? Why do you have this idea? Why are you so sure?*
T: *I saw it, when I was young, by using play dough.*
El: *Yes but, I get what you 're saying when you mix them red outweighs.*
T: *Yes.*
El: *But now that this isn't playdough, how, how you are so sure?*
B: *Are we going to add it?*
El: *We are going to add it. How you are so sure that it's going to go everywhere, Tom?*
T: *Red is red. Vinegar is vinegar.*

(Children add the red color.)

T: *I found it. All the rescuers* (lids of markers) *died and shed their blood.*
El: *Oh! Could you believe this guy? A good idea, indeed. All the rescuers died and shed their blood.*
T: *Miss, the shark did it though.*

(...)

El: *So, why did you suggest we should try it?*
T: *To know.*
El: *Aaa. So, if I wanted to add something to it, how could I know that...*
T: *You should do it first.*

(...)

El: *Lily? Tell me your idea about... The one you told me about before because I got confused. If I want to add sugar, how can I be sure that what is going to happen?*
L: *You are going to see it on the computer or the tablet.*
El: *Or at the tablet, yes. Yes. So? Otherwise? So, if I have here, now, water. Here, now. And I don't have a computer or a tablet, how could I know? So, I have here the sugar. I hold it. Let's imagine that I have sugar now and I want to add it. How do I know if it's going to go everywhere? What can I do?*
L: *You will see it and you will understand.*

In the above excerpt, the process through which Tom and Lily built the trial skill for the dissolution phenomenon is shown. Tom hypothesized that if the red color falls into the lake, the lake will be red, because of previous experience with a mixture of playdough. During their scientific play, he stated that its substance is different (red is red, vinegar is vinegar). He used this idea many times to explain that we need to try a substance to know if it would be dissolved. He explained what happened during play as rescuers shed their blood after a shark attack in the imaginary situation, and

Image 5.3 Developing the skill of trial

red color was dissolved out the imaginary situation. He also tried to communicate his idea about the methodological tool for dissolution. Lily also mentioned simulations and watching videos of experiments on a computer or a tablet as a means of getting more information. Later, when Olivia asked how they could know if a substance would be dissolved without using the above, Lily suggested trying the substance and testing it (Image 5.3).

The evidence presented here makes visible that children manage to develop one more core idea in approaching the dissolution phenomenon: the trial skill. Through collaboration, discussion, wondering, hypothesizing, and utilizing previous observations children develop during scientific play this basic methodological tool. Following this fundamental science methodology managed to further explore the phenomenon beyond specific objects, procedures, and situations.

5.10 Discussion and Conclusions

The present study explored how scientific play created the conditions for the formation of the concept of dissolution by preschool children within early childhood educational settings. Taken together, the findings revealed that through a set of play-based activities children were oriented towards specific science learning outcomes. The way children developed their scientific play within this framework created the conditions for the children to form the concept of dissolution. Children's learning about the phenomenon became evident through the way they gradually approached critical aspects of the phenomenon. These aspects are mentioned in turn: (a) children started thinking about the phenomenon of dissolution in an abstract way focusing on properties of the matter such as moisture and consistency, (b) children's thinking about the phenomenon made a transition from a macroscopic to a microscopic level where substances still exist even not visible to the naked eye, (c) children managed to distinguish the substances into two main categories: substances that can be dissolved and others that cannot, crafting a narrative around the conservation of the substances in both case, and d) children developed the trial skill to test if a substance can be dissolved or not.

The findings also highlighted that within scientific play children unpacked their everyday knowledge and understandings about the phenomenon and made critical interrelations with their social and cultural reality using a wide set of cultural artifacts such as drawings to expand and deepen their understandings. Children then used these everyday concepts that were formed through everyday experiences and understandings e.g., hydration and dehydration to form aspects of the scientific concept of dissolution e.g., properties of the matter. Social interactions and participation within the collective scientific play also allowed children to approach the phenomenon collectively as a team and reach more advanced understandings. As children start forming the scientific concept of dissolution their thinking about the phenomenon became more abstract and they were able to conceptualize the phenomenon beyond specific objects, procedures, and situations as it was shown in the fourth vignette.

Imagination was critical during children's scientific play. Children's imagining made them think at a microscopic level imagining a substance so small that exists even though people's eyes cannot see it without a "tool". This could be seen for example, when Lily mentions that if her peers were smaller and could emerge themselves into the water, perhaps they could see the vinegar. These findings come in line with findings from Fleer (2017) that showcases how play can be a means for children to imagine being microscopic and form an understanding of the microscopic world.

It became also evident that using drawings allowed children to reflect on their learning experience and deepen their understandings of the phenomenon (Delserieys et al., 2017; Georgantopoulou et al., 2016; Fragkiadaki et al., 2019; Fragkiadaki et al., 2021a). What was evident is that in the process of creating their drawings children reflected on and expanded their collective science experience. The narratives that children crafted while drawing support children in deepening their understandings. This could be seen for example when Lily mentions in her drawing that vinegar is presented everywhere although her idea during play was that when something dissolves, it does not exist.

Early childhood teachers' pedagogical practices motivated children toward science learning and supported the development of their scientific thinking. Both early childhood teachers played with the children and while playing enhanced children's concept formation about the phenomenon by constantly supporting children to elaborate and explain their ideas, provide arguments, and form explanatory schemes compatible with the scientific model used in early childhood science education. What was important here was that the early childhood teachers engaged with children's play after having a teaching and learning plan based on a set of teaching goals related to the science concept. Early childhood teachers' engagement with children's play was aimed at improving and advancing children's learning experience along with their play experience as a whole.

The overall findings agree with the findings in Christidou's research (2006) as children appeared to focus on the dissolved substance to explain the phenomenon. What is different here is that no typical teaching and learning intervention about the phenomenon took place. It was shown that children formed the concept through their scientific play. Thus, the above findings are particularly important given that the learning process was achieved through a play-full process that was personally meaningful for the children as well as led by the children. What was also shown is the critical role of children's creativity in developing their scientific play. Thinking creatively in their play children managed to form and test new ideas, generate solutions, and advance their understandings of the materials and the phenomenon. Being creative in their scientific play allowed children to introduce new aspects to the imaginary scenario and this also deepened their wonder and expanded the inquiry over several days. At the same time, children's imagination appeared to promote creativity in play in a sense that allowed children to change the meanings of the objects and the materials they used and create new worlds of play and new opportunities for learning about the concept. The dialectical interrelation between imagination and creativity was made visible in children's scientific play.

The present study has only explored a case study of five children from one kindergarten in an urban area in Greece. Therefore, the sampling was limited in terms of the number of children that participated as well as the regional range of the participants. Given the small sample size, the findings might not be transferable to other pedagogical contexts in the same way. Future work will focus on expanding the sample and the amount of the generated data. Despite the above limitations, unique qualities, and the dynamics of learning in play-based settings were shown through the empirical data analysis. The study adds to the literature in the field by providing empirical documentation about how science learning happens during children's scientific play and how scientific play can be shaped and supported by early childhood teachers to address specific teachings goals with concrete and advanced learning outcomes. The findings suggest that learning goals based on subject matter knowledge are critical for shaping the overall scientific play experience of the child. The coherence, consistency, and balance between teaching-learning goals, subject matter knowledge, and play is the key asset for advancing learning outcomes in science in the early years. The study informs everyday educational practice and science Pedagogy by providing new insights into learning and development in science unpacking children's intellectual, social, emotional, and enactive needs. These remarks constitute fertile ground for future research works.

References

Adbo, K., & Vidal Carulla, C. (2019). Designing play-based learning activities in the preschool environment. *Chemistry Education Research and Practice, 20*(3), 542–553.

Akman, B., & Özgül, S. G. (2015). Role of play in teaching science in the early childhood years. In K. Cabe Trundle & M. Saçkes (Eds.), *Research in early childhood science education* (pp. 237–258). Springer.

Andersson, K., & Gullberg, A. (2014). What is science in preschool and what do teachers have to know to empower children? *Cultural Studies of Science Education, 9*(2), 275–296.

Australian Government, Department of Education, Skills and Employment. (2010). *Belonging, being & becoming – The early years learning framework for Australia*. Retrieved April 21, 2020, from https://bit.ly/2yCuGTb

Bulunuz, M. (2013). Teaching science through play in kindergarten: Does integrated play and science instruction build understanding*? European Early Childhood Education Research Journal, 21*(2), 226–249.

Christidou, V. (2006). Accounting for natural phenomena. Explanatory modes used by children. *International Journal of Learning, 12*(8), 21–28.

Delserieys, A., Impedovo, M. A., Fragkiadaki, G., & Kampeza, M. (2017). Using drawings to explore preschool children's ideas about shadow formation. *Review of Science, Mathematics and ICT Education, 11*(1), 55–69.

Elkonin, D. B. (1999). The development of play in preschoolers. *Journal of Russian and East European Psychology, 37*(6), 31–70.

Elkonin, D. B. (2005). The psychology of play. *Journal of Russian and East European Psychology, 43*(1), 11–21.

Fesseha, E., & Pyle, A. (2016). Conceptualising play-based learning from kindergarten teachers' perspectives. *International Journal of Early Years Education, 24*(3), 361–377.

Fleer, M. (2006). Meaning-making science: Exploring the sociocultural dimensions of early childhood teacher education. In K. Appleton (Ed.), *Elementary science teacher education. International perspectives on contemporary issues and practice* (pp. 107–124). Lawrence Erlbaum Associates.

Fleer, M. (2009). Supporting scientific conceptual consciousness of learning in 'a roundabout way' in play-based contexts. *International Journal of Science Education, 31*(8), 1069–1089.

Fleer, M. (2011). 'Conceptual play': Foregrounding imagination and cognition during concept formation in early years education. *Contemporary Issues in Early Childhood, 12*(3), 224–240.

Fleer, M. (2017). *Play in the early years*. Cambridge University Press.

Fleer, M. (2018). Conceptual playworlds: The role of imagination in play and learning. *Early Years, 3*, 1–12.

Fleer, M., & Pramling, N. (2014). *A cultural-historical study of children learning science*. Springer.

Fleer, M., Fragkiadaki, G., & Rai, P. (2020a). STEM begins in infancy: Conceptual PlayWorlds to support new practices for professionals and families. *International Journal of Birth and Parent Education, 7*(4), 27–31.

Fleer, M., Fragkiadaki, G., & Rai, P. (2020b). Methodological challenges of studying children in a living laboratory: Case example of conceptual PlayLab. *Cultural-Historical Psychology, 16*(3), 47–59. https://doi.org/10.17759/chp.2020160306

Fragkiadaki, G., Fleer, M., & Ravanis, K. (2019). A cultural-historical study of the development of children's scientific thinking about clouds in everyday life. *Research in Science Education, 49*(6), 1523–1545. https://doi.org/10.1007/s11165-017-9665-8

Fragkiadaki, G., Fleer, M., & Ravanis, K. (2021a). Understanding the complexity of young children' s learning and development in science: A twofold methodological model building on constructivism and cultural- historical strengths. *Learning Culture and Social Interaction, 28*, 10046. https://doi.org/10.1016/j.lcsi.2020.100461

Fragkiadaki, G., Fleer, M., & Rai, P. (2021b). Innovation in early childhood and primary education. In *Educational innovation in society 5.0 era: Challenges and opportunities* (pp. 7–10). Routledge.

Georgantopoulou, A., Fragkiadaki, G., & Ravanis, K. (2016). Clouds as natural entities in preschool children's thought. *Educational Journal of the University of Patras, 3*(2), 114–128.

Greek Ministry of National Education and Religious Affairs – Greek Pedagogical Institute. (2002). *Cross-thematic curriculum framework for the kindergarten and curriculum for activities' development*. Greek Pedagogical Institute. Retrieved April 21, 2020, from http://www.pi-schools.gr/programs/depps

Greek Ministry of National Education and Religious Affairs- Greek Pedagogical Institute. (2011). *Kindergarten Curriculum*. Greek Pedagogical Institute.

Hakkarainen, P. (2008). The challenges and possibilities of a narrative learning approach in the Finnish early childhood education system. *International Journal of Educational Research, 47*(5), 292–300.

Holding, B. (1987). *Investigation of school children's understanding of the process of dissolving with special reference to the conservation of matter and the development of atomistic ideas* (PhD thesis). University of Leeds, Leeds.

Irish Government. (2009). *Early childhood curriculum framework*. Retrieved April 21, 2020, from https://bit.ly/2VbzkyY

Kravtsov, G. G., & Kravtsova, E. E. (2010). Play in LS Vygotsky's nonclassical psychology. *Journal of Russian & East European Psychology, 48*(4), 25–41.

Lynch, M. (2015). More play, please: The perspective of kindergarten teachers on play in the classroom. *American Journal of Play, 7*(3), 347–370.

Miller, E., & Almon, J. (2009). *Crisis in kindergarten: Why children need to play in school*. Alliance for Childhood.

O'Connor, G., Fragkiadaki, G., Fleer, M., & Rai, P. (2021). Early childhood science education from 0 to 6: A literature review. *Education Sciences, 11*(4), 178. https://doi.org/10.3390/educsci11104017

Panagiotaki, M. A., & Ravanis, K. (2014). What would happen if we strew sugar in water or oil? Predictions and drawings of pre-schoolers. *International Journal of Research in Education Methodology, 5*(2), 579–585.

Pea, R. D. (1993). Learning scientific concepts through material and social activities: Conversational analysis meets conceptual change. *Educational Psychologist, 28*(3), 265–277.

Piaget, J., & Inhelder, B. (1974). *The child's construction of quantities*. Routledge and Kegan Paul.

Rai, R., Fleer, M., & Fragkiadaki, G. (2021). Theorising digital tools: Mutual constitution of the person and digital in a conceptual PlayWorld. *Human Arenas, 1–18*. https://doi.org/10.1007/s42087-020-00178-8

Remountaki, E.-L., Fragkiadaki, G., & Ravanis, K. (2017). The approach of solid in liquid dissolution in early childhood education settings: A socio-cultural approach. *European Journal of Education Studies, 3*(6), 303–318.

Rosen, A., & Rosin, P. (1993). Now you see it, now you don't. The pre-school child's conception of invisible particles in the context of dissolving. *Developmental Psychology, 29*(2), 300–311.

Roth, W.-M., Goulart, M. I. M., & Plakitsi, K. (2013). *Science education during early childhood. A cultural-historical perspective*. Springer.

Sacks, H. (1995). *Lectures on conversation*. Blackwell.

Tu, T. (2006). Preschool science environment: What is available in a preschool classroom? *Early Childhood Education Journal, 33*, 245–251.

UK Government, Early years foundation stage statutory framework. (2017). *Statutory framework for the early years foundation stage. Setting the standards for learning, development and care for children from birth to five*. Retrieved April 21, 2020, from https://bit.ly/3bTHcfd

Vartiainen, J., & Kumpulainen, K. (2020). Playing with science: Manifestation of scientific play in early science inquiry. *European Early Childhood Education Research Journal, 28*(4), 490–503.

Vygotsky, L. S. (1966). Play and its role in the mental development of the child. *Voprosy Psikhologii, 12*(6), 62–76. https://doi.org/10.2753/RPO1061-040505036

Vygotsky, L. S. (1987). Imagination and its development in childhood. In R. W. Rieber & A. S. Carton (Eds.), *The collected works of L.S. Vygotsky* (Vol. 1, pp. 339–349). Plenum Press.

Vygotsky, L. S. (2004). Imagination and creativity in childhood. *Journal of Russian and East European Psychology, 42*(1), 7–97. https://doi.org/10.2753/RPO1061-0405280184

Eirini-Lida Remountaki works as an early childhood teacher. She graduated with honors from the Department of Early Childhood Education and Educational Sciences of the University of Patras in Greece. She has a MSc in Educational Sciences with specialization in Natural Sciences, Mathematics and ICT Didactics from the University of Patras.

Glykeria Fragkiadaki is an Assistant Professor of early childhood science education at the Faculty of Education, Aristotle University of Thessaloniki, Greece. Prior to this appointment, she held the position of Senior Research Fellow in the Conceptual PlayLab at Monash University, Australia, leading part of Australia's first National Programmatic Study of Conceptual Play in Science, Engineering, and Technologies. Glykeria's work builds upon the tradition of cultural-historical theory. Her research is focused on young children's conceptual learning and development in science in dialectical interrelation with the child's social and cultural reality. The concepts of play, imagination, and creativity are central to her research work. Apart from her academic background, she has also extensive teaching experience as an Early Childhood Teacher and Director, has acted as OMEP's Patras Local Committee Secretary in Greece, and has been a tutor to Professional Development Programs for Early Childhood Teachers. Glykeria has published over 30 research papers in international journals and 20 book chapters. She has experience as an Editor in collected volumes and is also an Associate Editor at the Learning, Culture, and Social Interaction Journal. She has presented her research work widely at international and national conferences and has also been invited as a keynote and plenary speaker at academic events worldwide.

Konstantinos Ravanis is Professor in Physics Education at the University of Patras in Greece. He did two types of undergraduate studies, in Physics and in Educational Sciences at the University of Patras in Greece, completed his postgraduate studies in Science Education in the Université Paris VII-Denis Diderot and received his PhD degree in Physics Education from the Department of Educational Sciences and Early Childhood Education of the University of Patras. In 1988 K. Ravanis was appointed "Special Scientist", and then he was elected through at all levels of the academic hierarchy to reach the degree of Professor (2004) in the Department of Educational Sciences and Early Childhood Education of the University of Patras, specializing in "Didactics of Physics". He was an invited Professor in Université de Provence, Aix-Marseille Université, Universidad de Buenos Aires, Université de Bretagne Occidentale, University of Nicosia and he had a research fellowship from the Jean Piaget Archives at the University of Geneva. K. Ravanis is the author of about 230 publications in scientific journals, collective editions and conference proceedings and he has been the supervisor of 14 Ph.D. thesis. He is the author of three (4) books and he was the editor of two (2) collective scientific editions. K. Ravanis was director of the Division of Theoretical and Applied Pedagogies (2003–2008) and has been Chairman in the Department of Educational Sciences and Early Childhood Education (1999–2001, 2005–2006), Vice-Rector (2006–2010) and member of the Council Foundation (2014) of the University of Patras.

Chapter 6
'On the Way to Science...' Development of the Scientific Method in the Early Years

Eleni Kolokouri and Katerina Plakitsi

6.1 Science Education in the Early Years: Concepts and Processes

Science Education in the early years, has been considered a significant learning domain and has been connected with scientific inquiry as well as the development of skills and attitudes. According to the cross-disciplinary approach in the early years' curricula, Science Education includes Physics, Biology, Chemistry, Environmental Education and Education for Sustainable Development. Science Education in the early years is directly connected with the exploration of authentic learning environments, as well as working with the scientific method, using science process skills such as observation, classification, communication, etc. in an effort to understand the world around us. Furthermore, there is a strong relationship between science and society in the early years. As learners make their first steps in the world of education matters such as ethics, culture, informal learning and social changes are strongly connected with scientific learning (Plakitsi, 2008).

Investing on the scientific interest of students contributes to the creation of a basis for the promotion of scientific knowledge from the early years. Many researchers (Vosniadou, 2019; Kampourakis, 2018; Ravanis, 2017; Fleer, 2015; Roth, 2011; Wells, 1994) believe that Science Education must start in the early years as at this age, learners construct structures for understanding scientific concepts and

E. Kolokouri (✉)
Department of Early Childhood Education, University of Ioannina, Ioannina, Greece
e-mail: ekolokouri@uoi.gr

K. Plakitsi
Department of Early Childhood Education, School of Education, University of Ioannina, Ioannina, Greece
e-mail: kplakits@uoi.gr

K. Plakitsi, S. Barma (eds.), *Sociocultural Approaches to STEM Education*,
Sociocultural Explorations of Science Education 21,
https://doi.org/10.1007/978-3-031-44377-0_6

moreover, develop a lifelong interest in science. In relation with the processes of approaching the properties of physical objects and materials, the phenomena and the scientific concepts, four frameworks of learning and development in the early years summarize the different trends in Science Education: the empiricist, the Piagetian, the socio-cognitive and the socio-cultural or cultural-historical approach (Ravanis, 2021).

Early years Science Education curricula provide the opportunity to learn about the characteristics of scientific concepts, encourage the use of the scientific methods and science process skills, develop positive attitudes and values to science and the environment, all of which contribute to responsible citizenship. In this context, teacher and learner collaborate to achieve a learning outcome connected with real life situations. During the design of educational material, ideas and previous knowledge of learners are taken into account and are finally transformed to scientific knowledge (Ravanis et al., 2013; Plakitsi, 2008). Within this frame, knowledge is related to the theories, laws and principles of each individual subject, didactic transformation and design of educational material so as to fit early year learners' needs. The role of the teacher, is that of a mediator and facilitator who creates learning environments adjusted to the learners' needs, develops the dynamics of the learners' groups, and provides them the appropriate tools and methods to reach scientific knowledge (Kolokouri et al., 2012). Within this frame, the approach of scientific concepts and phenomena in the early years is made through observations, recordings, drawings, formulations of hypotheses or even development of argumentation. During the preparation and design of the educational material, teachers have to take into account the personal development of pupils, the development of life skills as well as positive attitudes towards nature and the environment, all of which contribute to scientific literacy and responsible citizenship. The early years curriculum in Greece provides pupils a holistic view of the world around them (Birbili & Alexandra, 2020). At this level, all subjects are dealt through a horizontal linking and not as independent fields of study and furthermore, they are connected with social sciences. Furthermore, it sets five frames of learning for the early grades, that is games, routines, everyday-life situations, explorations, organized learning activities in which all early-grade pupils should equally participate. Learning in this frame is transferred from the individual to the learning community and is connected with the sociocultural backgrounds of the learners.

The development of concepts in the early years is an ongoing process dialectically connected with moments of change and as a result they cannot be self-identical, they are always different (Roth et al., 2013). In other words, they are a cultural-historical product of the wider community, transmitted through time to the learners by instruction. Concepts lead to actions and allow the observer to make sense of a subject's actions (Blunden, 2013). According to Vygotsky (1993), the development of concepts takes the cultural and historical environment of the learners as concepts are units of a culture, from which they may be acquired by an individual. Furthermore, he claimed that the level of development of scientific concepts forms a zone of proximal development of everyday concepts. There has been a differentiation between everyday and scientific concepts as one the one hand, they are both linked dialectically to actions (Bakhtin, 1986), there is a significant difference between the two types of concepts on the other. Everyday concepts are learnt through participation to everyday activities while scientific concepts are learnt through instruction in specialized settings (Wells, 1994).

Science Education in the early years' curricula is connected to authentic learning situations in formal and informal environments using the scientific method.[1] In this sense, early years learners are engaged in scientific inquiry following the scientific method to construct their knowledge as well as a variety of science process skills. Science process skills are skills that can address to the complexity of learning in the early years as they are connected with culture and development. According to Cole et al. (2010), culture is a vehicle to make sense of the world while development includes the use of cultural tools such as language or symbols which reinforce learning. Science Education in the early years is a complex issue as it aims to make basic scientific concepts and phenomena understandable on the one hand and deal with adapting the scientific content to the experiential way of teaching for early learners on the other. Another issue is the relation between teaching and learning and what early learners comprehend and remember from participating in the teaching and learning process (Åkerblom et al., 2019).

Science process skills facilitate Science Education in the early years as they encourage active learning, appropriate to many science disciplines and reflect the way that scientists work (NARST, 1990). The basic science process skills are described below:

Observing: collect information about an item or a phenomenon by using the senses.
Inferring: use data or information to draw a conclusion about a phenomenon.
Measuring: use measures or estimates to describe the dimensions of an item.
Communicating: a verbal or nonverbal action to describe an action, item or event.
Classifying: create groups or categories of certain items or events based on criteria.
Predicting: use evidence to describe the outcome of a future event.
Formulating hypotheses: describe the expected outcome of an experiment.
Controlling variables: identify variables that can affect results in an experiment, by manipulating only the independent variable.
Defining operationally: define ways to measure a variable in an experiment.
Interpreting data: organize data and draw conclusions from it.
Experimenting: conduct experiments.
Formulating models: create a model as a representation of a natural phenomenon or concept.

These skills address to the early learners' natural curiosity and their tendency to ask questions and in this sense, engage them in authentic learning situations in which

[1] The scientific method in the early years describes a systematic model to approach scientific concepts and includes science process skills such as observation, questioning, predicting, experimenting, summarizing, and sharing results (Gerde et al., 2013). It supports scientific inquiry through a holistic approach both in learning in a school classroom and throughout daily activities in the early learners' lifeworlds.

learners approach scientific concepts through investigation, are encouraged student to share their questioning, while offering them space for scaffolding in their zone of proximal development (Jirout & Zimmerman, 2015). Thus, science process skills facilitate the developmental trajectory of learning scientific concepts as they support following the scientific method in structured activities mediated by artifacts and put learning in a socio-cultural and historical basis.

6.2 Cultural Historical Activity Theory (CHAT) as a Framework for Science Education

Activity Theory has origins in the classical German philosophy (from Kant to Hegel), in the writings of Marx and Engels, and in the Russian studies in psychology of Vygotsky, Leontiev and Luria. The sociocultural frame of Activity theory provides a context of human activity and links the individual to the social level (Engeström, 2005). The basic principles of Activity Theory include object-orientedness, the hierarchical structure of activity, internalization/externalization, tool mediation, development, and contradictions as a source of change and development (Engeström, 2015; Kaptelinin & Nardi, 2017). Within this frame, the central questions that define the object in the course of learning are the selection of the subjects of the learning process, the motivation towards learning, the content, the key actions of learning as well as the outcome (Plakitsi, 2013). Knowledge is considered as part of object-oriented and artifact-mediated activity (Vygotsky, 1978).

The unit of analysis is the activity which includes the person or group who is acting towards an object, following certain rules and the dynamic relationships that develop within the activity system (Engeström, 2016). Furthermore, learners interact with one another as well as with tools and means into the learning community and work on the construction of knowledge with outcomes that are scientifically accurate (Engeström, 2005). Within the cultural-historical framework of CHAT, human interactions with the environment are socially defined and contribute to the comprehension of the human mind (Blunden, 2013). In a school classroom teacher and students co-operate to reach the object of learning, they both use tools which are on the top of the activity system and they explore about learning concepts and their connections with everyday life; subjects become engaged in learning activities and use physical or mental tools in order to deal with scientific concepts. The role of the teacher is that of a mediator who does not just teach concepts but provides tools to learners in order to reach knowledge. Learning is expanded in the wider community which includes family and social organizations all of which act in various ways within an activity system. Each learning task is a system that makes progress through space and time (historicity). The internal four-level contradictions that appear within an activity system (Engeström, 1987) lead to progress and furthermore, contribute to re-organizing the process of problem-solving and comprehension of scientific concepts.

"Contradictions are not the same as problems or conflicts. Contradictions are historically accumulating structural tensions within and between activity systems...Activities are open systems...Such contradictions generate disturbances and conflicts, but also innovative attempts to change the activity." (Engeström, 2001, p. 137).

The object is a fundamental concept in CHAT (Blunden, 2013). In fact, it generates an activity system as it provides motivation towards the desired aim as well as the outcomes of the activity. Science as a learning object is not considered as an isolated task of a curriculum but it is incorporated in general education, which aims at the development of the individual through the development of skills as well as a positive attitude towards creativity and progress on a personal and a social level.

In this process tools and artifacts play a significant role in accordance with Vygotsky's initial insight that tools mediate activities and as a result create connections between human and objects as well as other people (Wells, 1994).

6.3 On the Way to the Scientific Method with the Aid of Cartoons

Cartoons and animations have been widely used by education researchers in different domains of teaching and have been designed to teach specific concepts in an effective way (Doran, 2019; Naylor & Keogh, 2013; Bowkett, 2011). Ted-ed has produced a series of animation in which they combine graphics and storytelling so as to produce memorable and entertaining learning content. Naylor and Keogh (2013), support teaching and learning scientific concepts by using concept cartoons. Thus, they offer an alternative method of science teaching which involves discussion, investigation and motivation for learners of all educational levels. Furthermore, the use of cartoons and animations for education in general and more specifically for Science Education seems to have had a growing interest. There has been an increase in the number of articles that mention animation in major science education journals in 2019. According to Unsworth (2020), '*The number of articles mentioning animation in major science education journals in 2019 was double the number in 2010. In Research in Science Education the increase was from 9 to 24 and in the International Journal of Science Education from 20 to 40*' (p. 3).

The use of cartoons in the early years' education is connected with everyday situations which make pupils lacking in confidence more likely to participate in science activities. Moreover, it is an easy way to visualize and explain concepts no matter how complex they are as complex information is depicted with illustrations and then these are animated to describe a process. In this sense, cartoons aid the creation of mental representations of concepts and also deal with difficult cognitive processes. Humor, exaggeration, symbols, emotions are elements that provide learners with very interesting types of knowledge presented in a familiar context. Furthermore, media representations of teaching using cartoons and animations capture pupils' attention and can be very effective as they are very familiar with their everyday life. Cartoons can support teaching as they become a strategy to elicit pupils' ideas, challenge their thinking and develop their understanding. In this sense, they engage early learners in exploring a variety of scientific concepts, in developing science process skills, in creative thinking and providing solutions to problems without restraint. Teachers can create a simulation-based intelligent learning environment for their pupils than can promote change in their professional practice (Kolokouri, 2016). This chapter presents part of a research study in which an early years' science curriculum was developed in two parts, to teach floating and sinking and light properties with the use of cartoons. The purpose of the study was set under a CHAT perspective and accordingly seeks to:

- Design and analyze Science Education activities in the CHAT framework with emphasis on the interactions that take place in the activity systems while teaching.
- Use cartoons as a mediating tool to help early grade pupils gain experience about scientific concepts such as properties of light, shadows and colors.
- Provide pupils the opportunity to obtain skills of the scientific method and life skills.
- Connect knowledge with everyday life and develop metacognitive skills.

The chapter puts emphasis on the study of the science process skills that early learners can practice with the use of cartoons and addresses the need to connect learning with real life situations and this way reinforce the sociocultural aspect of Science Education in the early years.

The science curriculum consisted of two parts and in the first part, a popular cartoon was used to teach floating and sinking through educational drama strategies such as receiving a letter from the cartoon hero, teacher in role, role play, developing freeze frames, role on the wall, conscience alley and thought tracking. The basic idea was that the cartoon hero needed to find out which items sink or float in water and the pupils with the teacher. Pupils exchanged roles in order to find the solution of problem concerning floating and sinking concepts. They defined the place and time and through the drama strategies and a series of science education activities of experiments they reached the solution of the problem and sent the cartoon hero all the information they collected about floating and sinking (Kolokouri & Plakitsi, 2013). The second part of the curriculum concerns a narrative about light, colors and shadows entitled 'Colors from the past' which makes connections of scientific concepts with History of Science and scientists. Then, the narrative was turned to a twenty-minute animation about light, colors and shadows in the program scratch (http://scratch.mit.edu/) and a series of science activities about light, colors and shadows were designed. The animation was divided in five episodes and each time pupils watched the episode and were involved in didactically transformed activities such as discussion about the episode, role playing, organizing experiments, making lists and drawing results of experiments. Pupils participated in problem solving situations and interacted with each other as well as with the teacher (Kolokouri, 2016).

The didactic strategies that were used in both parts of the science curriculum followed the methodological framework of CHAT (group work, use of instrumental and conceptual tools in object-oriented classroom activities, interactions between subjects, contradictions). Furthermore, the curriculum was connected with the five frames of learning in the early grades (games, routines, everyday-life situations, explorations, organized learning activities) in which all early-grade pupils should equally participate. Following Vygotsky's ideas about the development of imagination through childhood play (1999), all five frames are connected with play as a means of learning, building relationships and practicing roles in different situations. Within this frame, spontaneous learning is encouraged mostly in the first four frames while in the fifth frame the teacher acts as the source of learning through the organized learning activities. In the following table (Table 6.1), a cluster analysis of the two parts of the science curriculum shows the activities that have been grouped together because of the characteristics they have in common regarding their connection with the five frames of learning. This was done in the beginning phases of the analysis to help see the connections between the classroom activities and the frames of learning of the National Curriculum for Early Childhood.

The methodology used in this study was based on the analysis of Activity Systems by the view of Engeström (2005) and the cultural- historical approach in the early years development by Fleer and Ridgway (2014). Within the frame of CHAT, learning, knowledge and expertise are distributed in the community through learners' participation. Learning is a process of social interaction and takes place through collaboration with other people so as to develop skills and abilities. Central roles are collaboration and language as a tool to strengthen the identity of the individual (Vygotsky, 1978; Leontiev, 1979; Nardi, 1996). This notion puts emphasis on the complexity of group work, the increasing levels of awareness and the consciousness of higher intellectual functions such as problem solving and reasoning in the teaching approach (Wertsch, 1985).

In this study, emphasis was put on: developing conceptual tools to understand dialogue, multiple perspectives, and networks of interacting activity systems and

Table 6.1 Connection of the science curriculum with the 5 frames of learning

	Games	Routines	Everyday life situations explorations	Organized learning activities
Part 1	Spongebob's games	Painting of bikini bottom	The letter	First contact with materials
	Frozen images	Discussion in circle	Teacher in role	Experiments
		Evaluation	Role on the wall	Telephone conversation and experiments
			Conscience alley	Prediction board
Part 2	Rays of light	Expressing ideas about colors	Narration: Color visions from the past	Sources of light
	Course of light	Colors activities	Life of a scientist	Color experiments
			The letter	Creation of shadows

introducing teaching scientific concepts by using cartoons to develop scientific thinking. As a result, the study focuses on the use of CHAT as a framework for designing and analyzing science education activities for the early grades in developing science process skills as well as studying the interactions within the learning community.

The design and implementation of the science curriculum followed the basic parts of SCOPES (Systems of activity, Contradictions, Outcomes, Praxis, Expansive learning, Science education), a methodological tool for Science Education in early childhood and primary education based on CHAT (Kolokouri & Kornelaki, 2019). Within this frame, the curriculum involved the activity systems of the pre-school classrooms that it was implemented, faced the contradictions that occurred as a source of change and development, focused on the outcomes as it was expected to become a paradigm for designing educational material for the early grades and included praxis through participatory methods in both parts. It was developed in an expansive cycle of learning (Engeström, 2015), which consists of a sequence of learning actions. The theoretical grounds of expansive learning are summarized in eight principles (Engeström & Sannino, 2010, 2012):

1. The separation of the action of the activity with the introduction of the division of labor by Leontiev (2009).
2. The concept of zone of proximal development as a fundamental element of the theory of expansive learning (Engeström, 2000).
3. The theory of expansive learning as an application of activity theory is based on object orientation (Engeström, 2000).
4. Activity Theory is a dialectical theory and the concept of contradiction plays a very important role (Ilyenkov, 1977).
5. Expansive learning is based on the dialectical method of the abstract to the concrete developed by Davydov (1999).
6. The mediation of action through cultural tools and signs is based on Vygotsky's human psychological functioning (Vygotsky, 2003).
7. The contribution of Bateson's theory about the three levels of learning (Bateson, 2000).
8. The idea of Bakhtin (1986) for the heteroglossia (voicedness).

The majority of these principles were followed in the design and implementation of the curriculum as it was object-oriented, focused on the division of labor within the groups of learners, used strong cultural tools such as cartoons, dealing with contradictions as a source of development and promote learning through social interaction. The series of learning actions used in this curriculum are presented below as the steps of in an expansive cycle (Engeström, 1999):

Questioning: In what way can cartoons be used so as to deal with scientific concepts in the early years? Which science process skills ca be developed with the use of cartoons?

Analyzing the situation: This part included the analysis and comprehension of existing methods and practices of cartoons in early years education. Analysis was conducted through the educational curricula both at a local and an international level so as to develop educational material.

Modeling: This part concerned the development of a workshop within the frame of a university course in which the first part of the curriculum concerning floating and sinking was designed and pre-service teachers used it during their internship.

Examining-Implementing the model: Further development of the first part of the curriculum and development of the second part which concerned light, colors and shadows. Implementation of the curriculum in 4 pre-primary school classrooms.

Reflecting and evaluating: Collection and analysis of all data from both interventions.

Consolidating the new practice: Making final amendments in the curriculum and examining the case that this approach would also fit for teaching different scientific concepts. Emphasis was put on the use of tools that will be appropriate for learning communities in different educational settings as well as on the contradictions that occurred during the curriculum implementation.

6.4 Data Analysis and Results

The Science Curriculum was implemented in four kindergarten school classrooms of 25 pupils each, 5 and 6 years old (Table 6.2).

The duration of the curriculum implementation was 10 weeks. The 4 schools were situated in the city of Ioannina, Greece and the teachers belong to the collaborating network of schools with the Department of Early Childhood Education. Every year, the specific schools welcome students of the Department of Early Childhood Education of the University of Ioannina both for their internship and for additional research activities in Science Education, in Environmental Education and in Education for Sustainable development in the early years. During the implementation of the curriculum, 20 lessons were conducted by the educational researcher lasting 30–55 minutes, depending on the number of activities of each teaching unit and the interest of the students. In any case, the participation of the class teacher was optional.

Research data were collected by observations, field notes, video recordings, interviews and classroom materials. Research data were analyzed through the creation of different projects in the Nvivo9 research software. The collection of

Table 6.2 Sample of data

Schools of Ioannina	Number of pupils	Boys	Girls	Classroom teachers
1st	25	14	11	1
2nd	20	8	12	1
3rd	23	13	10	1
4th	25	12	13	2
Total	93	47	46	5

data included 236 video files, 179 drawings, 2 word files of the 2 parts of the Science curriculum and 195 photos. All data were articulated and connected with the methodology and the research questions. Nvivo9 research software was used to classify, sort and arrange all the collected data and examine the relationships as well as combine analysis with the theoretical framework of CHAT.

In the word frequency dialogue below (Fig. 6.1) we can see the most frequently words used while analyzing. The words ‘interactions, pupils, group, communication, hypotheses, experiments, object’ are of high importance as they show the course of action taken within the activity system in order to reach the object. Thus, pupils interact within the learning community and use the scientific method in order to approach the scientific concepts.

In the word frequency query in Fig. 6.2, conceptual relations that concern the learning procedure are illustrated. Pupils realize that scientific concepts can be developed in different institutional settings of the present and the past, as a result of collaborative action, critical thinking, problem solving and argumentation. In this sense, internal activities, such as pupils’ understanding of the scientific concepts are shaped with external activities and they both unify to form knowledge structures.

The figure below (Fig. 6.3) is illustrating the science process skills that pupils practiced during the implementation of the science curriculum. Using cartoon characters and role-playing pupils visualized time and space relations, made hypotheses in order to approach the scientific concepts. Furthermore, they organized experiments, made predictions and verified them at the end. Communication was the most prevalent and was combined with almost every other skill. Communication

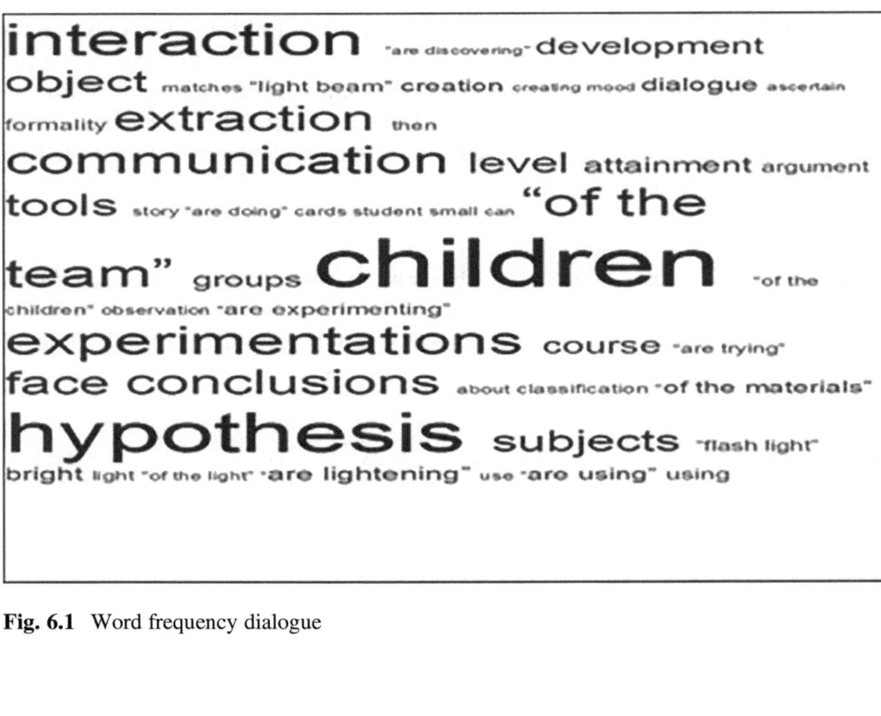

Fig. 6.1 Word frequency dialogue

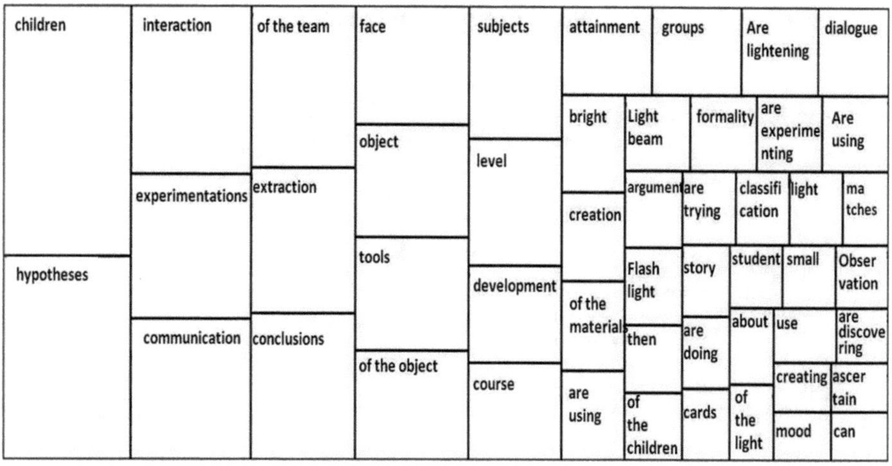

Fig. 6.2 Conceptual relations

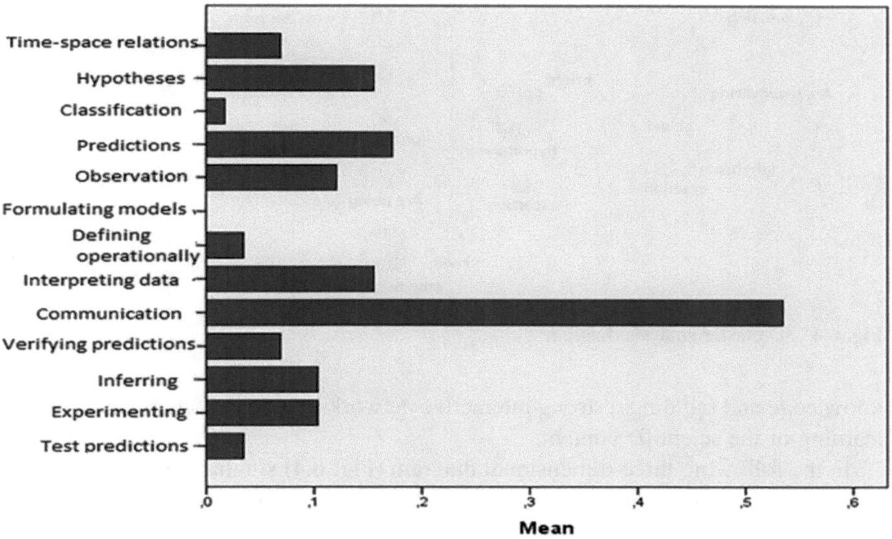

Fig. 6.3 Science process skills

is an essential skill for early-year pupils as they use a variety of means to describe an action, object or event. During the implementation of the curriculum communication took many forms and was present at every moment of action. Pupils communicated in order to share their observations or predictions and tried to make themselves understood in an effective way. Furthermore, dealing with scientific concepts with the aid of cartoon characters and role-playing involved forms of communication which contributed to better understanding of science, connecting with prior

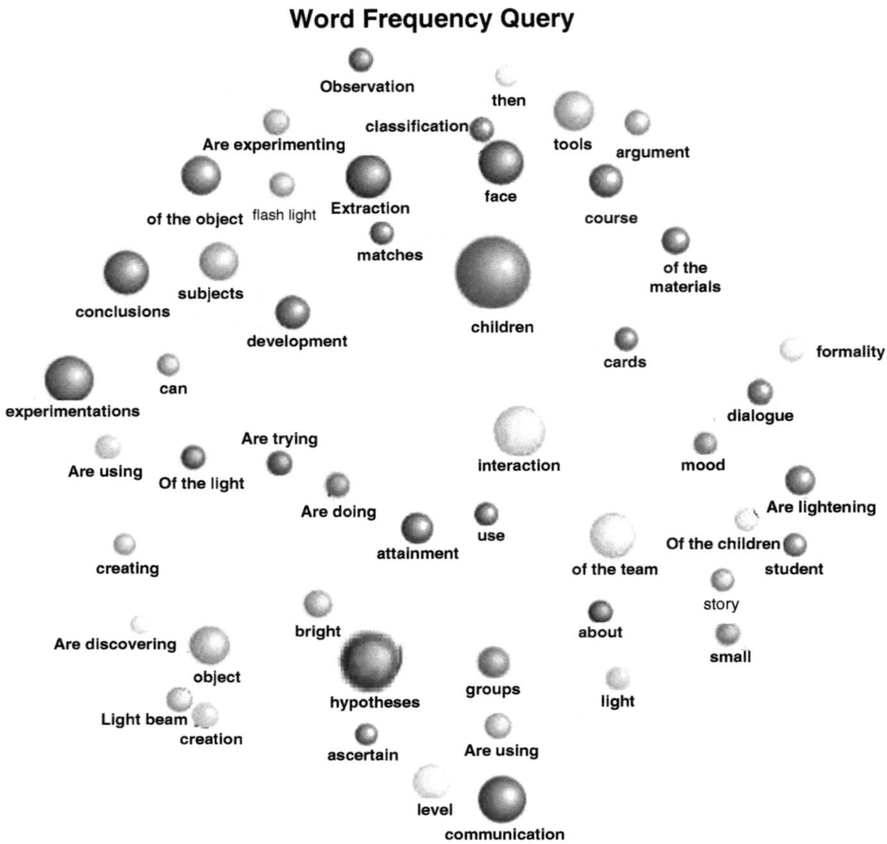

Fig. 6.4 3D cluster analysis diagram

knowledge and building a strong interactive network in order to achieve meaningful learning of the scientific content.

In the following three-dimensional diagram (Fig. 6.4) similar items are clustered together and form groups of the most important concepts of the didactical intervention such as types of interactions (group, course, dialogue) and science process skills that pupils follow (children, hypotheses classifications, conclusions). The connections in these clusters show the development of the essential skills for early grade pupils while they participate in an action, object or event.

During the implementation of the curriculum science process skills were present at almost every moment of action. Pupils communicated in order to share their observations or predictions made hypotheses and tried to predict before the experiments and drew conclusions. Furthermore, pupils tried to make themselves clear and effective and interact with the other person to make them understand their point

Figs. 6.5 and 6.6 Pupils conduct experiments following the scientific method with the aid of the cartoon hero

of view. The cartoon hero was present in all the activities, leading the pupils' work on the science process skills (Figs. 6.5 and 6.6).

Through the application of the tree clustering algorithm with the addition of Clementine the conditions under which students are led to the acquisition of knowledge by drawing conclusions were identified (Fig. 6.7). It is obvious that there is a significant degree of interdependence between the skill of drawing conclusions verification of predictions and interpretation. Drawing conclusions depends directly on the interpretation skill and indirectly on the prediction verification skill. We therefore conclude that science process skills aid pupils to drawing conclusions related to scientific concepts and make interpretations of data directly related to the scientific method.

Pupils described scientific concepts providing examples of their logical thinking and connected them with everyday life situations. This was a result of both group work and whole classroom discussion, in which knowledge was constructed in relation to what pupils already know with the shared classroom experience.

The following extract (Kolokouri, 2016), which is taken from a class discussion, occurred immediately after a game in pupils faced different situations connected with the presence of light. Pupils exchange their ideas and experience about light sources. Then, they classify materials to those that produce light and to those who do not (Fig. 6.8a, b).

T:	*Where do you think light comes from?*
P1:	*From the light bulb but there is light in the room anyway.*
T:	*Where does this light come from if not from the bulb?*
P1:	*From the window I suppose.*
P2:	*Light comes from the stars and the moon.*
P3:	*The fire gives out light and the candles as well.*
P4:	*I have seen light in the fireflies*
P5:	*The lighthouse, there is one in my mum's village, we go during the summer there....*
T:	*What is the use of a lighthouse?*
P5:	*I know that, it enables the ships to see at night!*

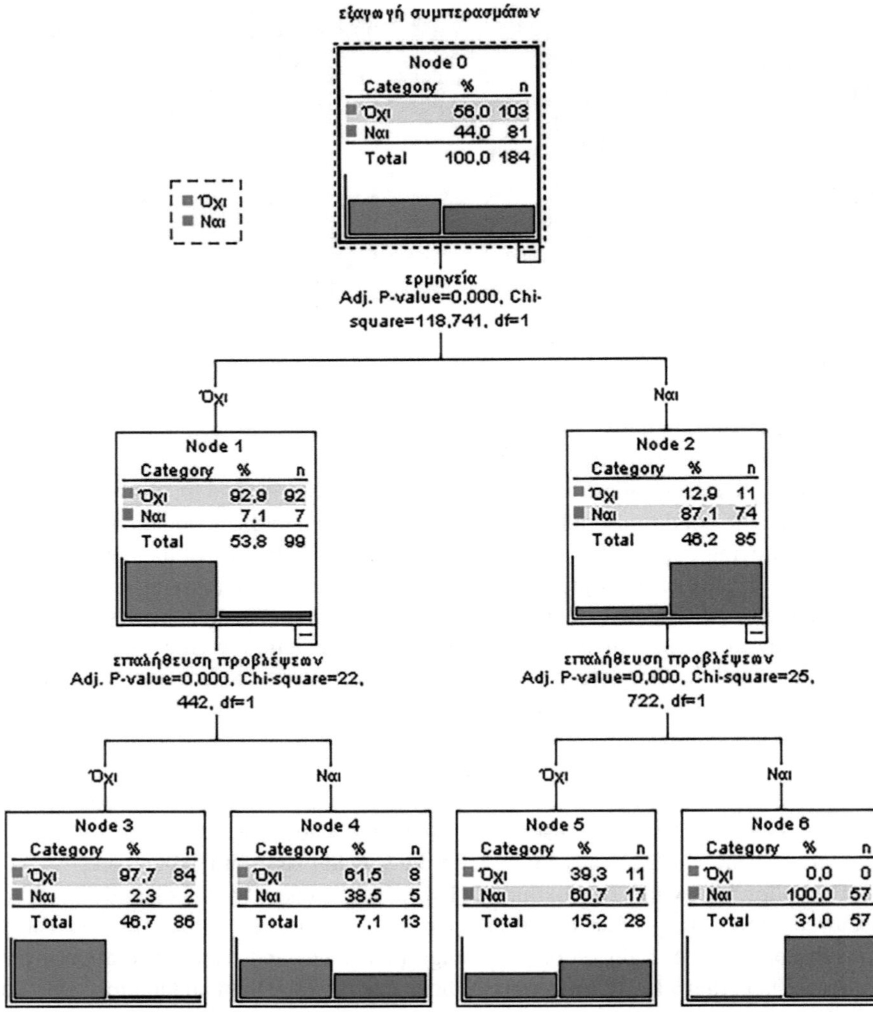

Fig. 6.7 Correlation of science process skills

P2:	*Oops, we can't see now it is very cloudy, where is the sun?*
T:	*What shall we do then?*
P2:	*We can switch on the light, there is light in the electricity poles that bring light to our homes and schools...*
T:	*Can you see now? (T. switches on the light)*
P2:	*Yes, we can.*
T:	*And now? (T. switches on the light).*
P2:	*No, we can't! Well, not very well*

Fig. 6.8 (**a**, **b**) Classification of light sources

P3: *I can see very clearly.*
P2: *No, you can't!*
P4: *Miss, miss, we can see, the light is here.*
T: *Where does it come from?*
P4 (looking around): *there it is, it comes from the window, from the sun.*

In this example, science process skills were used in an oral exchange so as to make further connections and co-construct scientific concepts. The teacher was able to explore pupils' ideas about the presence of light to gain information to design new activities and at the same time to leave them space to develop scientific thinking within their individual zones of proximal development.

6.5 Discussion and Conclusions

The present study concerned the development of the scientific method as well as the science process skills in the early years through a science curriculum with cartoon characters in CHAT framework. CHAT focuses on the connection of school instruction with everyday life and provides artifacts and approaches for analyzing collective activity, interactions within a community of practice and structural change and development.

The development of the curriculum followed an expansive learning cycle (Engeström, 1987; Engeström & Sannino, 2012; Plakitsi et al., 2018). In the course of the development, the subjects acting in the activity systems were the pupils of the pre-elementary schools and the teacher-researcher. Their actions were influenced both by the school management and their sociocultural background. They used physical and mental tools to achieve the object which were adapted to the learners' needs. The object of the pre-service teachers was to design educational material for the early years' education. The outcome included the possibility that the curriculum could become a paradigm for designing science educational material based on cartoons. The community was expanded to include the pupils' teachers on the classroom, family members and university students during their internship in certain cases. The rules were defined within the learning community with the cartoon characters being prevalent most of the time. Finally, the division of labor referred to group work, collaboration and sharing of responsibilities. Contradictions that occurred concerned mainly the complexity of the role of the teacher-researcher during the implementation of the curriculum, the rules and the use of tools within the community of early learners. The resolution of contradictions in each step of the expansive cycle led to practice change and development of the activity systems of learners.

Through the curriculum, early years learners developed the scientific way of thinking and science process skills, such as observation, hypothesis and experimental testing, gathering-utilization of information, analysis and interpretation of data as well as drawing conclusions. Furthermore, they connected their learning with everyday life situations as part of their cultural - historical and social context as we can see in the dialogues about classification of light sources.

Through the whole process it seems that all three dimensions that Miller recognizes in scientific literacy are approached: (1) understanding the scientific method and approach through the nature of science; (2) understanding basic scientific concepts and (3) awareness of the impact of science and technology on society (Miller, 1983).

The curriculum has effectively contributed to the approach of all three dimensions and to the creation of conditions for the development of scientific literate citizens who have the opportunity to deal problem solving situations.

Facione and Facione (2007) argues that critical thinking develops mainly through interpretation, analysis, evaluation of processes and drawing conclusions. In the didactic interventions of the curriculum, the skills of classification and categorization were directly related to the interpretation of facts and natural phenomena. The examination and exchange of ideas, the formulation of hypotheses, the search for arguments led to processes of analysis that strengthen the critical thinking of students. Finally, the process of drawing conclusions and evaluating the procedures included the examination of cases and alternatives and the formulation of views, the development of which led to the conclusions.

From data analysis, we observe that communication is the process with the highest frequency of occurrence, which appears in all the snapshots of both didactic interventions. Communication including verbal and non-verbal forms, is one of the basic skills in the early years' curricula. In line with creative and critical thinking, the strengthening of personal identity and autonomy and the development of citizenship-related skills fulfil the main purpose of the curriculum development.

Pupils communicated in a verbal and non-verbal way during their collaboration in both didactic interventions. Due to its collaborative nature, communication is linked to all other science process skills. Through communication they organized their action, made assumptions, gathered materials for their experiments, decided on the use of new tools and formed the rules within the learning community. In their contact with the cartoon heroes, communication was evident in the role-playing activities and in the problem-solving situations.

Experimenting by the use of the scientific method as proposed in this chapter enhanced the students' scientific way of thinking. For example, in the second part of the curriculum concerning light, shadows and colors, pupils were able through experiments to determine their position in relation to the white cloth and the light source so as to create shadows. In another case, they determined the position of their body in relation to the projector and the whiteboard to create the shadow of their hand. In order to determine position they had to stand in relation to the cloth and the light source to create their shadow, the students made assumptions, experiment and drew conclusions.

The following are some examples from the qualitative analysis of the results which show the development of science process skills:

- Pupils create shadows by moving different parts of their body in front of the light source (hypotheses, experimentation, drawing conclusions).
- Pupils raise the cards with the 'see' sign each time the hero of the story they hear can see (creating cognitive conflicts, hypotheses, developing arguments, drawing conclusions).
- Pupils experiment with the rotation of the earth around itself and around the sun (hypotheses, experimentation, drawing conclusions, communication)
- After experimenting, pupils classify items into those that produce light and those that do not by placing them on the corresponding tabs (observation, dialogue development, classification, communication).
- Pupils (Subjects) use cd (Tools) and observe that if they hold them to the light the rainbow is formed inside them (hypotheses, experimentation, drawing conclusions, interaction of group members).
- Pupils, in groups of 2–3, conduct Newton's color experiment with colors using a glass prism after watching a relevant animation episode (communication, hypotheses, experiments).

Furthermore, data analysis has shown a degree of asymmetry in the development of science process skills. Thus, not all science process skills develop in the same way, they develop in accordance with the different activities of the curriculum. Strong correlations were also observed between certain skills which show a high level of interdependence. More specifically, there is a strong positive correlation between conducting experiments and verifying predictions, between verifying predictions and drawing conclusions, between hypothesis and observation, and between interpreting and drawing conclusions. Both the strong positive correlations and the moderate intensity correlations that occur show the important contribution of the curriculum to the development of science process skills.

Another interesting aspect was that through the curriculum, pupils became familiar with some episodes in the history of science which helped them to organize their own scientific work and progress. For example, in the second part of the curriculum pupils organized experiments about light and colors and tried to follow the way Newton presented his discoveries to the scientific society. Furthermore, pupils described scientific concepts providing examples of their logical thinking and connected them with everyday life. History of science played an important role as it presents the scientific concepts with different interpretations through different periods of time. The connection of scientific explorations and experiments in the school classroom with scientists and scientific facts underlines the cultural dimension of scientific knowledge. In this sense, scientific knowledge is considered a result of human actions in the past, connected with real life situations and affecting the contemporary world in which pupils live (Kolokouri & Plakitsi, 2016).

The limitations and restrictions of this study are related to the complexity of teaching and learning in the early years. Science teaching needs to develop reasoning abilities, epistemological beliefs and skills of early years' learners. The development of these skills and ways of reasoning should be a constituent part of science curricula. Moreover, science teaching should focus on supporting learners to realize their alternative ideas based on experience and overcome their misconceptions (Vosniadou, 2019). Towards this direction, we propose further research mainly focusing in the following issues:

- What is required for the teacher and the learners to achieve the complex activity of learning?
- How does a teacher consciously and unconsciously conduct her/himself in and toward this trajectory?
- What are the conditions that permit the emergence of a curriculum based on scientific concepts?
- If we consider concept formation as crucially dependent on cultural mediation what is the role of cultural artifacts, including signs?

There is a need to re-think the scope of science education, environmental education and education for sustainable development, restructure the context of scientific learning and regenerate methods and practices of expansion and connection with society. Expansive learning puts the priority on communities as learners who create and transform culture and finally form theoretical concepts (Engeström & Sannino, 2010; Stetsenko, 2017). Through collective activity that involves change, learners construct new knowledge and put it in practice in new situations. Systems of activity are transformed though collective practices beyond the walls of a classroom (Engeström, 2019) and create an open and sustainable scientific community. Science education needs to address the complexity of issues that humanity of the contemporary world faces, rather than focus on traditional school science (Fensham, 2012). Through the acquisition of knowledge, the development of skills, positive attitudes and values, the learner becomes able to interpret human activity in space and time as a product of interactions that takes place in a situated and socio-cultural environment. Through a Science Education curriculum from the early years the learner acquires a global perception of life and strengthens his position as a future citizen.

References

Åkerblom, A., Součková, D., & Pramling, D. (2019). Preschool children's conceptions of water, molecule, and chemistry before and after participating in a playfully dramatized early childhood education activity. *Cultural Studies of Science Education, 14*, 879–895. https://doi.org/10.1007/s11422-018-9894-9

Bakhtin, M. (1986). *Speech genres and other late essays* (pp. 60–102). University of Texas Press.

Bateson, G. (2000) *Steps to an ecology of mind: Collected essays in anthropology, psychiatry, evolution and epistemology* (revised edition) University of Chicago Press. https://doi.org/10.7208/chicago/9780226924601.001.0001.

Birbili, M., & Alexandra, M. (2020). Early childhood teachers' relationship with the official curriculum: The mediating role of professional and policy contexts. *Education Inquiry, 11*(2), 110–125. https://doi.org/10.1080/20004508.2019.1687080

Blunden, A. (2013, July 19). *Cultural-historical activity theory glossary of terms*. Retrieved from: https://www.ethicalpolitics.org/ablunden/pdfs/Glossary_of_Cultural_Historical_Activity.pdf

Bowkett, S. (2011). *Using comic art to improve speaking, reading and writing*. Routledge, Taylor & Francis. https://doi.org/10.4324/9780203809440

Cole, M., Hakkarainen, P., & Bredikyte, M. (2010). Culture and early childhood learning. In R. E. Tremblay, M. Boivin, Peters R., V. De (Eds.), *Encyclopedia on early childhood development* [online]. Montreal, Quebec: Centre of Excellence for Early Childhood Development (CEECD) and the Strategic Knowledge Cluster on Early Child Development (SKC-ECD); 2010:1–6. Available at: http://www.child-encyclopedia.com/documents/Cole-HakkarainenBredikyteANGxp.pdf. Accessed [20-01-2022].

Davydov, V. (1999). The content and unsolved problems of activity theory. In Y. Engeström, R. Miettinen, & R. Punamäki (Eds.), *Perspectives on activity theory* (pp. 39–52). Cambridge University Press. https://doi.org/10.1017/CBO9780511812774.004

Doran, Y. (2019). Encyclopedia on early childhood development. Building knowledge through images in physics. *Visual Communication, 18*(2), 251–277. https://doi.org/10.1177/1470357218759825

Engeström, Y. (1987). *Learning by expanding*. Orienta – Konsultit.

Engeström, Y. (1999). Activity theory and individual and social transformation. In Y. Engeström, R. Miettinen, & R.-L. Punamääki (Eds.), *Perspectives on activity theory* (pp. 19–38). Cambridge University Press. https://doi.org/10.1017/CBO9780511812774

Engeström, Y. (2000). From individual action to collective activity and back: Developmental work research as an interventionist methodology. In P. Luff, J. Hindmarsh, & C. Heath (Eds.), *Workplace studies* (pp. 150–166). Cambridge University Press. https://doi.org/10.1017/CBO9780511628122.008

Engeström, Y. (2001). Expansive learning at work: Toward an activity theoretical reconceptualization. *Journal of Education and Work, 14*(1), 133–156. https://doi.org/10.1080/13639080020028747

Engeström, Y. (2005). *Developmental work research: Expanding activity theory in practice.* Lehmanns Media.

Engeström, Y. (2015). *Learning by expanding: An activity-theoretical approach to developmental research* (2nd ed.). Cambridge University Press. https://doi.org/10.1017/CBO9781139814744

Engeström, Y. (2016). Foreword: Making use of activity theory in educational research. In D. Gedera & J. Williams (Eds.), *Activity theory in education* (pp. vii–ix). Sense Publishers.

Engeström, Y. (2019). Expansive learning and concept formation: Cycles, germ cells and formative interventions. In *Iscar Regional Conference, 19–24 March 2019,* Ioannina, Greece.

Engeström, Y., & Sannino, A. (2010). Studies of expansive learning: Foundations, findings and future challenges. *Educational Research Review, 5*, 1–24. https://doi.org/10.1016/j.edurev.2009.12.002

Engeström, Y., & Sannino, A. (2012). Whatever happened to process theories of learning? *Learning, Culture and Social Interaction, 1*(1), 45–56. https://doi.org/10.1016/j.lcsi.2012.03.002

Facione, P., & Facione, N. (2007). Talking critical thinking. *The Magazine of Higher Learning, 39*, 38–45.

Fensham, P. J. (2012). Preparing citizens for a complex world: The grand challenge of teaching socio-scientific issues in science education. In A. Zeyer & R. Kyburz-Graber (Eds.), *Science|Environment|Health. Towards a renewed pedagogy for Science Education* (pp. 7–29). Springer. https://doi.org/10.1007/978-90-481-3949-1_2

Fleer, M. (2015). A cultural-historical model of early childhood science education. In M. Fleer & N. Pramling (Eds.), *A cultural historical study of children learning science* (pp. 199–213). Springer. https://doi.org/10.1007/978-94-017-9370-4_13

Fleer, M., & Ridgway, A. (2014). *Visual methodologies and digital tools for researching with young children*. Springer International Publishing. https://doi.org/10.1007/978-3-319-01469-2

Gerde, H. K., Schachter, R. E., & Wasik, B. A. (2013). Using the scientific method to guide learning: An integrated approach to early childhood curriculum. *Early Childhood Education Journal, 41*, 315–323. https://doi.org/10.1007/s10643-013-0579-4

Ilyenkov, E. (1977, January 22). *Dialectical logic, essays on its history and theory*. Progress Publishers. Retrieved from: http://aworldtowin.net/documents/Ilyenkov_Dialectical_Logic.pdf

Jirout, J., & Zimmerman, C. (2015). Development of science process skills in the early childhood years. In K. C. Trundle & M. Sackes (Eds.), *Research in early childhood science education* (pp. 143–165). Springer. https://doi.org/10.1007/978-94-017-9505-0_7

Kampourakis, K. (2018). On the meaning of concepts in science education. *Science & Education, 27*, 591–592. https://doi.org/10.1007/s11191-018-0004-x

Kaptelinin, V., & Nardi, B. (2017). Activity theory as a framework for human-technology interaction research. *Mind Culture and Activity, 25*(1), 1–3. https://doi.org/10.1080/10749039.2016.1267227

Kolokouri, E. (2016). *Cultural Historical Activity Theory (CHAT) in Formal and Informal Science Education in the Early Grades (5–9 years old). The case of cartoons*. University of Ioannina.

Kolokouri, E., & Kornelaki A. C. (2019). Introducing a new socio-cultural tool for science education in first grades: SCOPES. In *Conference proceedings ISCAR 2019*: *Crisis in contexts*, Ioannina, Greece, (pp. 87–101).

Kolokouri, E., & Plakitsi, K. (2013). A cultural historical scene of natural sciences for early learners: A chat scene. In K. Plakitsi (Ed.), *Activity theory in formal and informal education* (pp. 197–208). Sense Publishers. https://doi.org/10.1007/978-94-6091-317-4_8

Kolokouri, E., & Plakitsi, K. (2016). A CHAT approach of light and colors in science teaching for the early grades. *World Journal of Education, 6*(4), 1–13. https://doi.org/10.5430/wje.v6n4p1

Kolokouri, E., Theodoraki, C., & Plakitsi, K. (2012). A cultural historical activity theory approach in natural sciences education laboratory lessons towards reforming teachers' training. *World Journal of Education, 2*(2), 23–40. https://doi.org/10.5430/wje.v2n2p23

Leontiev, A. (1979). The problem of activity in psychology. In J. Wertsch (Ed.), *The concept of activity in soviet psychology*. Armonk, M.E. Sharpe.

Leontiev, A. N. (2009). *Activity and consciousness*. Marxists. Internet Archive.

Miller, J. D. (1983). Scientific literacy: A conceptual and empirical review. *Daedalus, 112*(2), 29–48.

Nardi, B. A. (1996). Activity theory and human-computer interaction. In B. A. Nardi (Ed.), *Context and consciousness: activity theory and human-computer interaction* (pp. 69–103). MIT Press. https://doi.org/10.7551/mitpress/2137.001.0001

NARST (1990, March 21). *The science process skills*. Retrieved from: https://narst.org/research-matters/science-process-skills

Naylor, S., & Keogh, B. (2013). Concept cartoons: What have we learnt? *Journal of Turkish Science Education, 10*(1), 3–11.

Plakitsi, K. (2008). *Didactics of natural sciences in pre-school and early school age: Modern trends and perspectives*. Patakis. (in Greek).

Plakitsi, K. (2013). *Activity theory in formal and informal science education*. Sense Publishers. https://doi.org/10.1007/978-94-6091-317-4

Plakitsi, K., Stamoulis, E., Theodoraki, X., Kolokouri, E., Nanni, E., & Kornelaki, A. (2018). *Cultural-historical activity theory and science education: A new dimension in STEAM education*. Gutenberg. (in Greek).

Ravanis, K. (2017). Early childhood science education: State of the art and perspectives. *Journal of Baltic Science Education, 16*(3), 284–288. https://doi.org/10.33225/jbse/17.16.284

Ravanis, K. (2021). The physical sciences in early childhood education: Theoretical frameworks, strategies and activities. *Journal of Physics: Conference Series, 1796*. https://doi.org/10.1088/1742-6596/1796/1/012092

Ravanis, K., Christidou, V., & Hatzinikita, V. (2013). Enhancing conceptual change in preschool children's representations of light: A socio-cognitive approach. *Research in Science Education, 43*(6), 2257–2276. https://doi.org/10.1007/s11165-013-9356-z

Roth, W.-M. (2011). Science in/for early childhood: More than lip service. In K. Plakitsi (Ed.), *Sociocultural and sociocognitive approaches in the didactics of natural sciences in early childhood*. Patakis.

Roth, W.-M., Goulart, M. I. M., & Plakitsi, K. (2013). *Science during early childhood: A cultural-historical perspective*. Springer. https://doi.org/10.1007/978-94-007-5186-6

Stetsenko, A. (2017). Science education and transformative activist stance: Activism as a quest for becoming via authentic-authorial contribution to communal practices. In L. Bryan & K. Tobin (Eds.), *13 Questions: Reframing education's conversation: science* (pp. 33–47). Peter Lang.

Unsworth, L. (2020). A Multidisciplinary perspective on animation design and use in science education. In L. Unsworth (Ed.), *Learning from animations in science education. Innovations in science education and technology* (pp. 3–22). Springer. https://doi.org/10.1007/978-3-030-56,047-8_1

Vosniadou, S. (2019). The development of students' understanding of science. *Frontiers in Education, 4*, 32. https://doi.org/10.3389/feduc.2019.00032

Vygotsky, L. S. (1978). *Mind in society*. Harvard University Press.

Vygotsky, L. S. (1993). *The collected works of L. S. Vygotsky* (The fundamentals of defectology) (Vol. 2). Kluwer Academic Publishers.

Vygotsky, L. S. (1999). *The collected works of L.S. Vygotsky, vol. 6* (R.W. Rieber, ed.). Kluwer.

Vygotsky, L. S. (2003). Imagination and creativity in childhood. *Journal of Russian and East European Psychology, 42*(1), 7–97. https://doi.org/10.1080/10610405.2004.11059210

Wells, G. (1994). *Learning and teaching scientific concepts: Vygotsky's ideas revisited.* Paper presented, Vygotsky and the Human Sciences, Conference, Moscow.

Wertsch, J. V. (1985). *Vygotsky and the social formation of mind*. Harvard University Press.

Eleni Kolokouri works as a Laboratory Teaching Staff in the Department of Early Childhood, University of Ioannina, Greece. She has a PhD in Science Education in Early Childhood (5–9 years old) and a master's diploma. Her main research interests concern approaches of Cultural Historical Activity Theory in Education as well as education for Sustainability. She is a member of the international research group @Formal and Informal Science Education Group (@FISE group). More specifically, she conducts research on Science Education in early grades under a cultural historical perspective, the use of innovative tools in teaching procedure as well as Environmental Education, Education for Sustainability and STEAM Education. Her approach is focused on Cultural Historical Activity Theory and its use as a theoretical framework as well as a tool of analysis. She has experience in adult training and has worked as a researcher and trainer in several national and European training courses. Her research work has synergy with regional and international congresses and journals.

Katerina Plakitsi is a member of the Governing Council of the University of Ioannina. She is a full professor of Science Education with two bachelor's degrees in Physics and Pedagogy, a master's diploma, and a PhD in Science Education. Her main researching interests are Science Education, Formal and Informal Science Education, and Cultural Historical Activity Theory applied in Science

Education. She has written many books in Greek and English while she has published in different international academic journals. She is scientific coordinator in many European Projects and supervises some PhD scholars in Science Education. She has written many schools environmental science textbooks, and she coordinated a science curriculum reform in Greece for the contemporary Education. She co-founded the interdisciplinary master's program "Environmental Sciences and Education for Sustainability" in collaboration with the Faculty of Medicine and the Department of Biological Applications and Technology at the University of Ioannina. She is also the co-founder of the ISCAR-STEM Thematic Section and the principal investigator of the @formal and informal science education group (@fise group). She is the editor-in-chief of the international bilingual journal *Science Education: Research and Praxis*. She had been Head of the Early Childhood Department at the University of Ioannina. Katerina Plakitsi is President of the International Society for Cultural Historical Activity and Research (https://www.iscar.org/).

Part III
Instrument Producing Activity and the Role of Techno-Creative Activities in STEM Education

Chapter 7
We Have Problems! Analysis of Collaborative Problem Solving in an International Educational Robotics Challenge

Margarida Romero and Sylvie Barma

7.1 Collaborative Problem Solving in Educational Robotics

In this chapter, we start by introducing collaborative problem solving (CPS) in the context of educational robotics challenges to support not only the CPS competency but also, computational thinking in each context requiring a high degree of engagement for the children participating in the international educational robotic challenge. Afterwards, we analyze the process of collaborative problem solving (CPS) based on the PISA framework. In this challenge, children are secured at the socioemotional level through the collective nature of their activity, but at the same time they are challenged cognitively and in their cooperative capacities. In this type of complex activity, different types of contractions arise during the activity. Contradictions presupposes a dual existence between two alternative competing problem-solving strategies to produce a solution meeting the task objectives (Barma et al., 2017). Contradictions are considered necessary to induce change and demand qualitatively new instruments for their resolution (Engeström, 1987). In activity theory, contradictions are dialectical, at the source of self and collective development, and central to the theory of expansive learning (*ibid*). Most importantly, contradictions are the driving force of transformation. The object of an activity is always internally contradictory. It is these internal contradictions that make the object a moving, motivating and future-generating target.

M. Romero (✉)
Université Côte d'Azur, Nice, France
e-mail: margarida.romero@univ-cotedazur.fr

S. Barma
Department of Teaching and Learning Studies, Laval University, Quebec, QC, Canada
e-mail: sylvie.barma@fse.ulaval.ca

K. Plakitsi, S. Barma (eds.), *Sociocultural Approaches to STEM Education*, Sociocultural Explorations of Science Education 21,
https://doi.org/10.1007/978-3-031-44377-0_7

The contractions arising during a collaborative problem-solving activity will engage the learners and the teachers in a more efficient articulation of their inner resources and the external mediating artefacts and instruments, but also in relation to the organization of the other participants of the activity. Problem-solving activities are a fertile context for contradictions allowing to create new solutions towards an expansive learning cycle. The different contradictory states in the collaborative problem-solving activity emerges from a conflict that aims to provoke qualitative leaps in the students such that they conceptualize new rules for engaging in the activity and new ways of organizing their work leading to the actions they will put in place. This ill-defined problem-solving situation should be developed in a positive climate for engaging the learners in a difficult challenge (Stinkeste et al., 2021). We develop an analysis of the educational robotics challenges within a socio-cultural approach. This R2T2 challenge has the intention of reassuring the children through a joint activity allowing them to delineate a zone of proximal development (ZPD) and engage in problem solving at the different moments of the robotic challenge. Nevertheless, in the use case described in this chapter, we observed children being overly distressed at certain moments. We also observed that the ZPD was totally lost at a certain moment within the technical problems, situating them beyond their capacity to control their actions. We analyze this distress, and the way adults act as socioemotional support and activity regulators to support different actions which allow the children to overcome the problem and reach the object of the activity (Fig. 7.1).

After describing the development of the R2T2 Richter educational robotics challenge, the chapter discusses the way external help has helped children to overcome the challenges outside their zone of proximal development. Debriefing on the activity was very important in order to permit the institutionalization of the different concepts developed during the activity but also to situate the difficulties experienced by the children in a way that could be correctly managed: allowing them to learn, even though this too stressful situation. The teachers and facilitators in the activity were required to develop a process of restitution of the lived experience as one that can be a reflective challenging experience giving them cues for new complex problem-solving situations.

Collaborative problem solving through educational robotics challenges Collaborative solving of complex problems is a key competency today. To develop this type

Fig. 7.1 Team engaged in problem solving during the R2T2 Richter educational robotics challenge

of competency, it is important to be able to design activities of a certain level of complexity that can bring into play a collaborative resolution of complex problems in contexts with a certain degree of authenticity. In this context, the challenges of cooperative-type educational robotics present a problem-situation of great engagement in which a team of students is mobilized to program the robots pursuing a common goal. The R2T2 type educational robotics challenges developed by Francesco Mondada's team at EPFL provide a favorable context for the development of collaborative resolution of complex problems.

The R2T2 challenge is a learning activity based on challenges (challenge-based education) with the aim of developing problem-solving competency. Several studies highlight collaborative problem solving as an essential competency in the twenty-first century (Häkkinen et al., 2016; Hesse et al., 2015). Problem-based learning is seen as a context for learning rather than an approach for teaching (Arsac et al., 1991; Charnay, 1992; Lajoie & Bednarz, 2012). It is an open activity with an open structure (ill defined) on which participants are engaged to produce a creative solution to a goal that is to be developed in positive interdependence (Johnson & Johnson, 1989). The engagement within the framework of a challenge places us in a dynamic of cooperative game with time pressure close to an escape game situation with the differences that in the challenge R2T2 the movements are not carried out by the students themselves in a confined space, but by robots confined in a room located at a distance in EPFL Lausanne. The R2T2 challenge is also developing under a thematic approach (subject-based learning) which helps develop the potential for inter disciplinary work (Darbellay et al., 2019) while highlighting the inquiry-based learning approach.

The R2T2 challenge engages in a problem-solving situation with robots in an ill-defined task. For these reasons, the activity supports not only prob. lem solving but also two additional components of computational thinking: formal systems (code literacy) and physical systems (hardware literacy) as represented in Fig. 7.2. Computational thinking is a set of socio-cognitive and metacognitive strategies engaged

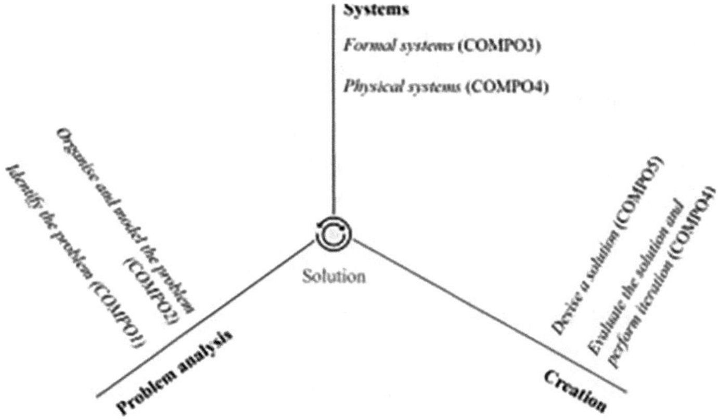

Fig. 7.2 Components of the computational thinking competency

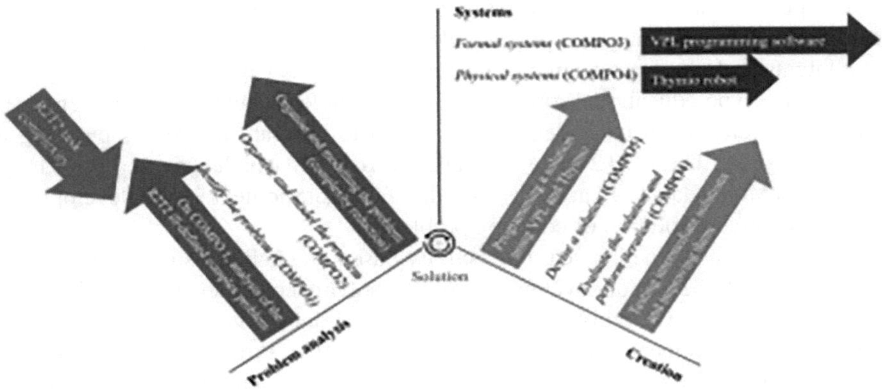

Fig. 7.3 Computational thinking in the R2T challenge

in the analysis and modeling of (complex) problem-situations, and the creation and test of digital artefacts to provide a creative solution to the problem-situation (Romero et al., 2017).

Computational Thinking has been initially defined by Wing (2009) as the capacity to solve problems using computer concepts and procedures. As we can see in Fig. 7.3, there are parts of the components of computational thinking that are part of the problem-solving competency and there are two specific components which are specific to computational thinking on the systems axis integrating formal systems, comprising code literacy (COMPO3) and physical systems (COMPO4) comprising hardware literacy. On the axis of the problem analysis is the identification of the problem (COMPO1) and the organization and modeling of the problem (COMPO2). Finally, on the creation axis, the creation of a solution (COMPO5) and the process for evaluating this solution are presented, as well as the commitment to an iterative improvement process (COMPO6).

On the R2T2 challenge the learners start facing the complexity of the ill-Collaborative problem solving in ER 5 defined problem situation (COMPO1) and organizing the activity (COMPO2). They are required to understand the Visual Programming Language (VPL) software (COMPO3) and the Thymio robot (COMPO4) features in order to mobilize both these software and hardware knowledges on the device of their solution (COMPO5) and the interactive improvement of the intermediate solutions allowing them to solve the problem (COMPO6).

7.2 Situational Pedagogy for Problem Solving

The learning activities that are presented in class are often calibrated at a relatively low level of difficulty which allows them to develop the learning objectives targeted for all students, considering their prior knowledge and their degree of autonomy.

However, to develop problem-solving competencies, it will be important to be able to engage students in more unstructured problem situations and with a higher level of complexity (Hesse et al., 2015). Thus, the challenges of pedagogical robotics present fairly high levels of complexity which precisely aim to allow students to confront problem situations so that they can develop the knowledge and competencies necessary to develop a collaborative approach to solving complex problems. In a spirit of situational pedagogy, students can modulate their previous representations and construct new ones when faced with problematic situations. These problem situations are both the motive and the means of learning because they allow the appropriation of knowledge and the development of competencies. Thus, "a teacher who wants to provoke a qualitative leap in the functioning of his students, to make them competent, will seek to develop a reality which resists them, which unbalances them, in a climate of confidence" (Beckers, 2011, p. 41). This imbalance therefore seems to be an important concept in a constructivist and social constructivist learning model. Minier (1998) insists on the fact that it is "necessary to take into account the capacities of reflexivity and conscious action when trying to understand the processes linked to cognitive, emotional and social regulation in the context of a learning situation" (p. 267). Faced with learning situations that are meaningful to him, the student can regulate his approach to appropriating knowledge, know-how and interpersonal competencies. (Roth & Lee, 2004), believe that for the situations presented to students not to trap them in a race for performance, they must: allow for a variety of activities in the way students will engage; emphasize a more democratic approach where they can effectively make decisions related to their lives and interests to foster a long-term commitment to resolving socio technical controversies they will face. Instead of favoring a disciplinary science, teachers have an advantage in encouraging situations that allow the negotiation of different forms of knowledge regarding significant problems (such as that of the arrival of a tsunami) while they emerge from the experience of the community in which the pupil fits (Barma, 2008). By promoting meaningful student participation in solving a complex problem, this entry-level approach would allow students to recognize the nature of knowledge as it can be experienced and interpreted in the community. We could also speak here of a real cognition located and distributed since the activities that emerge are inspired and come directly from what is experienced in the environment. For example, by trying to measure the length of the trip to the hospital in the event of injuries in the tsunami, students get to work on models before starting programming. It is about offering students a different approach to science and technology education. Thus, Roth and Lee (2004) propose that "rather than privileging disciplinary science, we must favor situations which allow the negotiation of different forms of knowledge adapted to particular problems as they arise in life, daily life of a community" (p. 287). Moreover, such a vision implies that teachers develop these situations in a spirit of technoscientific literacy which considers not only the individual dimension of this literacy but also its collective dimension (Fourez, 2002).

7.3 Analysis of the R2T2 Educational Robotics Challenge

As part of the second edition of the R2T2 "Operation Richter" international educational robotics challenge, we engaged students from different teams located internationally in a challenge to save the city from a natural threat: a tsunami. The robots that the students moved in the city model are in Lausanne. In the room, a total of 17 Thymio educational robots are placed in a play mat simulating the city. The student teams are spread over two continents: in France, there are teams in Nice and Montpellier, another team is in Martinique, another in Guadeloupe and Mexico. The mission is to reach the different rescue zones assigned to each team before the Tsunami approaches. The robots are remotely programmed to coordinate and activate the city's protection systems to secure four strategic points before the Tsunami arrives. The experience feedback makes it possible to observe a high level of commitment from all the teams within the framework of these project-type activities. Before analyzing the challenge, we present situational pedagogy related to problem solving and then analyze the development of the R2T2 task.

The challenge enabled the hiring of five teams located in five different locations: Nice and Montpellier in France, Guadeloupe and Martinique in the Caribbean and Mexico in North America. We analyze collaborative problem solving based on the analysis of the team's progress in Nice. The observation is carried out by four observers from various notetaking and video recording. The challenge analysis considers the PISA (2015) problem-solving model, whose four specific steps are: (1) explore and understand, (2) represent and formulate, (3) plan and execute, (4) monitor and reflect. The assignment of roles freezes the type of student activity, however, for fixed problem-solving activity the roles have limited the flexibility of autonomy needed.

7.4 Analysis of the Progress of the R2T2 Challenge

In Nice, the students of the Hubschool21 school divided into teams for the final preparations. There is the team responsible for communication with Lausanne, the YouTube chat team as well as the team responsible for checking, alerting and anticipating movements and finally that for programming related to travel. Each team is made up of four students aged between 8 and 15, highlighting the heterogeneity of the groups (see Fig. 7.4). Four expert observers joined each team for this R2T2 Richter challenge.

Fig. 7.4 The Hubschool21 team as part of the R2T2 Richter educational robotics challenge

7.5 The R2T2 Challenge, as a Collective Activity with a High Degree of Engagement

To put the students in a situation, a presentation video was created by the EPFL team. The challenge is presented in the form of a disaster scenario where a tsunami will sweep over the city in 90 min. This video is viewed by all the teams participating in the challenge. As the teams begin their work, they begin with a step of understanding the challenge. According to the problem solving competency developed by the international program PISA (OECD, 2017), the first step in perspective is that of exploring and understanding the problem to be solved. Anticipation work was carried out within each group. The students prepare for teamwork, they show joy, enthusiasm, and fun. A.

dynamism operates immediately. Positive energy and shared pleasure help overcome difficulties and reach an optimal solution. Anticipation of movements; which direction will the Thymio follow and how can we avoid it going in an unwanted direction or bumping into other Thymios? Many questions stir the children. They also plan to cooperate with other teams through Chat to carry out their mission. An establishment of a shared and formulated understanding and representation is implemented according to the second stage of the PISA Framework Problem Solving Competency. A reflection was made by a student by bringing back a layer on the miniature map providing an anticipation of the route to reach the hospital (Fig. 7.5). The children took to heart the mission offered to them.

A few minutes before the start of the R2T2 Richter challenge, we checked the connection with Lausanne. Feedback on the connection is positive, the Thymio assigned to Nice is on.

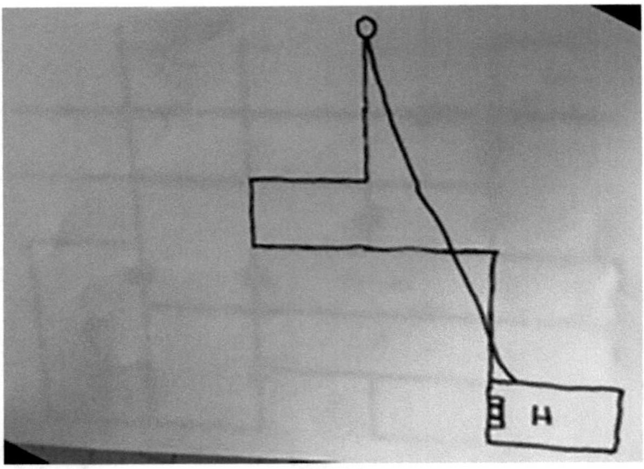

Fig. 7.5 Modeling of the challenge performed by one of the students

7.6 Problem Solving During the Challenge

Common understanding of the problem in a problem-solving process requires the effective engagement of an individual and understanding of the situation. In the context of the R2T2 challenge we must consider the common understanding of the problem situation which must then allow us to engage in the problem solving. Some older students have explained the mission to other younger ones in a different way to identify the problem situation for which the process and solution is not known in advance. The reformulation made it easier for the different students to assimilate the challenge. This ability to develop a shared understanding and to work in a coordinated manner with several people for a common goal determines the first and second component of collaborative competence.

First problem in perspective. We have noticed that collective work is not very effective for some sub-teams. Some students take ownership of the problem at their own pace through reflection, locking themselves into individual rather than collective autonomy. The ability to identify a problem situation for which the process and solution is not known in advance to operate very slowly. The students are locked into their solutions and do not accept the proposals of others, which provokes the demoralization of some because of this lack of team cohesion. We have seen that within this group, good students have taken power by imposing their points of view and their rhythms on others who have ended up giving up work. A feeling of rejection pierced a child who burst into tears. Other children dropped out and eventually left the group to draw on the board and take up other activities.

Although the understanding of knowledge and competencies has been shared in some groups, the establishment of individual strengths as well as the restriction of other team members to organize tasks towards the common goal was not implemented in all groups. The children did not know how to best manage the difficulties of working as a team while respecting and seeking solutions. The third and fourth component of collaboration has not been fully completed. Therefore, the help of the supervisors proved essential.

7.7 Problem Analysis During Challenge R2T2

Once the R2T2 mission is launched, the timer is running. The organizers and the students were enthusiastic about the launch of the challenge. While others were starting their programming, the Nice team encountered a connection problem with Lausanne. Which leads to a second problem. The Thymio was well lit, however, no computer at our base managed to connect to Lausanne. The second component of the problem-solving competency is an exploration of a variety of solutions considering the problem-situation. A test of possible solutions through several tests of connections from four different computers with our IP and port represents the third component.

Each team did its best to find solutions or external resources; the common goal was to successfully displace the Thymio. The use of varied strategies and the dynamism of students in exploring a variety of new solutions (Fig. 7.6).

7.8 Remediation to Succeed the Challenge

After several attempts, a few minutes before the arrival of the Tsunami, the engineer was present in Lausanne to cooperate with the students to advance the Thymio by programming it for them. For their part, the children, through Skype, guide her to indicate the path to take. Many emotions pierced the students in the face of this situation. Feelings of defeat, disappointment, anger and even irritability were present from the start of the challenge. Once the Thymio reached its point of arrival, the

Fig. 7.6 Students engaged in collaborative problem solving

Fig. 7.7 Agreeing on the solution following remediation with the help of an adult

students were relieved and happy about this victory, which would not have happened without the intervention of the engineer on site in Lausanne. The mission was successfully completed even though the computer coding was not handled by the children (Fig. 7.7). What is crucial and that we have been able to observe is an iterative co-regulation of this intermediate solution which led to an optimal solution despite the technical problems encountered. At the heart of this techno-creative activity, the adoption of a flexible operation is essential to achieve the common objective of this mission.

7.9 Discussion

The intensity of the challenge was very strong. The pupils moved out of their proximal development zone the moment the connection was lost. The students were no longer in control of the task, and they could no longer implement their action plan. The duration of this difficulty spanned over 30 min. This discouraged the students and disturbed or removed them from the task. This difficulty is more pronounced as they watch other teams succeed without being able to regain control of the situation. The team in Lausanne was involved in the entire challenge without being able to re-establish the connection, this waiting time was a difficult experience for the students. The collective challenge with the Mexican and Martinican teams came to an end without the students in Nice being able to complete theirs. The awareness of the tsunami-based narrative caused some students to infer that their failure would have resulted in the deaths of their city's citizens in such a situation. Despite the iterative co-regulation efforts, technical issues prevented teams from being able to search and share with external resources. This tension between the commitment of the team and their reliance on external resources created a feeling of helplessness and discouragement. Once the challenge was finalized, the facilitators highlighted the learning achieved and the team in Lausanne was able to restore the situation allowing the students to end up being able to take on the challenge. The outburst of joy that followed the period of discouragement was of great intensity. The teachers reported to us a few weeks after the challenge that the students still discuss the moments of great intensity experienced during the R2T2 Richter. The situational learning of the students in a challenge of solving complex problems helped to develop an understanding of the difficulties related to the organization, both technological and the necessary coordination efforts. Mutual assistance within the framework of the project between the various partners is also an aspect highlighted by the students participating in this study. Although complex, R2T2 type activities have great potential for developing concrete experiences of collaborative problem-solving in context and understanding collaboration under a distributed and international approach. CPS activities have the potential to address real world problem solving (RWPS) challenges (Isaac et al., 2021) to support the Sustainable Development Goals (SDGs) agenda including climate crisis, antibacterial resistance, pandemics, and other global issues.

Acknowledgements We thank Morgane Chevalier, Didier Roy, Nathalie Methelie, Jorge Sanabria, Stéphanie Netto, Dayle David, Saint-Clair Lefevre and Sarra Abdelmoula for their feedback and suggestions.

References

Arsac, G., Germain, G., & Mante, M. (1991). *Problème ouvert et situation problème*. Université Claude Bernard Lyon I.

Barma, S. (2008). Vers une lecture systémique du contexte, des enjeux et des contraintes du renouvellement des pratiques en éducation aux sciences au secondaire au Québec. *CJNSE/RCJCÉ, 1*(1).

Barma, S., Laferrière, T., Lemieux, B., Massé-Morneau, J., & Vincent, M. C. (2017). Early stages in building hybrid activity between school and work: The case of PénArt. *Journal of Education and Work, 30*(6), 669–687.

Beckers, J. (2011). *Développer et évaluer des compétences à l'école: Vers plus d'efficacité et d'équité*. Labor Éducation.

Charnay, R. (1992). Problème ouvert, problème pour chercher. *Grand N, 51*, 77–83.

Darbellay, F., Louviot, M., & Moody, Z. (Eds.). (2019). *L'interdisciplinarité à l'école : Succès, résistance, diversité*. Éditions Alphil.

Engeström, Y. (1987). An activity-theoretical approach to developmental research. *Learning by expanding*.

Fourez, G. (2002). *La construction des sciences: Les logiques des inventions scientifiques*. De Boeck & Larcier.

Häkkinen, P., Järvelä, S., Mäkitalo-Siegl, K., Ahonen, A., Näykki, P., & Valtonen, T. (2016). Preparing teacher-students for twenty-first-century learning practices (PREP 21): A framework for enhancing collaborative problem-solving and strategic learning skills. *Teachers and Teaching*, 1–17. https://doi.org/10.1080/13540602.2016.1203772

Hesse, F., Care, E., Buder, J., Sassenberg, K., & Griffin, P. (2015). A framework for teachable collaborative problem-solving skills. In *Assessment and teaching of 21st century skills* (pp. 37–56). Springer.

Isaac, G., Barma, S., & Romero, M. (2021). Cultural historical activity theory, double stimulation, and conflicts of motives in education science: Where have we been?(2012–2021). *Revue internationale du CRIRES: innover dans la tradition de Vygotsky/CRI_SAS international Journal: Vygotsky's Heritage: Innovation in Education, 5*(2), 86–94.

Johnson, D. W., & Johnson, R. T. (1989). *Cooperation and competition: Theory and research*. Interaction Book Company.

Lajoie, C., & Bednarz, N. (2012). Évolution de la résolution de problèmes en enseignement des mathématiques au Québec: Un parcours sur cent ans des programmes et documents pédagogiques. *Canadian Journal of Science, Mathematics and Technology Education, 12*(2), 178–213.

Minier, P. (1998). La métacognition selon une approche constructiviste sociale de l'apprentissage. In L. Lafortune, P. Mongeau, & R. Pallascio (Eds.), *Métacognition et compétences réflexives en éducation* (pp. 261–280). Éditions Logiques.

OECD. (2017). *PISA 2015 results (Volume V) : Collaborative problem solving*. OECD. https://doi.org/10.1787/9789264285521-en.

Romero, M., Lepage, A., & Lille, B. (2017). Computational thinking development through creative programming in higher education. *International Journal of Educational Technology in Higher Education, 14*(1), 42.

Roth, W.-M., & Lee, S. (2004). Science education as/for participation in the community. *Science Education, 88*(2), 263–291.

Stinkeste, C., Napala, A., & Romero, M. (2021). Impact of team climate on co-creativity in STEAM education in primary education. *Creative Education, 12*(8), 1977–1994.

Wing, J. (2009). Computational thinking. *Journal of Computing Sciences in Colleges, 24*(6), 6–7.

Margarida Romero is full professor at Université Côte d'Azur in France and an associate professor at Université Laval in Canada. Her research focuses on the study of transversal competencies, particularly in relation to computational thinking and creative problem solving.

Sylvie Barma is a full professor of Science Education at the Faculty of Education at Laval University since 2008. After teaching high school science for 20 years and contributing to the development of the new Quebec Science and Technology curriculum, she obtained a Ph.D. in Science Education.

Chapter 8
Creativity in Early Years Science Education Through the Exploitation of Robotics in the Sustainable School

Maria Topoliati, Katerina Plakitsi, and Fani Stylianidou

8.1 Introduction

In recent years, there has been an increasing focus on the interest of Science, Technology, Engineering, Arts, Mathematics and Robotics in early childhood research. This study focuses on the creative approach of Science Education by preschool students in the context of their participation in the applied Erasmus+ Project, CEYS: Creativity in Early Years Science Education.

The project aims to promote the use of creative approaches in early years science teaching (Ravanis, 1999) and brings together five distinguished partners from four countries across Europe and many teachers who are contacted by others in order to interact and work together in developing a bilateral pedagogical cooperation. The project also specifies that inquiry-based and creative approaches to learning and teaching have some features in common. These pedagogical synergies are identified as including: play and exploration; motivation and affect; dialogue and collaboration; questioning and curiosity; problem-solving and agency; reflection and reasoning; teacher scaffolding and involvement; assessment for learning (CEYS, 2017a, b).

During the planning and implementation phases, action research and field research are applied (Cohen et al., 2008). This interdisciplinary approach is based on the interaction of STEAM education and educational robotics with the national curriculums (Ministry of Education, 2011), focusing on the sustainable development. The dominant concern and the main research topic of this study is to investigate whether the socio-cultural approach of STEAM education and educational

M. Topoliati (✉) · K. Plakitsi
Department of Early Childhood Education, University of Ioannina, Ioannina, Greece
e-mail: m.topoliati@uoi.gr; maria.topoliati@gmail.com

F. Stylianidou
State Scholarships Foundation (I.K.Y.)/Hellenic Erasmus+ National Agency, Athens, Greece

K. Plakitsi, S. Barma (eds.), *Sociocultural Approaches to STEM Education*, Sociocultural Explorations of Science Education 21,
https://doi.org/10.1007/978-3-031-44377-0_8

robotics (Bee-bots, LegoWeDo 2.0) facilitate the creativity in science education through the empirical acquisition of knowledge (Shaheen, 2010), reinforce the adoption of positive attitudes towards science and approach all the learning areas of the national curriculums related to early years education (Ministry of Education, Pedagogical Institute, 2002), emphasizing on the sustainable development goals (Dimitriou, 2009).

In the second phase of this research, the educational intervention is extended through the implementation of two innovative educational projects: "Open Schools for Open Societies (OSOS, 2017)" and "Reflecting for Change (R4C, 2020)" (Erasmus+). The openness of the school in the local community and particularly the fostering of connection to the daily life of children contributes a significant value to the project (OSOS, 2017). The school involves policy makers and education organisations in an open and regular dialogue with the goal of increasing policy coherence and to benefit from stakeholders' experience and broad networks. In this context, Technology Enhanced Learning (TEL) opportunities are integrated in the school policy promoting inclusive, cross-disciplinary and differentiated learning opportunities, in order to ensure equity and equal opportunities to all students (R4C, 2020).

Based on the Activity Theory and the belief that learning is the result of interaction, the specific concern of this project is the expansion of the learning environment outside the classroom (Sannino et al., 2016). During the developmental and experiential phase of this pilot educational intervention is applied a formal, informal and non-formal type of education, in which the natural, social and cultural environment is utilized as a primary source of knowledge (Plakitsi, 2011). Parents and external stakeholders were involved in the collaborative action planning of the project and participated actively in indoor and outdooring STEAM activities. In this way, the gap between the kindergarten and the extended community is bridged, supporting students to approach the sustainable educations' goals, to develop all the learning areas of the curriculum related to early childhood and to cultivate twenty-first century skills (National Research Council, 2012) through the exploitation of science and innovation.

Twelve children from one classroom of the rural area of Klimatias Kindergarten in Greece, aged between four to six years old participated for two school years in the research. In order to be confirmed the effectiveness of the applied project and the existence of its positive correlation with the degree of the active involvement of the children is implemented collection of data methods. The process of the assessment includes qualitative and quantitative data, such as participatory and non-participatory observation of the teacher, children's reflections as a whole group, children's self-assessment, semi-structured interviews and questionnaires of parents and collaborative teachers. In more details, it could be mentioned that the results of the collected data from the first school year concerned four students, eight parents and twenty-four teachers, while in the second school year participated eight students, sixteen parents and twenty-four teachers (Table 8.1). The learning outcomes are disseminated by the pupils and the teacher to the local and extended community.

Table 8.1 Data collection

DATA COLLECTION		
Kindergarten of Klimatias, Ioannina, Greece	School Years	
Type of Data	**2018-2019**	**2019-2020**
Children's questionnaires	4	8
Children's semi-structured interviews	4	8
Collaborative teacher's questionnaires	24	24
Parents' semi-structured interviews	8	16
Children's written material	60	90
Photographs	180	240
Video recordings	32 minutes	68 minutes

8.2 Theoretical and Methodological Framework

According to the notion of development as a dialectical movement and interdependency between the individual mind and the surrounding world (Vygotsky, 1978), the theoretical and methodological framework of this research is based on the principal element of Cultural Historical Activity Theory (CHAT) and the belief that learning is the result of interaction (Plakitsi et al., 2018).

Subject of the activity system is the individual or group of the learning community who involved in the implementation phase. In this activity the subjects are the students and the teacher.

The **object** of activity concerns the formulation and delimitation of the objectives of the didactic interventions and the objectives of the teacher/researcher. At this stage, the goals of the didactic interventions and the goals of the teacher researcher are formulated and delimited.

As mediating and methodological **tools** are considered the expansion of the learning environment and the exploitation of educational robotics and STEAM education in the implementation phase of the activities.

The **rules** are explicit and implicit norms that regulate actions and interactions within the system.

The **community** refers to participants in an activity system who share the same object. In this case, the interaction is between the students, the teachers, the collaborative schools and the extended learning community.

The **division of labor** involves the division of tasks and roles among members of the community and the divisions of power and status.

The final stage of the **outcomes** concerns the evaluation of the results of the implemented project. This study assesses the effective and interdisciplinary approacht of the national curriculums related to early childhood and the cultivation of twenty-first century skills through the exploitation of STEAM education and the expansion of the learning environment outside the classroom.

Particularly, emphasizing on selective learning areas of the sustainable education, the project pursues to promote the quality of education providing the students with lifelong learning opportunities, to encourage the adoption of healthy lifestyles and well-being, to accommodate everyone to create sustainable places and communities with green and culturally inspiring living conditions and to reinforce the means of implementation the global partnership for sustainable development (Sonter & Kemp, 2021).

In this framework, the learning environment in the classroom is enriched with supporting science material and educational robotics systems, while the natural, social and cultural environment is utilized as a primary source of knowledge. This formal, informal and non-formal type of education (Asghar, 2012) includes applied indoor and outdooring activities aiming to the approachment of the sustainable educations' goals and to the development of all the learning areas of the curriculum related to early childhood using the mediating educational tools.

The learning activities are organized following the same structure, which may be presented in diagrammatic form (Fig. 8.1).

Accordingly, the implemented project is analyzed in the main parts of SCOPES, which is an important methodological tool that contributes to plan and analyze science activities:

- **S**ystems of activities: the sociocultural system of the activity
- **C**ontradictions: the conflicts that arise in the Activity system and contribute to its evolution.

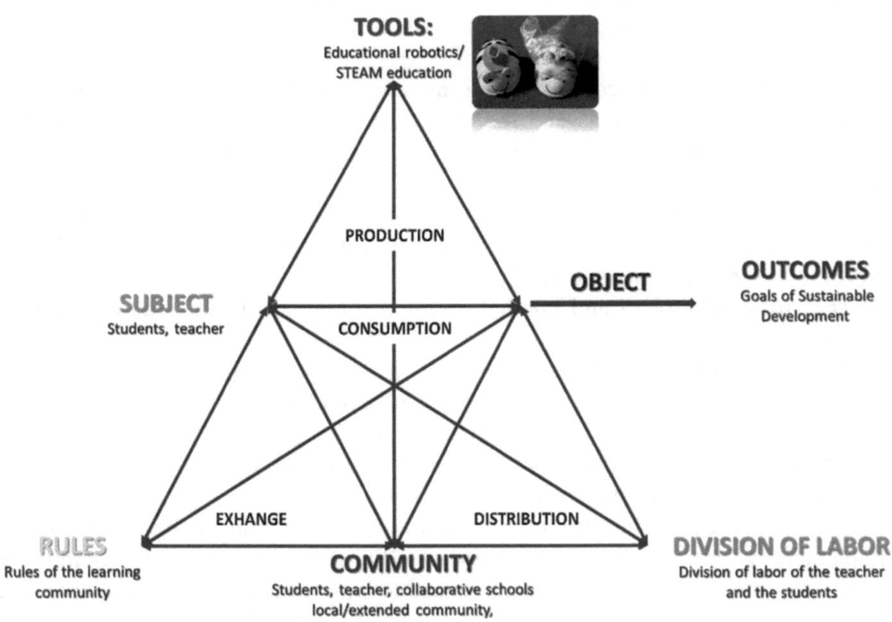

Fig. 8.1 Activity theory model. (Engeström, 1987)

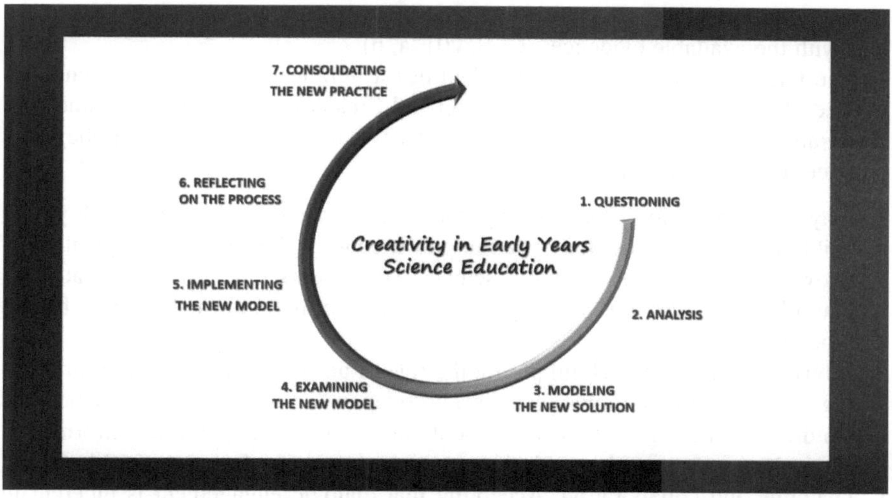

Fig. 8.2 Sequence of learning actions in an expansive learning cycle

- **O**utcomes: the results arising from the relationship between the individual subject and their object, which is transformed through collaborative activity.
- **P**raxis: exploitation of participatory methods during the developmental phase, which combine theory and practice.
- **E**xpansive learning distinguishes itself by its focus on learning within and between activities in society at large, beyond the confines of school. Expansive learning is the creative type of learning in which learners join their forces to literally create something novel, essentially learning something that does not yet exist (Engeström, 2015) (Fig. 8.2).
- **S**cience education: approach to concepts and phenomena of the Natural Sciences.

8.3 The Conceptual Framework

This effort is based on the conceptual framework of CEYS, which is built on definitions of creativity and inquiry drawn from reviews of literature in creativity education and science education conducted in the *Creative Little Scientists* project (CLS, 2012). The CLS focuses on the enabling creativity through Science and Mathematics in preschool and first years of primary education.

In this framework, creativity is identified as "a purposive imaginative activity generating outcomes that are original and valuable in relation to the learner", drawn from the National Advisory Committee on Creative and Cultural Education (NACCCE, 1999) report. Subsequently, the creativity in science and mathematics is defined as 'generating ideas and strategies as individual or community, reasoning critically between these and producing plausible explanations and strategies consistent with the available evidence' (CLS, 2014a, b).

The CEYS continuation project, identifies a number of pedagogical synergies between Inquiry Based Science Education and Creativity that provide a framework for examination of opportunities for creativity and inquiry in both policy and practice (CEYS, 2017a, b):

- Play and exploration. Recognizing that experimentation is inherent in all young children's activity, this study is aiming to encourage the empirical acquisition of knowledge through observation, science experiments with everyday materials, playful coding activities and implementation of formal, informal and non formal type of education.
- Motivation and affect. Highlighting the role of aesthetic engagement in promoting affective and emotional responses to science and mathematics activities, the children interact with the local socio-cultural environment, gathering information about the past, comparing with the present and planning for a sustainable future.
- Dialogue and collaboration. Accepting that dialogic engagement is inherent in everyday creativity in the classroom, the children are enabled to interact with collaborative schools and external educational institutions.
- Problem solving and agency. Recognizing that through scaffolding the learning environment of the children can be provided with shared, meaningful, physical experiences and opportunities to develop their own questions as well as ideas about scientifically relevant concepts. The students identify environmental issues and provide solutions as active citizens, involving local authorities and community.
- Questioning and curiosity. Recognizing that children are naturally curious about their world and enjoy exploring their surroundings, the teacher employs open ended questions, and promotes speculation by modelling their own curiosity.
- Reflection and reasoning. Emphasizing the importance of metacognitive processes, reflective awareness and deliberate control of cognitive activities, still developing in young children but incorporated into early years science and mathematics practice, this project attempts to connect the acquired and experiential knowledge with the daily life of the students.
- Teacher scaffolding and involvement. In this research, the teacher mediates the learning to meet children's needs, gives motivation for learning and contribute to forming positive attitudes of children toward Science and Sustainable development.
- Assessment for learning. Supporting and encouraging children's active engagement in learning through their active participation in the action planning, implementation and evaluation of the project (CEYS, 2017a, b).

8.4 The Development Phase

During the development phase of this action research, the teacher organizes indoor and outdoor education activities aimed at the empirical study of science and the approachment of the sustainable development goals through the exploitation of the educational robotics. The pedagogical role of the teacher is facilitator, inspirational and motivator. As action researcher stimulates pupils' curiosity about phenomena and events around the world, energizes their feeling and contributes to forming positive attitudes toward science.

In order to achieve the educational objectives, teacher concentrates on the enrichment of the *Corner of Science* in the classroom with supporting materials, programable floor robots, three dimensional simulations and installation of relevant software on the computer. Simultaneously, pursues the openness of the school to the local natural, historical and sociocultural environment and requests the collaboration with other schools, universities, authorities and experts.

It is worth to be mentioned that parents are implicated actively in the action planning, the implementation and the evaluation phases of the activities. Particularly, they participate as members of the school's wider pedagogical team and they submit proposed activities through brainstorming, which are taken into consideration and incorporated in the learning process. The effective parental communication takes place through face-to-face meetings at school, online discussions on the Webex by Cisco educational platform and collaborative work in the same on line document in real time.

At this point, a digital conceptual map of the proposed activities is modelled through democratic procedures (Topoliati, 2015). The use of an opensource software, the *Kidspiration*, in which coexist speech and image, allows users to visualize their ideas and facilitates the first stage of reading and writing process of the children. It is also used during the formative and the final assessment, to investigate the interdisciplinary knowledge and the skills that students have acquired during the implementation of the project and to assess the adoption of positive attitudes and behaviors.

In this context, the teacher relies on the suggestions of the pedagogical team and connects them with the learning modules of the national curriculums. The organized and emerging playful activities of the action planning are focused on the creative exploration of simple physical phenomena (Roth et al., 2013) and the approachment of selective goals of the sustainable development, including those related to quality of education, climate change, environmental degradation, peace and justice (Liarakou & Flogaiti, 2007), in order to achieve a better and more sustainable future for all. The research framework is completed with the process of disseminating the learning outcomes of the project, while the evaluation of the qualitative and quantitative evidence contributes to the school improvement through the analysis of the strengths and weaknesses (Fig. 8.3).

The applied educational intervention is extended through the implementation of two innovative educational projects: "Open Schools for Open Societies" and

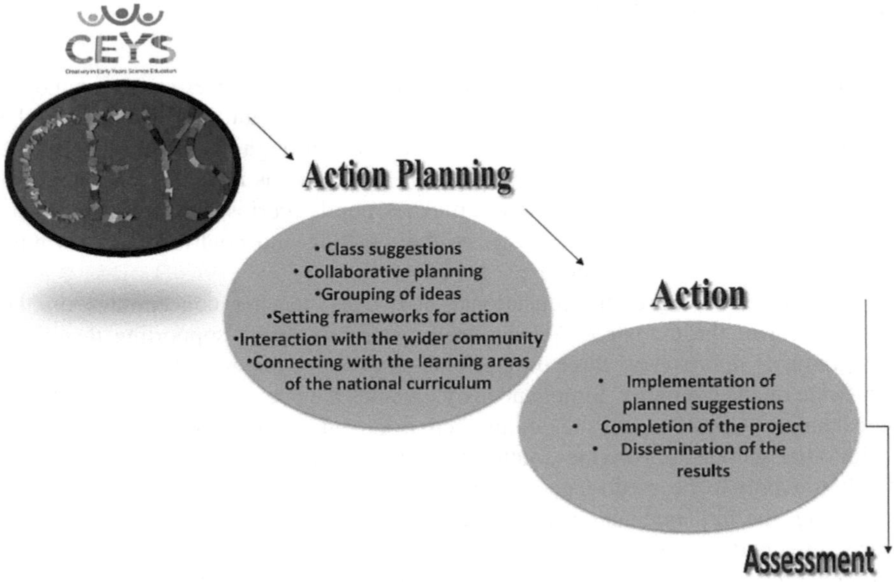

Fig. 8.3 Description of the action planning

"Reflecting for Change" (Erasmus+). Particularly, in the Open School, families, local and extended community, authorities, collaborative schools, universities, policy makers and experts are implicated in the action planning, in the implementation phase and in the evaluation of the activities. This openness of the school in the local and wider community has many advantages, while the students get the opportunity to learn inside but also outside school and can rely on the support of people in the field. Their projects meet real needs and draw upon local expertise and experience. Moreover, learning in and together with the real world makes learning more meaningful and motivates students, cultivating important education skills, such as critical thinking, creativity, problem solving, communication (OSOS, 2017). Additionally, according the "Reflecting for Change" project, the Technology Enhanced Learning is integrated in the school policy, promoting inclusive, cross-disciplinary and differentiated learning opportunities (R4C, 2020).

8.5 The Implementation Phase

Based on the Activity Theory, the principal element of CHAT and the belief that learning is the result of interaction, the main focus of this project is put on the application of formal, informal and non-formal teaching through the expansion of the learning environment outside the classroom (Plakitsi, 2012). During this learning process, the physical, social and cultural aspects of the environment is used as a primary source of knowledge. This learning expansion is considered as a process of acquiring and creating empirical knowledge and concepts and a process that leads to the formation of theoretical knowledge and concepts (Engeström & Sannino, 2010).

It is worth to point out the extended openness and the co-operations of the school with external partners. Indicatively, we mention its participation in the Unesco Associated Schools Network (ASPnet), in the Eco-School and Sustainable Networks, in the Scientix, eTwinning and EU Code Week Communities, the interaction with the Department of Preschool Education of the University of Ioannina, the local authorities, the Foundation for Environmental Education, the collaboration with the Communication Officer for Greece and Cyprus of the United Nations Regional Information Centre (UNRIC), the Holy Metropolis of Ioannina, the Ephorate of Antiquities of Ioannina (Archaeological and Byzantine Museums), the National Observatory of Athens, the European Council for Nuclear Research (Cern), the European Gravitational Observatory, the educational sector of the European Space Agency, the Research and Development Department of Ellinogermaniki Agogi and numerous educational institutions that inspire our effort.

Initially, the "Little Creative Scientists" (Fig. 8.4) of the Kindergarten "are adopted" by the Ephorate of Antiquities of Ioannina and participate in a cultural heritage project of substantial acquaintance with the local post-Byzantine park of Klimatias (Ministry of Culture, 2014), the Archaeological, the Byzantine Museums and the castle of Ioannina.

The students gather information through outdoor activities about their natural and socio-cultural local environment in an experiential and contextual way. According to the opinion that "there are no inappropriate weather conditions, but inappropriate clothing", the learning modules of the curriculum, related to pre-school education, are approached by organizing interactive activities outside the classroom in

Fig. 8.4 The little creative scientists of Klimatias Kindergarten

Fig. 8.5 Exploring the natural and socio-cultural local environment

collaboration with other Schools of the eTwinning Community. Among others, they observe the post-byzantine monuments, the waterfall and the hydroelectric plant of the local area, understanding experientially the utilization of the water energy. Subsequently, the young students engage in a dialogue with the archaeologists (Kokkotas & Plakitsi, 2005), clarifying and comparing concepts such as "Past", "Present" and "Future" (Fig. 8.5).

After the field study, the children present their experiences through indoor activities with creative ways. Firstly, they try to explore the power of the water, representing the waterfall, through the implementation of simple experiments (Fig. 8.6).

Secondly, they create a three-dimensional mapping with the reuse of recycled materials, in which is reflected their path, the acquired knowledge and the monuments they have visited. Thirdly, they are encouraged to exploit this model by creating their first algorithm and coding robotic systems, such as Bee-bot and LegoWeDo 2.0.

The Bee-Bot, is a programable floor robot and an excellent educational tool to introduce children to some of the basic concepts of coding, ideal for early-stage programming. It moves accurately in steps of 15 cm with the forward, backward, left, and right buttons, it turns in 90° degrees and remembers up to 40 steps. It can also be incorporated the pause button for momentarily pauses (Fig. 8.7).

Fig. 8.6 Implementation of simple experiments

Fig. 8.7 Exploitation of the Bee-Bot, the educational floor robot

While the Bee-bot is programmed to move from one destination to another, the children cultivate their computational thinking, improve their spatial development and identify challenges by proposing solutions, promoting the learning process, through the use of modelling, algorithms and decomposition. In this context the children participate in the Panhellenic Educational Robotics Competition creating a short narrative film in which the robot visits the monuments following the path drawn in the field study (https://www.youtube.com/watch?v=3wSgGf2xYkM&t=1s). During their preparation for the competition, the students, with the support of the teacher understand science, programming and automation, learn to think like engineers, develop their problem-solving skills and expand their creativity (WRO Hellas) (Fig. 8.8).

Accordingly, the parents are encouraged to create with their children floor tracks, to implement relevant coding activities and present them in the class plenary. In this

Fig. 8.8 Coding activities with the exploitation of the Bee bot

Fig. 8.9 Effective parental communication

effort they are supported by the students, who disseminate their knowledge (Fig. 8.9). It is worth to be mentioned that during the pandemic period, the effective cooperation between school and families was continued through synchronous distance learning. Particularly, utilizing the educational platform Webex by Cisco, the teacher communicates with the students and their parents through online discussions and they use cooperatively the computer program Google Earth, that renders a three-dimensional representation of Earth. Subsequently, they take an aerial-view satellite image from their path and make it interactive, through the exploitation of the web platform of ThingLink, https://www.thinglink.com/scene/1438844118888349698.

Following the same process, students create a space map and explore our solar system with the help of Paxi, the European Space Agency's education mascot (ESA, 2014). In this activity, the challenge is to code the floor robots (Bee-bot, E.a.R.L., Botley) in order to move them to the planets. The children, using cards with the symbols of the robot buttons, put them in the order they plan to solve the problem, test and debug their solution. According the computational thinking and the concept of the 'decomposition', they can be facilitated breaking the problem down into smaller parts (Fig. 8.10).

The "Little Creative Scientists", based on the implemented activities, decide to protect their local natural and cultural environment and make their world a better place to live. They communicate with the Communication Officer for Greece and Cyprus, United Nations Regional Information Centre (UNRIC), who visits the school and informs the students about the goals of sustainable development.

In this framework, they create an Ecological-Sustainable Committee and produce an Eco-Code, as statement that represents the school's commitment to sustainability. Among others, they decide to recycle, reduce and reuse their trash, they visit the environmental school of the Holy Metropolis of Ioannina and encourage their parents and the wider community to participate in this action (Fig. 8.11). In addition, they create a short film approaching the sustainable development goals, using technology to disseminate their messages. The actor of the "film" is the Active Citizen of the independent international organization, Action Aid, who travels around the world and encourages the active citizenship and global understanding of the students (https://youtu.be/jPUaqaNxQDI).

Fig. 8.10 Computational thinking: Programming space exploration

Fig. 8.11 Environmental school of the Holy Metropolis of Ioannina

Fig. 8.12 Dissemination of the results to the students of the Department of Preschool Education of the University of Ioannina

Eventually, the children present the results of their efforts to the students of the Department of Preschool Education of the University of Ioannina (Fig. 8.12), to the Festival of Science and Sustainability of the University of Ioannina, to the Festival of European School Radio, on the municipal radio of Ioannina, to the student conference of the national thematic network "Sustainable City: The city as a field of education for sustainability", to national and international educational conferences.

The multiple benefits of using robotics and STEAM education are being disseminated to teachers and colleagues from other schools who are encouraged to create their algorithm and participate in the pilot Preschool Robotics Competition (WRO Hellas) and the Europe Code Week (CodeWeek4All) action which aims to bring coding and digital literacy to everybody in a fun and engaging way.

8.6 Discussion and Conclusions

The research process is completed by evaluating and disseminating the learning outcomes of the project. The results are consistent with the corresponding research question, as the socio-cultural approach of STEAM education and educational robotics facilitates preschool children to approach the learning area of sustainable development successfully. Simultaneously, the applied project contributes to the spontaneous and effortless involvement of the students and promotes their active citizenship. The progress of the research includes qualitative and quantitative collection data methods, such as participatory and non participatory observation, semi-structured interviews and questionnaires of students, parents and collaborating teachers, observational notes. Conceptual mapping, photographs and audio and video recordings, written material from the students are being investigated in the individual and classroom portfolios children's reflections on their learning. In addition to these evaluation tools, a rubric is used as an alternative assessment method, which emphasizes strengths and identifies areas that need improvement.

According to the results, educational robotics and STEAM education are introduced as powerful and flexible learning tools that involve students actively in the learning module of sustainable development through authentic problem-solving activities. Children learn to work in teams, to communicate and collaborate effectively, in order to find suitable solutions to everyday problems through the use of robotics and science.

Although, the planned and emerging activities emphasize on STEAM education and robotics, they are connected with all the learning modules of the curriculum, relating to early childhood. Indicatively, we mention that:

- personal and social development is achieved through the collaboration of students and their interaction with the wider social environment.
- sustainable education is fostered through the acquisition of active citizenship and the approachment of selective sustainable development goals.
- science is linked to simple experimental activities and are related to the daily life of children.
- algorithmic thinking is promoted through the robot programming
- use of technology is developed by concept mapping (Kidspiration), digital mapping of their area (Google map), communication with partner schools and distance learning with the parents and other educational institutions.
- written and oral speech is improved as their vocabulary is enriched.

- arts by creating the three-dimensional robotic model from useless/recyclable materials.
- physical development is actualized with outdoor activities during field study.
- critical and creative thinking (E. de Bono, 1993) is developed, as children engage in solving problems that arise in an innovative unexpected way.

The analysis of the collected data of this case study demonstrates that a high level of student engagement was observed. The educational robotics and STEAM education, as methodological tools, motivate them to participate more during all phases. The active participation of students, the authenticity of the material that was produced and their enthusiasm during the implementation of this pedagogical intervention for the last two schoolyears, confirm the strong emotional involvement of student with relation to the object of their attention. When the children were asked if they like the presence of robots in their classroom, the most of them answered positively. Only a small percentage of students gave neutral answer about the presence of robotics learning corner (Fig. 8.13).

Simultaneously, the collaborating pre-primary teachers who participated in the pilot educational project confirm that educational robotics and STEAM education facilitate the approach of the 17 goals of sustainable development and contribute to the access of all learning areas of the curriculum for kindergarten. Most of the teachers believe that "The interaction with educational robotics and STEAM education enhance the curriculum, help students to learn problem-solving and can be

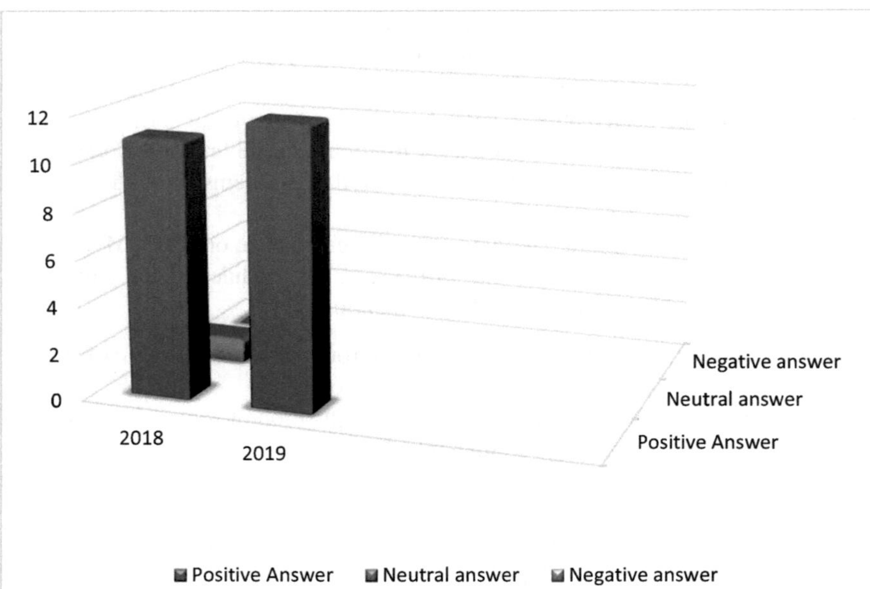

Fig. 8.13 Views of the students about the presence of robotics in the classroom

considered as powerful learning tools" (Fig. 8.14). When they asked if they would recommend to other colleagues the exploitation of educational robotics they responded positively (Fig. 8.15).

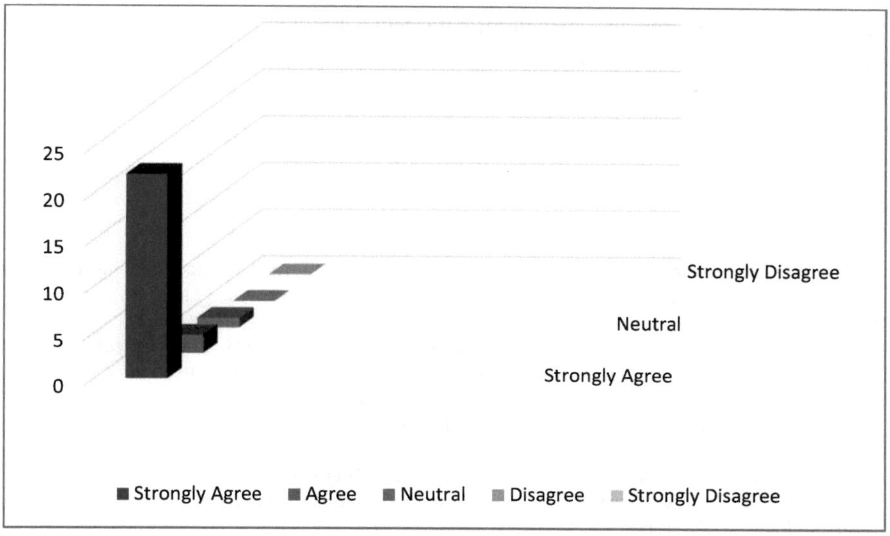

Fig. 8.14 Views of cooperating teachers on the exploitation of educational robotics and STEAM education

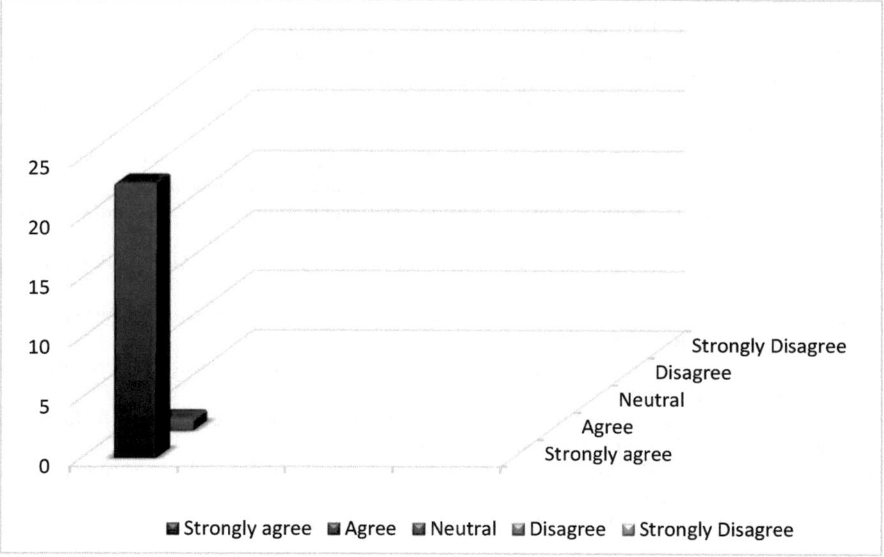

Fig. 8.15 Views of cooperating teachers about possible recommendation of educational robotics and STEAM education to other colleagues

It is essential to note that through the implementation of this pilot project the school was awarded as Sustainable, Ecological and ICT oriented School, as honored school with National and European Quality labels from the eTwinning community, with the European Label of Code Week Schools, with award from the European Space Agency for the implementation of coding activities and as "Ambassador School of the 17 Sustainable Development Goals". Finally, the teacher was awarded from the United Nations, Educational, Scientific and Cultural Organization (Hellenic National Commission for UNESCO).

References

Asghar, A. (2012). Informal science contexts: Implications for formal science learning. *Learning Landscapes, 5*(2), 55–72.

Bono, E. (1993). *Teach your child how to think.* Penguin Books.

CEYS (Creativity in Early Years Science Education). (2017a). *Report O3-A4 CEYS training guide and scenarios of use.* Authors: Esme Glauert, Jillian Trevethan – UCL Institute of Education. Available at: http://www.ceys-project.eu/sites/default/files/03_A4_CEYS%20Training%20Guide%20and%20Scenarios%20of%20Use_FINAL_EN_0.pdf

CEYS (Creativity in Early Years Science Education). (2017b). *Curriculum development methodology report.* Authors: Teresa Cremin, Tatjana Dragovic and Jessica Baines-Holmes. Available at: http://www.ceys-project.eu/sites/default/files/O2_A1_Curriculum_development_methodology_FINAL.pdf

CLS (Creative Little Scientists). (2012) *Conceptual framework.* Deliverable D2.2. EU Project (FP7 Contract: SIS-CP-2011-289081 – Project Coordinator: Ellinogermaniki Agogi, Greece). Leading Authors: A. Craft, T. Cremin, J. Clack, A. Compton, J. Johnston. Available at: http://www.creative-little-scientists.eu/sites/default/files/CLS_Conceptual_Framework_FINAL.pdf

CLS (Creative Little Scientists). (2014a). "*Comparative Report.*" Deliverable D3.4. EU Project (FP7) (Coordinator: Ellinogermaniki Agogi, Greece). Leading Authors: S. Havu-Nuutinen, D. Rossis and F. Stylianidou.. Available at: http://www.creative-little-scientists.eu/sites/default/files/D3_4_Comparative_Report_FINAL.pdf

CLS (Creative Little Scientists). (2014b). "*Set of Recommendations to Policy Makers and Stakeholders.*" Deliverable 6.6. EU Project (FP7) Coordinator: Ellinogermaniki Agogi, Greece. Lead Authors: D. Rossis and F. Stylianidou. Available at: http://www.creative-little-scientists.eu/sites/default/files/Recommendations_to_Policy_Makers_and_Stakeholders.pdf

Cohen, L., Manion, L., & Morrison, K. (2008). *The methodology of educational research.* Metaichmio.

Dimitriou, A. (2009). *Environmental education: Environment, sustainability. Theoretical and pedagogical approaches.* Focus.

Engeström, Y. (1987). *Learning by expanding. An activity-theoretical approach to developmental research.* Orienta-Konsultit.

Engeström, Y. (2015). *Learning by expanding: An activity-theoretical approach to developmental research* (2nd ed.). Cambridge University Press.

Engeström, Y., & Sannino, A. (2010). Studies of expansive learning: Foundations, findings and future challenges. *Educational Research Review, 5*, 1–24.

European Space Agency (ESA). (2014). *Explore the universe.* Available at: https://esamultimedia.esa.int/docs/edu/PaxiFunBook.pdf

Kokkotas, P., & Plakitsi, K. (2005). *Museum pedagogy and science education. Theory and action.* Patakis.

Liarakou, G., & Flogaiti, E. (2007). *From environmental education to sustainable development education.* Nissos.

Ministry of Culture. (2014). *Post-Byzantine Monuments of Klimatia*. Theodoridis Graphic Arts.

Ministry of Education. (2011). New school (21st century)-new curriculum-horizontal act, NSRF 2007–2013

Ministry of Education, Pedagogical Institute. (2002). Interdisciplinary curriculum framework for kindergarten and activity planning and development programs, Athens

NACCCE. (1999). *All our futures: Creativity, culture and education, national advisory committee on creative and cultural education*. DFEE.

National Research Council. (2012). *Education for life and work: Developing transferable knowledge and skills in the 21st century*. The National Academies Press. https://doi.org/10.17226/13398

OSOS (Open Schools for Open Societies). (2017). Open Schooling Model. Author(s) Sotiriou, S., Cherouvis S. Contributor(s): Mauer M., Bogner F.X., Bagiati A., Sarma S.E., Zygouritsas N., Giannakopoulou, A., Mylopoulos G., Fermeli G., Kiriakidi, E., Van Laar, E., M., Guenaga, M., Verboon, F., Parente, R. Di Martino, F., Gaist, O., Bloch, N., Oron, E., Ben-Horin, Y. Reviewer (s) Bogner F. X., Sotiriou, S. Available at: https://www.openschools.eu/wp-content/uploads/2018/01/D2.1-Open-Schooling-Model.pdf

Plakitsi, K. (2011). *Activity theory in formal and informal science education*. Sense Publishers.

Plakitsi, K. (2012). *Socio-cognitive and socio-cultural approaches to the teaching of the natural sciences in preschool and early school age*. Patakis.

Plakitsi, K., Stamoulis, E., Theodoraki, X., Kolokouri, E., Nanni, E., & Kornelaki, A. (2018). *Activity theory and science education: A new dimension in STEAM education*. Gutenberg.

Ravanis, K. (1999). *Science in preschool education: Teaching and cognitive approach*. Gutenberg.

R4C (Reflecting for Change). (2020). *Implementation plan*. Author(s) Nikos Zygouritsas, Sofoklis Sotiriou. Reviewer(s) Menalaos Sotiriou. Available at: https://reflecting4change.eu/wp-content/uploads/2023/10/D3.1-Implementation-Plan.pdf

Roth, W.-M., Goulart, M. I. M., & Plakitsi, K. (2013). *Science during early childhood: A cultural-historical perspective*. Springer.

Sannino, A., Engeström, Y., & Lemos, M. (2016). Formative interventions for expansive learning and transformative agency. *Journal of the Learning Sciences, 25*(4), 599–633. https://doi.org/10.1080/10508406.2016.1204547

Shaheen, R. (2010). *Creative education*. Scientific. *Research, 1*(3), 166–169. https://doi.org/10.4236/ce.2010.13026

Sonter, L., & Kemp, S. (2021). Connecting the UNESCO sustainable development goals with Australian early childhood education policy to transform practice. *International Journal of Early Childhood Environmental Education, 9*(1), 11. North American Association for Environmental Education.

Topoliati, M. (2015). Democracy and participation in kindergarten. In *Collective volume of the sustainable Greek school* (pp. 43–47). Elliniki Etairia. Society for the Environment and Culture Heritage.

Vygotsky, L. S. (1978). *Mind in society: The development of higher psychological processes*. Harvard University Press.

WRO Hellas. https://wrohellas.gr/

Maria Topoliati is a PhD candidate of the University of Ioannina with two bachelor's degrees in Pedagogy of Early Childhood Education and Philosophy, Pedagogy and Psychology. She has completed Postgraduate Training at the University of Ioannina. She has Diploma in musical instrument teaching, Degree of Advanced Theoretical Music, Certification of knowledge of the Braille system, Certification of skills and knowledge in ICT and Certification of training in Orff-Schulwerk and Dalcroze Eurythmics methods. She has participated as trainer in the program "Distance Learning Techniques through Moodle" and as mentor for the teachers of the Greek Sustainable Schools. She is an ambassador of Scientix Community, of the Erasmus+ project CEYS: Creativity in Early Years Childhood Education, and national ambassador of educational robotics.

She has worked as a music teacher in primary and elementary education. She works as an early childhood teacher since 2005. She is a member of the Activity Theory in Formal and Informal Science Education (@FISE) Researching Group. Her researching interests are focused on the sustainable development, STEAM education and robotics in early childhood and Cultural Historical Activity Theory applied in Early Years Science Education. She presents innovative educational and scientific projects in several national and international conferences. Maria Topoliati was awarded for her educational contribution by UNESCO Hellas and the Foundation for Environmental Education (FEE Global).

Katerina Plakitsi is a member of the Governing Council of the University of Ioannina. She is a full professor of Science Education with two bachelor's degrees in Physics and Pedagogy, a master's diploma, and a PhD in Science Education. Her main researching interests are Science Education, Formal and Informal Science Education, and Cultural Historical Activity Theory applied in Science Education. She has written many books in Greek and English while she has published in different international academic journals. She is scientific coordinator in many European Projects and supervises some PhD scholars in Science Education. She has written many schools environmental science textbooks, and she coordinated a science curriculum reform in Greece for the contemporary Education. She co-founded the interdisciplinary master's program "Environmental Sciences and Education for Sustainability" in collaboration with the Faculty of Medicine and the Department of Biological Applications and Technology at the University of Ioannina. She is also the co-founder of the ISCAR-STEM Thematic Section and the principal investigator of the @formal and informal science education group (@fise group). She is the editor-in-chief of the international bilingual journal *Science Education: Research and Praxis*. She had been Head of the Early Childhood Department at the University of Ioannina. Katerina Plakitsi is President of the International Society for Cultural Historical Activity and Research (https://www.iscar.org/).

Fani Stylianidou is a Senior Project Manager in the Hellenic Erasmus+ National Agency, Higher Education sector. She has a wide research experience in the field of science education, focusing on students' learning and teachers' professional development. In Ellinogermaniki Agogi, she coordinated the EU/FP7 project Creative Little Scientists and the Erasmus+ project Creativity in Early Years Science Education. She has a Physics degree from the University of Athens and an MA and PhD in Science Education from the Institute of Education, UCL. She has taught in the University of Aegean and been the Deputy Director of Science Learning Centre London at the Institute of Education, UCL. She has contributed as consultant to the OECD in education policy reviews.

Chapter 9
Science Education Program "Thunderbolt Hunt:" Practicing Scientific Method in the Archaeological Museum of Ioannina

Athina-Christina Kornelaki and Katerina Plakitsi

Over the last 20 years, science museums focus more and more on the dissemination of scientific knowledge considering it as a path to scientifically literate societies and participatory citizenship. The Science Center World Summit in Tokyo (2017) affirmed this view by recognizing the Sustainable Development Goals, endorsed by the United Nations (2015) as principal priority towards global sustainability and prosperity (Kornelaki & Plakitsi, 2018). The European Commission (2015) often incites synergies between formal and non-formal education in order to provoke students' engagement and to increase their interest about science (Mujtaba et al., 2018).

The goals envisioned by these international bodies with regard to science education can be promoted successfully using educational materials available in museums for organizing science programs. In terms of the synergy between school and museum, some obstacles are highlighted in the literature connected with the lack of resources and inclusion for pre-school students as well as the inequalities for those who come from agricultural communities and those with low socio-economic background (Falk et al., 2014). Within this research an effort is made to bridge the aforementioned gap. First, by the design of educational programs appropriate for pre-school and early-grade primary students. Their inner curiosity for the world around them as well as their excitement about discovering and exploring it at this age can be harnessed to promote science education effectively (Kornelaki & Plakitsi, 2020). Secondly, this research suggests that the implementation of the educational programs in museums of general interest creates opportunities for provincial schools and widely to the regions of convergence to participate in science education

A.-C. Kornelaki (✉) · K. Plakitsi
Department of Early Childhood Education, University of Ioannina, Ioannina, Greece
e-mail: akornelaki@uoi.gr; kplakits@uoi.gr

K. Plakitsi, S. Barma (eds.), *Sociocultural Approaches to STEM Education*,
Sociocultural Explorations of Science Education 21,
https://doi.org/10.1007/978-3-031-44377-0_9

programs in their cities. Epirus region is one of the regions of convergence according to EU meaning that it is considered a less developed region financially in the European Union. Moreover, Ioannina city, as the center of Epirus region, serves as a cultural destination of the provincial schools having an impact across the region. In Greece, science museums and centers are very few and are located in big cities. As a result, school visits especially for younger students is extremely difficult if not impossible.

This study explores the potential in cultural-historical activity theory (CHAT) and its methodologies to not only design and organize inclusive science education program for young children, but also to study its implementation. CHAT, as a socio-cultural approach is transferred to non-formal settings, such as museums of general interest, where it is embedded within a wider learning community. Within this learning community, students, carrying their own socio-cultural background, interact with other students, with the museum's personnel, even with their parents. This helps them get to know aspects of their culture. Besides, they practice scientific method processes, while at the same time they meet their historical, social and cultural background (Plakitsi, 2013). Museum constitutes an authentic learning community which is distinguished for the mediation of alternative tools that exceed the boundaries of formal education and offer students opportunities to adopt positive stances towards science education and museum. This is made possible because students learn in an environment with reduced hierarchy, student-centered, where they work collaboratively, interacting with each other, conducting their inquiry and making decisions about their teammates' role in the group. Finally, this approach, through non-formal and informal learning environments seeks to open science to society, via social agents that promote scientific knowledge and feature not only interpersonal, but also cultural and historical aspects of it (Foot, 2014). This is very well connected with the approach of science education in early grades towards students' scientific literacy. It is worth noting that especially in early grades the focus in the field of science education lies not on the scientific content (concepts, laws etc.) but on the scientific method process. The scientific method processes or science process skills are connected with science learning through a prism, according to which, science education is considered the cultivation of children's rational thinking and understanding of the world around them. The scientific method processes are observation, classification, measuring, predictions, hypotheses, interpreting and drawing conclusions, communication (Plakitsi, 2008). Hence, our focus on science education is to equip students with a set of skills that will support them on learning how to learn science and solve everyday problems (Suryanti & Lede, 2018).

The focus of the present research is the dissemination of the scientific knowledge within the Archaeological Museum of Ioannina. While there is plenty literature about disseminating scientific knowledge through science museums or centers, the literature about doing that in museums of general interest is very limited (Kornelaki & Plakitsi, 2022). There is some rigorous work from an interdisciplinary interpretative perspective of museum exhibits and collections that involves the construction of a mobile structure placed in the collection area, which provides the public with methodological tools to examine scientific hypotheses connected with the original archaeological object-exhibit (Georgopoulou et al., 2021). The present research is based on the visitor's/educator's free interpretation of the museum's exhibits and collections (Hooper-Greenhill, 2000) and the design of educational programs which does not require structural interventions in the museum area. Educators can observe and study a museum's collection, identify the bridges that connect the exhibits with the curriculum of science education and start building up on the idea of the educational program which corresponds to their students' interests and level. The bridges can be explicit, by observing the exhibit or they can be implicit, coming from the exhibits prior use or its testimony (Kornelaki & Plakitsi, 2022).

The present research attempted to exploit, on one hand, the benefits offered by the aforementioned learning environments with the reduced hierarchy, the strong cultural tools and the attractive settings as alternative learning communities through their educational programs, while on the other, to introduce to these environments scientific concepts cultivating scientific method processes. These objectives are achieved exploiting the principles of CHAT, which provide tools to design as well as analyse the educational program. The features of the educational programs introduced originate in the field of science education, cultivate scientific method processes, adopt collaborative and experiential learning, the process of scaffolding and fading make good use of students' prior experiences, create connections with students' everyday lives and offer alternative play-based learning experiences. The features of the educational programs are depicted in the conceptual map below (Fig. 9.1).

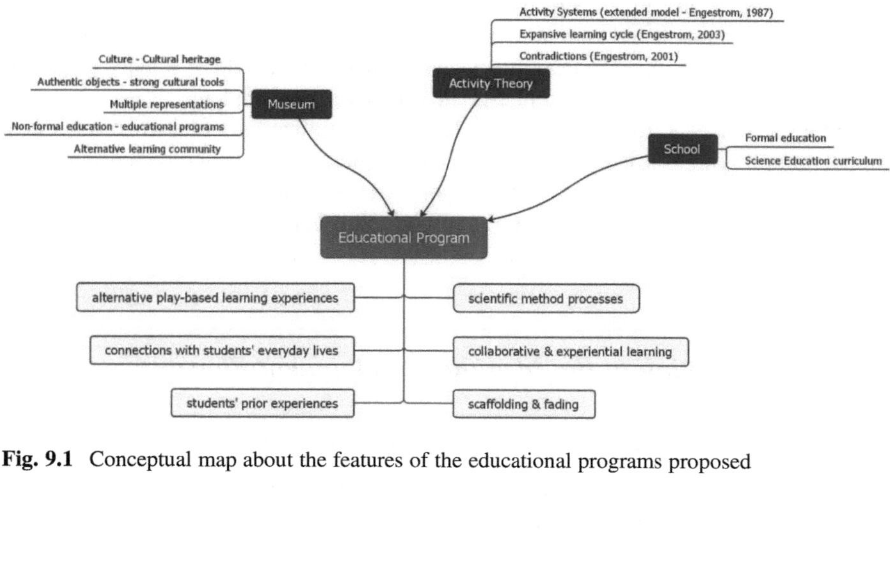

Fig. 9.1 Conceptual map about the features of the educational programs proposed

9.1 The Design Framework SciEPIMGI

The design framework introduced in the present study, SciEPIMGI (Scientific Educational Programs In Museums of General Interest), is based on two principal axes methodologically. The first is the general/widespread design methodology (Radice, 2014), while the second is the methodological tool for Science Education in early childhood and primary education based on CHAT, SCOPES (Systems of activity, Contradictions, Outcomes, Praxis, Expansive learning, Science education) (Kolokouri & Kornelaki, 2019).

The first is the general methodology followed by the museums for the development of museum exhibits and collections as well as educational programs and materials, which are based on participatory design and promote audience's engagement (Radice, 2014). The general/widespread methodology, according to the literature (Ambrose & Paine, 2018), consists of four phases which are (1) conception of the idea or the development phase, (2) design, (3) implementation, and (4) evaluation.

The second is SCOPES, which seeks to expand the scope of science education integrating "scientific and humanistic knowledge through participatory methods" (Kolokouri & Kornelaki, 2019). It offers a prism with which learning is holistically analyzed and dealt as a complex process, focusing on the parts of the tool (Systems of activity, Contradictions, Outcomes, Praxis, Expansive learning, Science education). In the beginning, the Systems of activity involved are identified as well as the Contradictions emerging through the constant interactions occurring within the systems' components, between them and/or between the different systems. All this process leads to a desired Outcome for students, which is achieved through Praxis, *the conscious, purposive practical experience* (Blunden, 2013). The aforementioned process is placed and studied in an Expansive learning cycle, which not only leads to the design, implementation and evaluation of a new model, but supports the shift from the individual to the collective, which in turn, reconceptualizes the object and the motive of the activity system transforming it and enriching it with new knowledge and practices (Engeström, 1999). Finally, the expansive learning cycle is situated in the field of science education and especially for pre-school and first-grade primary school students.

Important aspects from both methodologies were utilized in order to create educational programs appropriate to be implemented in museums of general interest while at the same time cultivate scientific method processes for students of early grades.

The basic steps of the design framework SciEPIMGI and what each phase includes are briefly presented below.

9.1.1 Conception of the Idea/the Development Phase

The development phase includes the following considerations:

1. The rationale for the design of the educational program
2. The objective/-s of the educational program
3. The target group

4. Bridges between the exhibits and the concepts/topics of Science Education – Connection with the Science Education curriculum

9.1.2 The Design Phase

The stages of the design phase adapted to the design of science education programs for museums of general interest are:

1. The selection of the museum exhibits that will be included in the educational program studying the cultural-historical value they hold.
2. The connection of the selected exhibits with the topic/-s of Science Education that the educational program deals with.
3. The development of the plot, the story that runs through the educational program and serves the smooth succession of the individual actions and the maintenance of the students' interest and engagement. Cultural and historical information about the exhibits compose the plot.
4. The design of the individual actions of the educational program.
5. Choosing the pedagogical methods and educational techniques
6. Determining instructor's role in an effort to give him a mediative, supportive role.
7. Choosing the evaluation method/-s

9.1.3 The Implementation Phase

The implementation phase includes the following stages:

1. Production planning: The final budget and time schedule of the project are determined, the approval of the museum director is given, and the dissemination of the educational programs starts, issuing press releases etc. All the necessary materials for the educational program are decided and prepared (materials for the experiments, drawing equipment, games, character dolls and anything else that may need to be prepared for the educational program's needs).
2. Production: All the above are gathered or constructed, if necessary, and are placed in the museum collection according to educational program's needs.
3. Operational stages: The maintenance of the materials produced as well as the renewal of the expendable materials for the future applications.

9.1.4 The Evaluation Phase

The widespread methodology proposes all the typical evaluation methods (front-end evaluation, formative evaluation, remedial evaluation, summative evaluation). The designed framework SciEPIMGI proposes reflection under the prism of CHAT focusing on active and interactive learning. There is a lot of discussion around the evaluation of learning in museum areas and opinions vary. Some experts in the respective literature argue that short-term metrics which focus on recalling facts and concepts are not very sufficient (Mujtaba et al., 2018). The same authors point out that essential learning in museums is very likely to become apparent later and may occur a few weeks to a few years after the first or next visits (Mujtaba et al., 2018). This is the reason why CHAT provides the lens through which learning is expressed through students' interactions during the learning process, while they experience those interactions. Through these lenses the following may be studied.

(a) The scientific method processes during the implementation of the educational program.
(b) The interactions occur among the students, between the students and the instructor and vice versa during the implementation of the educational program.
(c) The structural components of the activity systems and the interactions between them utilizing Engeström's extended model of the theory (Engeström, 1999).
(d) The contradictions occurred during the implementation.

The present chapter presents only the results connected with the scientific method processes due to space limitation. Hence, not all the phases of the design framework are presented in the chapter nor other research questions from those mentioned above which are presented in older publications or will be presented in future work.

9.2 The Educational Program "Thunderbolt Hunt" in the Archaeological Museum of Ioannina

Following the steps of the design framework, an educational program was developed, entitled "Thunderbolt hunt". "Thunderbolt hunt" is designed for students from 4 to 9 years old i.e., for pre-school and first-grade primary students. It is connected with the respective science education curricula and more specifically, with the scientific concept of air and its properties. The aim of the educational program "Thunderbolt hunt" was for students to perform experiments about air and its properties, utilizing mythological elements and personalities as well as museum's exhibits and their testimonies. More specifically, the objectives of the educational program were to cultivate scientific method processes, get to know some information about the exhibits and their use, participate in play-based actions and cooperate with their classmates, in order to achieve the goal of the educational program, which was the finding of Zeus' thunderbolt.[1] Overall, what was important was for students to

[1] According to Greek mythology, Zeus was the father of both gods and men, and he is depicted holding a thunderbolt as a symbol of his authority. He was considered a god of the sky and the weather, and it was believed that he was the sender of thunder, lightning etc. and his weapon was his thunder, so losing it was a stain to his reputation.

Aeolus was the father of six sons and six daughters and as a respecting ruler of his kingdom, Aeolia, he was favoured by Zeus who made him the king of the winds. Aeolus is usually depicted carrying a big sack which is full of the winds. He could control the winds releasing them only after Zeus' command.

gain a positive experience performing experiments from the field of science education in the Archaeological Museum of Ioannina.

The educational program "Thunderbolt hunt" consists of 7 individual actions and lasts 90 min. It includes different kinds of actions which play a different role in the program. There are activities that cultivate scientific method (2 & 4), a game (6), activities for which drama in education is used (5 & 7) and the plot, which introduces students to a problem-solving situation (3) (Kornelaki & Plakitsi, 2020). The individual actions are briefly described in the table below (Table 9.1).

Table 9.1 The actions of the Educational Program "Thunderbolt Hunt" (Kornelaki & Plakitsi, 2018, p. 91)

No	Actions	Short description
1	Group formation and discussion about museum exhibits	Students are welcomed in the collection room "Dodoni". They are taking one card which illustrates an exhibit. The cards form the three groups. A discussion takes place about the cards and what students think they represent.
2	Search for museum exhibits – The common element	The first task for the groups is to use the tools given (magnifiers, torches etc.) in order to find the exhibit which is illustrated in the group's cards and observe it. When all the groups find their exhibit, they describe it to the rest of the groups and altogether try to figure out the common element which is the thunderbolt.
3	How was the thunderbolt lost? – Narrative	This activity constitutes a narrative about Dodoni's oracle, which is pictured on the wall, and explains how Zeus lost his thunderbolt when Aeolus set his winds free from his sack without warning Zeus. Now Aeolus is accused, threatened for his life and ordered by Zeus to find thunderbolt. Aeolus turns to students for help.
4	Experiments on air	The common element of activity 2 gives students the pass for the next task, which is the experiments about air and its properties. Students do the experiments by using the materials given (balloons, straws, syringes, plastic bottles with or without a hole etc.), test their predictions, communicate their findings, draw conclusions and generally gather data in order to help Aeolus by giving him advice on where and how to find the thunderbolt.
5	Role on the wall	Students draw or write their advice and give them to Aeolus. He is pictured on a big paper and students glue their ideas on his head, so he can think and choose the best idea to find the thunderbolt.
6	Zeus' winged thunderbolt	While he is fast as the wind, Aeolus shortly and secretly leaves pieces of the thunderbolt to the instructor and students must assemble the pieces of the puzzle to get the Zeus' thunderbolt.
7	Aeolus' sack	Aeolus surprises students with the last task which aims to decompress. Before students leave the museum, they are asked to gather Aeolus' winds and put them back into his sack which is left to the instructor. When all the winds are in the sack one or more students tie the sack with a band, so the winds will not escape.

9.2.1 The Concept of Air and Its Properties in the Educational Program "Thunderbolt Hunt"

Air is a key component of the atmosphere and therefore a structural element of nature and life itself. The movement of the air creates wind, which gave impetus to the evolution of civilization and was the driving force for many inventions and mechanisms that made human life easier. Although air is an integral part of students' lives, they find it difficult to grasp the existence of air since it is not visible. What the educational program introduces to students is the idea of existence of air around us, as well as in ostensibly empty spaces, the property of air about space occupation and that it takes the shape of its container, that air can be compressed and finally, that two objects cannot occupy the same space at the same time. The above properties were introduced with three experiments performed in the fourth action of the program. Before the implementation of the experiments, the instructor discussed with students their prior knowledge about air and their hypotheses and predictions about the experiments. The three experiments are described below.

9.2.1.1 First Experiment

The materials of the experiment are plastic bags and plastic bottles. In the first experiment students were squeezing an empty plastic bottle and capturing air with an empty plastic bag from different spots of the room. The goal was to experience that even though we are not able to see air, it exists among us in the room and every corner of this room, or every isolated spot is full of air. In this way, students were introduced to the existence of air, as well as the property of air to take the shape of the container it is in.

9.2.1.2 Second Experiment

The experiment requires syringes. In the second experiment students were challenged to pull the syringe's plunger and then try to push it back having syringe's opening closed with their finger. This experiment introduces the property of air to occupy space and that it can be compressed, because even though we are not able to push the plunger entirely, we are able to squeeze it at some point.

9.2.1.3 Third Experiment

The materials required for the experiment are balloons, plastic bottles, some with a hole on the bottom and some without a hole. The balloons are placed in the spout of every bottle facing inwards. In the third experiment students try to inflate the balloon inside the bottle. With this experiment students experience that when there is a hole on the bottle, the balloon inflates because the air existed in the empty bottle went out from the hole and left the space for the balloon. In the second case, when there is not a hole on the bottle, the balloon cannot be inflated at all since the air in the empty bottle occupied all the space and it does not allow the balloon to inflate. In this way, students are introduced to the property of air to occupy space and that two objects cannot occupy the same space at the same time (Kornelaki & Plakitsi, 2020).

9.3 Methodology

9.3.1 Participants

The educational program "Thunderbolt hunt" was implemented to schools that were interested to attend it and booked an available date for their students. In total, 8 different classes of schools attended the educational program in the Archaeological Museum of Ioannina the first year. Two of them were private schools, four were from the city of Ioannina and two from villages in the outskirts of the city. In numbers, 136 students, 6–8 years old and 12 teachers participated in the research.

9.3.2 Data Analysis

The data collection included video recordings from the implementation of the educational program in the museum, photographs from the implementation, students' drawings and researcher's fieldnotes during or right after the implementation of the educational program. In total, almost 13 h of video were analyzed, 136 students' drawings, 39 photographs and fieldnotes which enriched the data.

Due to the large amount of data, the qualitative data analysis software, NVivo 9 was utilized to support the complex analysis. The dialogues from the implementations of the educational program were recorded, transcribed, and coded in software NVivo 9. A multilevel data analysis was performed within the NVivo 9 environment using the coding process and following the evaluation phase of the SciEPIMGI framework. The coding process is carried out by creating nodes and serves the researcher to identify patterns and ideas in the data regarding the research questions. The node in simple words is a gathering point of references/excerpts from the data that present a common theme such as communication which is a scientific method process. A lot of nodes falling in a common theme, create a parent node, such as scientific method processes altogether.

The research question the present chapter focuses on, is which scientific method processes are practiced by the students during the implementation of the educational program "Thunderbolt hunt" and how. In this sense, scientific method processes serve as nodes in data analysis. The results of the analysis will constitute an indication related to another research question about the extent to which the museums of general interest such as the Archaeological Museum of Ioannina, are suitable learning environments to practise the scientific method processes and furthermore, whether the SciEPIMGI framework is suitable for such educational programs' design.

The analysis of the data goes beyond the mere identification of the scientific method processes in students' dialogues. What interests the research is to show the way with which the scientific method processes are practiced by the students in the museum's learning community focusing on active and interactive learning processes. This process captures what happens when it happens. In fact, all the quantitative results are accompanied by experts of data that support and frame the results under the prism of CHAT.

9.3.3 CHAT Frame in Relation to the Analysis

The connection of the study's methodology with CHAT lies on key points taken into account long before the analysis, during the design as well as the implementation of the program. The educational program considers students' prior knowledge and experiences about air. Before every action, discussion takes place where the students express themselves along with their misconceptions about air which are considered very valuable in the process. Discussion also follows the actions where students reflect on the process orally and/or by depicting their reflection in their drawings.

The instructor plays an essential role in the process. She is responsible to create a learning environment, where all different opinions are welcome and respected by all participants in the learning community. She is in the position to provoke interaction among the actors throughout discussions, collaboration, and feedback in the frame of the collective learning (Vygotsky, 1962). The interaction of students with peers or adults on working on assignments is connected with their zone of proximal development according to which, via cooperation, they can reach their possible development transcending their actual development (Veresov, 2017). Moreover, the interaction with their peers helps them confront themselves with their misconceptions and cross their own boundaries in the process of double stimulation (Sannino, 2015). Again, the instructor's role is to support students on their journey by scaffolding or fading when needed (Reiser & Tabak, 2014), allowing students to change their established ideas based on their vivid experiences rather than forcing them towards the desired direction.

The definition of the concept formation process is observed when students are building on their classmates' ideas, gradually approaching the desired concepts in the trajectory of the experiments (Kornelaki & Plakitsi, 2020). All the aforementioned elements are considered in the process of data analysis.

9.4 Results

9.4.1 Some Aggregate Results

In the following sections, the results from the analysis of the video recordings, the drawings and the fieldnotes will be presented. The coding process shows five scientific method processes, communication, observation, hypothesis & predictions, experiment, interpreting data. The order of the processes indicates the frequency according to which they appear in students' dialogues. The aforementioned order emerges from a word frequency query tag cloud applied in the data related to the scientific method processes in NVivo (Fig. 9.2).

The same query applied in the content of the parent node "scientific method processes", give the results depicted in Fig. 9.3. In the tag cloud extracted by NVivo, the size and the weight of the font indicates the frequency of the words used by the students. The most frequently used words are air, because, thunderbolt, balloon, can, bottle, groups, hole. Most of these words refer to the educational program's fourth action and to materials used during the experiments about air (balloon, bottle, hole, syringe, cap).

Text search queries help drawing upon details about the most frequent words in the data. Hence, the next step is to apply text search queries on the most frequently used words in the dialogues, in order to explore the connections and their interrelations with other words or phrases from the data analyzed and coded. The text search queries won't be displayed since the number of references connected with the words make them illegible. Therefore, only the comments from the text search queries will follow.

Fig. 9.2 Word frequency query tag cloud in students' dialogues related to scientific method

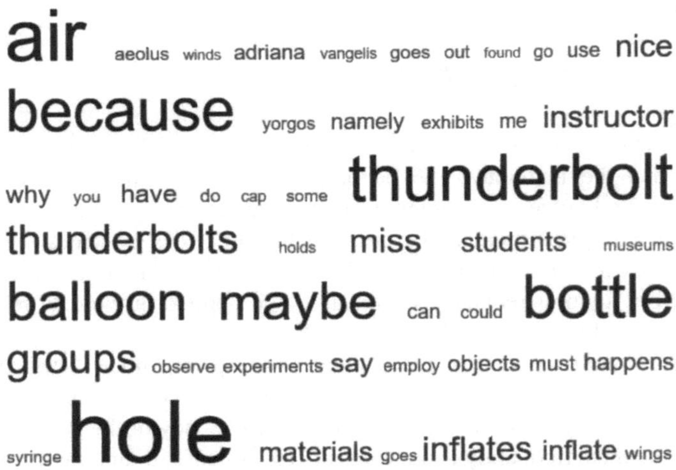

Fig. 9.3 Word frequency query tag cloud in the parent node "scientific method processes"

9.4.2 The Most Frequently Used Words by the Students

Starting from the word hole, which is the most frequently used, it is obvious that it is a dynamic and strong word in the data. Hole has a significant role in the research since it appears in the action (fourth) with the experiments and the acknowledgement of its presence is essential in order students to understand the reason why the balloon was either inflated or not in each case. Most of the references are related to this specific experiment with the plastic bottle, the balloon, and the hole and only some references are related with the experiment with syringes. In the first case, students mostly use the word while they are trying to successfully complete the experiments. Towards this direction, students use the materials they have in their disposal while at the same time, they use communication, oral speech and signs, as a tool during their collaboration (it inflates, look; it goes out from here; right here, look).

Moving on to the next most frequently used word by the students, air, and applying the text search query to find some connections and interrelations, most of the content seems connected with the word originates from the students' dialogues during the fourth action of the educational program, during which experiments were performed about the concept of air. There are a lot of references in materials used during the experimentation such as bottle, balloon, syringe as well as references of their use, either the use foreseen by the educational program, or the alternative uses students discovered while freely employing the materials. There are also references about students' observations while experimenting as well as interpretating these observations.

Moreover, we can identify references from students' dialogues from the fifth action, where students advise Aeolus to help him find Zeus' thunderbolt. Students make up ways, according to which Aeolus is asked to employ his winds and the air to find the missing thunderbolt. Most of the students encourage him to create a tornado, implicating his children, the winds and then thunderbolt would arise and become visible to Aeolus. Some students suggest that he should look for it in places where wind already exists, because they believe that wind has carried it away. So, wherever there is wind, Zeus' thunderbolt will be around. Finally, some students advise Aeolus to avoid wind because wind may carry himself away. At last, there is a group of references from the third action where students make predictions and hypotheses about how they can capture air.

Moving on to the next most frequently used word by the students, hole, and applying the text search query to find some connections and interrelations, it is obvious that it is a dynamic and strong word in the data. Hole has a significant role in the research since it appears in the action (fourth) with the experiments and the acknowledgement of its presence is essential in order students to understand the reason why the balloon was either inflated or not in each case. Most of the references are related to this specific experiment with the plastic bottle, the balloon, and the hole and only some references are related with the experiment with syringes. In the first case, students mostly use the word while they are trying to successfully complete the experiments. Towards this direction, students use the materials they have in their disposal while at the same time, they use communication, oral speech and signs, as a tool during their collaboration (it inflates, look; it goes out from here; right here, look).

9.4.3 The Instructor's Role

The instructor's involvement in the experiment process is ancillary. She appears when she judges that she can assist students to better understand the concepts and the phenomena posing questions for consideration or when she notices that during the process, obstacles appear that students cannot overcome by themselves (scaffolding), but she does not intervene when students collaborate and seem like they are discovering and interpreting the phenomena without her assistance (fading).

9.4.4 Examples of Scientific Method Processes in Students' Drawings

Correspondingly with their existence in students' dialogues, scientific method processes are present in students' drawings. It is noteworthy that when the students finished their drawings, the instructor asked them to describe what exactly each drawing depicts, in order to use these inputs for a more accurate analysis. Observing students' drawings, the process that prevails is experimenting. This is due to the use of tools by students while experimenting, which are depicted in their drawings. Hence, the scientific method processes are analyzed via the use of tools and students' descriptions of their drawings.

Nikia is a second-grade primary student, who drew, according to her description, "a straw from which air goes out, a cloud that rains and blows, my badge, the syringe that blows and air comes out and a bottle inside of which air goes out" (Picture 9.1). In Nikia's drawing all the experiments of the fourth action are included.

The same elements are depicted in Danae's drawing (Picture 9.2), who is Nikia's friend, and they were both members of the same group. Danae drew all the experiments just like Nikia, but what differs from Nikia's drawing is that Danae additionally depicted an experiment which was not designed for the educational program. She drew the straw close to the bottle's spout to depict her attempt to blow the balloon into the bottle by blowing the straw instead of the balloon. Before the instructor challenges the groups of students to try the experiments included in the educational program, students could freely manipulate tools and materials and use them the way they wanted. During this process students discovered a lot of alternative ways to use the tools in their disposal, such as using the straws to inflate the balloons into the bottles. What remarkable about the alternative experiments is that students considered them an essential part of the process and included them in their drawings.

George, a first-grade primary student, depicted in his drawing two experiments from the fourth action. In the first part of his drawing, we see "a human who inflates a

Picture 9.1 Nikia's drawing

Picture 9.2 Danae's drawing

balloon in the bottle". It is followed by "a syringe with which we are trying to inflate a balloon", and the last part of his drawing shows "a bottle with a hole, from which air comes out when we blow, and the balloon is inflated". In George's drawing (Picture 9.3), we can spot some scientific method processes such as observation, experimenting, but also interpreting, since in the third part of the drawing George implicitly explains why the balloon inflates in the bottle as air comes out of the hole while the balloon is inflated.

Finally, in George's drawing, as in Dane's drawing, an alternative experiment is depicted, according to which the syringe is used as a pump to inflate the balloon.

9.4.5 The Scientific Method Processes Through the Data

Cultivating scientific method processes in an alternative learning community such as in the museum, is considered a very important goal in regard to the design of educational programs, which deal with science education concepts into non-formal learning environments and museums of general interest. This is exactly why it is considered essential to study in-depth the scientific method processes identified in the implementations of the educational program "Thunderbolt hunt".

9.4.5.1 Communication

Communication exists in all the phases of the scientific method, since students use communication in order to express their observations and ideas, to form predictions

Picture 9.3 George's drawing

and hypotheses, to interpret data. Nevertheless, communication is not only seen as a scientific method process, but it is considered a significant skill students need to develop in order to express themselves. In this direction, it is very important educators to offer students opportunities to express their ideas and thoughts and science education is a fruitful field to do so.

Having that in mind, the results of Fig. 9.4 show the extent to which communication appeared in the individual actions (first to seventh) of the educational program, in relation with the quantity of references coded in the node "communication" in NVivo software. Communication appears in all the actions in greater or lesser degree. The most coded references appear in the second action, almost equally in the

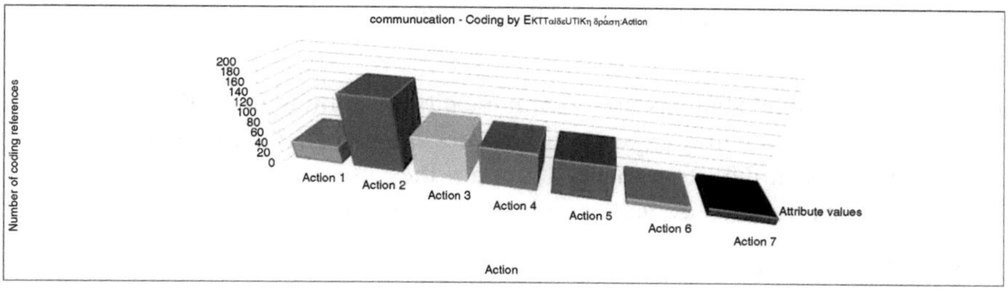

Fig. 9.4 Chart of communication in relation to the number of coding references in each action

third, fourth and fifth and less in the first and much less in the sixth and seventh actions.

Some examples of the references coded in the node communication are presented below. The excerpts are from the students' dialogues during the implementation of the educational program. Communication in these excerpts appears solely or accompanied with other scientific method processes.

In the following excerpt students use communication in the second action, when they are asked to find the exhibits on their cards in the museum's collections and observe the details of their group's exhibit.

- It was not just like in the picture, there was something else on top of it. In the picture the rock is half but in the collection is complete. (Konstantinos)
- Hmmm very interesting. So, what else did you see? What was on the rock? (instructor)
- An eagle who holds a thunderbolt. (Vasilis)

In the third action, the instructor narrates a story of how Zeus' thunderbolt was lost and why Aeolus undertook the mission of finding thunderbolt. After the narration, the instructor invites students to think ways of capturing air. Students communicate their predictions and hypotheses.

- In a glass and we close it. (Yorgos)
- Yorgos says in a glass, any other ideas? (instructor)
- In a vase. (Christos)
- In a vase. So, according to your ideas, I can take a bottle, close the cap and will have trapped air in it, right? (instructor)
- You cannot do that. (Marios)
- I cannot? Why not? (instructor)
- Because it will escape. (Marios)
- How will it escape? (instructor)
- Because it is air. (Marios)
- Even if I close the cap? (instructor)
- The air is transparent. (Marios)
- It is transparent, right. (instructor)
- Right. (Marios)

- Do you believe that air can escape through the cap? (instructor)
- Yes. (Alexandros)

In the excerpt above, students use communication to communicate their predictions and hypotheses before they experiment. In the same way, in the fourth action, where students implement the experiments about air, they use communication to describe the experiments' process as well as to interpret their observations.

- Miss, mine inflates, look. (Rafaela)
- Miss, it has a hole! (Odyssey)
- Oh, it has hole. And why does it inflate? (instructor)
- Because the air goes out. (Rafaela)

In the previous actions described above, communication is used accompanied with other scientific method processes such as observation, predictions and hypotheses, experimenting, interpreting. In the fifth, sixth and seventh action, communication is employed by students in order to express their ideas and their concerns with regard to the development of Aeolus' mission, to advise Aeolus, to collaborate so they can successfully complete the actions. In other words, in the last cases, communication is not connected with scientific concepts or processes, however, its use is worthwhile under the prism of the interactions occurred among students within the learning community.

Equally interesting is the use of communication as a tool for the instructor, who uses it to support students to better understand the concepts and phenomena and to move their thought a step further, to motivate them and keep their interest throughout the educational program.

9.4.5.2 Observation

During observation students use their senses, observe details of the objects, identify similarities and differences and use tools for systematic observation. In the educational program "Thunderbolt hunt", observation was the second most used among the scientific method processes. In Fig. 9.5 the number of references coded in the node "observation" is presented in relation to the individual actions in which it is appeared. Most of the references coded concern the second action of the program, less references are coded in the fourth action and even less in the first, third, sixth and fifth accordingly. The aforementioned results were anticipated since the goal of the second action was to explore the museum's collection, identify each group's exhibit and observe it in order to present it to the rest of the groups.

Below there are some examples of the references coded in the node.

- So, try to fill the balloon with air. (instructor)
- None of them inflate. (a lot of students together)
- My balloon inflates. (Stavroula)
- There it is. (Vangelis)
- Did you make it? (instructor)

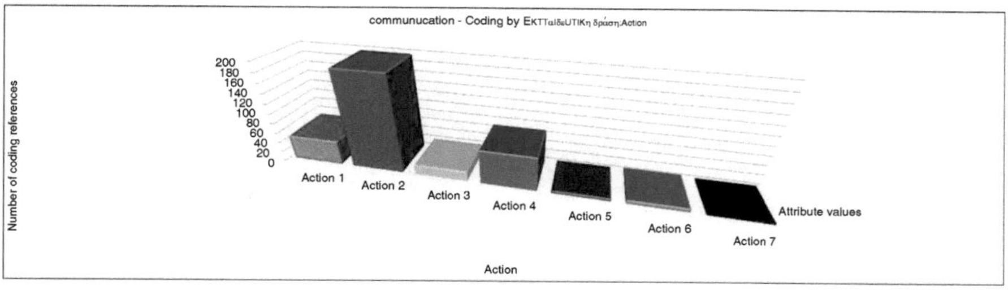

Fig. 9.5 Chart of observation in relation to the number of coding references in each action

- I cannot make it. (Marina)
- Are all the bottles identical? In some cases, the balloon inflates and in some other they don't. (instructor)
- They are all the same. (Angelica)
- Are they all the same? Do they all inflate? (instructor)
- No. (Angelica)
- This one inflates. (Nefeli)
- My balloon goes even lower in the bottle. Look where it is. (Vangelis)
- Hm, Vangelis' balloon inflates more, but why does this happen? (instructor)
- There is a hole over here. (Stavroula)
- Stavroula, where did you see the hole? (instructor)
- Here (showing). (Stavroula)
- Can I have your attention? Stavroula says that her bottle has a hole. Do your bottles have holes as well? (instructor)
- My bottle doesn't have. (Marina)
- My bottle has. (Theodor)
- Mine has too. (Vangelis)

This excerpt is from the fourth action of the educational program, in which students experiment. The co-existence of observation and communication as well as experiment is clear in the excerpt.

9.4.5.3 Predictions and Hypotheses

Making predictions and hypotheses is an important function of students' development in younger or older ages because they are connected with their everyday lives and by using them, they try to explain the phenomena they observe in the world around them. Despite the fact that predictions and hypotheses are two individual processes, in this research they have been grouped and studied as one process since in the data the two processes appear together in students' dialogues.

Getting the right answer is less important for the students of this age, what is important is the opportunity students have to start making connections of what it may happen employing their prior knowledge or experiences. In Fig. 9.6 the

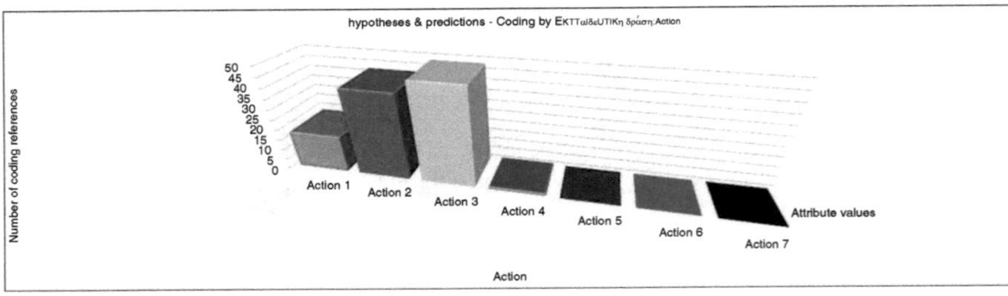

Fig. 9.6 Chart of hypotheses and predictions in relation to the number of coding references in each action

appearance of predictions and hypotheses in students' dialogues is presented, in relation to the individual actions of the educational program. Most of the references coded in the node appear in the third and second action. Much less references are coded in the first action and almost none in the fourth, fifth, sixth and seventh action.

An example of making predictions during the first action is depicted below:

- Where do you think we can find these (referring to the exhibits on the cards)? (instructor)
- In the desert. (Petros)
- In the desert? (instructor)
- A rock with an eagle on it. We can definitely find this in the desert. (Petros)

In the second action students make hypotheses about what is carved on the exhibits while they are observing them, and they make predictions about what was their use in the past:

- What do you think is this? (instructor)
- A belt. (Anna)
- A belt, very well, and what's on the belt? (instructor)
- A thunderbolt. (Ioanna and Lambrini)
- A thunderbolt, very nice! And what else is there close to the thunderbolt? (instructor)
- It has wings (showing on the exhibit). (Emmanuel)
- The thunderbolt has wings, so it's a winged thunderbolt, right? Very nice! (instructor)

In the third action students make predictions and hypotheses connected with the experiments about air, before they actually do the experiments of the fourth action. Students are challenged to think whether they are able to capture air, and if so, in which ways they can manage that:

- So, if I want to capture some air, how would I do that? Tell me your ideas. I want to take some air with me. Theodor? (instructor)
- You can ask Aeolus. (Theodor)

- Ask Aeolus, and what if he doesn't help me and I should do it by myself? Vangelis? (instructor)
- Then you should see in which direction the wind blows, towards North and collect some wind. (Vangelis)
- Hm, so you think I should wait the wind to blow... (instructor)
- So, please tell me, if I take an empty plastic bottle and close its cap, will there be any air inside? (instructor)
- No. (a lot of students together)
- You think not? (instructor)
- Yes! (Angelica)
- Yes, or no? What do you think? (instructor)
- Yes, yes. No, no. (altogether)
- Hm, it seems that we have a disagreement about this one. (instructor)

Just like in the previous cases, the process of predictions and hypotheses appears accompanied with the rest of the processes, showing a correlation.

9.4.5.4 Experimenting

Experimenting is a fundamental process of the scientific method. Students, after formulating their hypotheses, they test them implementing experiments. Experimenting requires the use of tools and materials and helps students testing their hypotheses, observing, discovering phenomena and concepts and on the next step, drawing conclusions. Exploring the process of experimenting in the implementation of the educational program, it is expected that it will appear in the fourth action, where students do the experiments. There are some references coded in the third action, according to Fig. 9.7, where students propose experiments for testing their hypotheses and with the instructor demonstrate some during the discussion about capturing air, using a plastic bag or an empty plastic bottle.

Such as in the previous cases, experimenting doesn't appear alone in students' dialogues, but accompanied by more processes while students communicate their observations during the experiments, which then interpret and form new hypotheses which will later on be tested through more experiments.

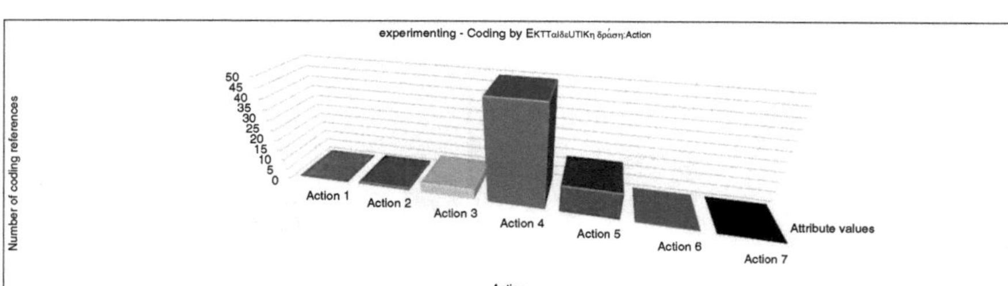

Fig. 9.7 Chart of experimenting in relation to the number of coding references in each action

In the next excerpt experimenting appears along with communication, observation, and interpretation:

- Did you try closing syringe's opening and pushing its plunger? (instructor)
- Can I try? (Vangelis)
- Miss, it can't be pushed. (Angelica)
- You can't close it huh? Why do you think it happens? (instructor)
- Miss, I can't close it either. (Marina)
- I know! It needs air to close the syringe. (Vangelis)
- Explain me what you mean by that. (instructor)
- Miss, it doesn't need air, air doesn't go out. (Nefeli)
- And what does that mean? What's in the syringe? (instructor)
- The air! (Nefeli and Marina)

9.4.5.5 Interpreting

Interpreting data is one of the final phases of the scientific method and helps students, after they have gathered scientific data from the phase of experimenting, and through observation, to interpret the phenomena under research. For students, interpreting has great significance, because on the one hand they understand the meaning of the variable and the connection of different variables between them while on the other hand, interpreting provides scientific value to data since the data alone does not value as much.

The process of interpreting is present in the educational program and more specifically, it mostly appears in the fourth action, where the students experiment. According to Fig. 9.8, interpretation also appears with much less references in the second action and even less in the third.

In the fourth action, students interpret the phenomena based on their experiments:

- Because the bottle has, doesn't have a hole and the air doesn't come out. And I can't inflate it because air has occupied all the space in the bottle. (Ektoras)
- Bravo! Did the rest of you listen what Ektoras said? That there is air inside the bottle and that is why the balloon can't be inflated. Please wait for your turn, Vasilis will tell us now. (instructor)

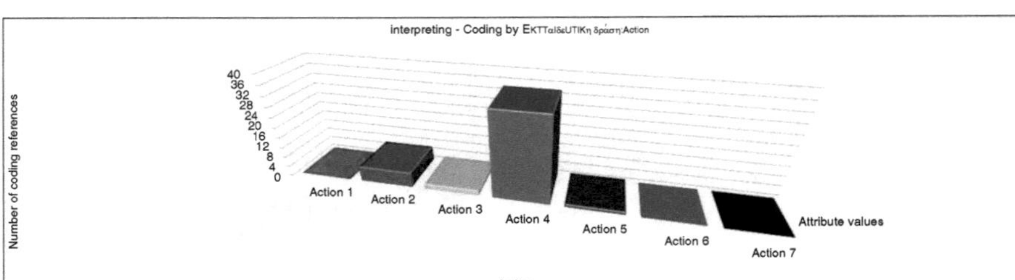

Fig. 9.8 Chart of interpreting in relation to the number of coding references in each action

- If we have the balloon and it (bottle) doesn't have a hole, air can't come out, if it doesn't . . . if it has a hole, the air will go out. (Vasilis)
- Hm. . . so, when the bottle has a hole, the air goes out and then . . . (instructor)
- If it doesn't, it doesn't go out. (Katerina)
- Right. Odyssey, what did you want to tell us earlier? (instructor)
- When we try inflating it, it doesn't, if there is a hole, we inflate it and it can go very low in the bottle because the air goes out from the hole. (Odyssey)
- Hm . . . very nice. So, do you think that air occupies space? (instructor)
- Yes. (Odyssey)

Like all the previous processes in students' dialogues, interpreting appears along with other processes such as communication, since students interpret using communication, observation and experimenting considering that these processes precede in relation with interpretation in the scientific method. In the following excerpt interpreting appears in the fourth action.

- Tell me Valia. (instructor)
- When it is, when it is a bottle that doesn't have a hole even though the balloon inflates a little bit, it doesn't properly inflate because there is no room since the air doesn't go out. (Valia)
- Please everybody, listen what Valia has to tell us. (instructor)
- When it is a bottle which has a hole, then the balloon inflates because there is space in the bottle. (Valia)
- Very well, tell us Elena. (instructor)
- When it doesn't have air, when it doesn't have a hole, it can't be inflated because it doesn't let it (air) inflate and when it has a hole, it can be inflated. (Elena)
- Nice! Maria?
- When a bottle has a hole, when you try to inflate the balloon, you can because the air goes out of the hole and it doesn't block the balloon to inflate while when it doesn't have (a hole) the air pushes the balloon upwards and it can't be inflated. (Maria)
- Very nice, Katerina? (instructor)
- Because when it doesn't have a hole, the bottle has air inside and it doesn't let the bottle (meaning the balloon) to go down and if it doesn't have a hole the air can't go out so that the balloon will inflate. (Katerina)
- So, according to what you tell me, inside the bottle there is air, right? (instructor)
- Yes! (a lot of students together)

9.4.5.6 Operational Definitions

As mentioned above, the scientific method processes discussed so far were foreseen from the design of the educational program and its actions. What is noteworthy, is that except from the expected processes, two more appear in the video recordings, operational definition, and measuring. These two processes are present in the data, but their frequency is very low, therefore they do not appear in the rest queries.

Operational definitions, as students' narratives to describe the world around them in their own words (Plakitsi, 2008) contribute to the formation of the self and consequently are precursors to the formation of concepts. Thus, students try to give their own definitions of air by giving it characteristics and connecting it with other phenomena and their personal experiences. Some examples of operational definitions for air in students' dialogues are listed below:

- The air is the clouds. (Ersa)
- In the clouds is a thunderbolt. (Elli)
- And when it rains, before it rains it blows. (Constantine)
- Thunderbolt strikes only when there are thunderstorms. (George)

The above excerpt comes from the same dialogue between second-grade primary students, trying to define the concept of air. It is obvious that students' experiences influence students' ideas about scientific concepts. Moreover, it is remarkable how students interact with each other in the frame of collaborative learning and complete each other's ideas and connect different concepts. Another student from a different class gives his own explanation for air:

- Where there is air there is rain, where there is rain there is thunderbolt. (Vassilis)

In a different implementation of the educational program, some first-grade students tried, according to the excerpt below, to define the concept of air.

- It blows. (a lot of students together)
- The air is, when it's cold it blows. (Ionas)
- When it's cold and it blows, very well. Anyone else? (instructor)
- It takes the leaves. (Nikolas)
- It takes the leaves, very nice. Yes? (instructor)
- In the sky. (Venia)
- In the sky, very nice. Tell me. (instructor)
- The more the wind blows the stronger it gets. (Achilles)
- Broom (imitating the sound of the thunderbolt) (Thomas)

In the above excerpt the interaction among the students is clear as well as their collaboration in defining the concept, of course from their everyday experiences.

9.4.5.7 Measuring

Another scientific method process that emerged from students' interaction but was not expected, was measuring. In the second action, in which students were searching for the exhibits in the museum's collection, two groups of students from different classes of the same school in the third grade of primary school, decided to measure the exhibit with the ruler and compare its size with the size of the exhibit on their cards, as seen in Picture 9.4.

Picture 9.4 Students measuring the exhibit on the display with the ruler

One of the groups even highlighted the difficulty they had as the distance of the exhibit from the glass of the showcase did not help them to have an accurate measurement:

- It was a little difficult to measure. (Marina)
- Hm it was hard to measure huh? (instructor)
- It was through the glass. (Marina)

The other group that tried to make the same measurement concluded that the exhibit shown on the group's card was smaller in size than the one in the display:

- Our exhibit was on the belt, where the weapons are placed and next to it was a similar one but it was different and we saw that it was smaller than here (showing the card). (Mary)

9.5 Discussion

Research's innovation lies on the cultivation of scientific method processes through educational programs which introduce scientific concepts and are implemented in informal settings such as the museums of general interest. The focus is not on students' right answers but in the cultivation of the scientific method and students' initiation in the scientific way of thinking in the Archaeological Museum of Ioannina. The latter was sought with the implementation of the educational program "Thunderbolt hunt", whose design and actions' succession supported students to get to know a new way of working and to familiarize themselves with the scientific method. Some student classes seemed more and others less familiar with the way of working adopted by the educational program. The truth is that for all students it was an unprecedented experience, a reason that lies, among other things, in the fact that it took part in the Archaeological Museum.

According to the results presented above, the scientific method processes cultivated through the program's actions were communication, observation, hypotheses and predictions, experimenting and interpreting data. These processes were expected to be cultivated since they were part of the program's expected learning outcomes while at the same time they were linked to the individual actions' objectives.

Communication not only is the most frequently appearing process, but also the process that appears in every individual action during the educational program. Not only argumentation and discussion in science education, but also the cultivation of language in general, are the central goals of the new curricula of formal education in Greece (Institute of Educational Policy, 2014) in both pre- and primary school. Communication is also of particular interest under a CHAT perspective, as it is synonymous with the concept of interaction (Plakitsi, 2008). According to Osborne (2010), students' participation in collaborative discussions and argumentation enhances their conceptual understanding and the cultivation of skills and competencies related to scientific reasoning.

Students, during the education program, were communicating with each other but also with the instructor, they were sharing their observations, formulating their hypotheses and predictions, experimenting and communicating their empirical data, interpreting phenomena, answering questions and justifying their answers. Students' communication is distinguished by both verbal and non-verbal expression. The adoption of the collaborative method throughout the program favored the cultivation of communication, which seems to co-exist with all the other processes of the scientific method.

Observation is a fundamental process in science education and was established according to Plakitsi (2008) as "an integral part of the scientific method and thought". Observation is a part of children's lives, who receive stimuli both inside their home and their close social environment as well as outside their family context, in the playground, in the school area, in the yard. That is why students are introduced to scientific observation from a very early age. During the educational program, students observed both with their senses (action 1 & 4) and with the use of simple instruments such as magnifying glasses (action 2). Observation, like communication, appears along with the other processes.

During the educational program students formulated hypotheses and predictions. The evolution of science is attributed to these processes as they are the reason for further investigation through the control of cases. It does not matter if they are true or false, it is more important for students to formulate hypotheses that can be tested (Plakitsi, 2008). Students made predictions and hypotheses basically when the instructor was encouraging them but during the experiments, they also tested hypotheses which arose from the students' contact with the tools and materials of the action. It is important that the students involved, through the process of predictions and hypotheses, their socio-cultural background as they relied on their previous knowledge and connected this process with experiences, observations and patterns from their everyday lives. This process does not seem to be independent as it appears with other processes.

Experimenting was the most enjoyable process for the students during the implementation of the educational program, judging from their reactions, but at the same time increased their difficulty of following the learning community's rules. This is due to the fact that the students had at their disposal materials to explore, to use, to try. Students' contact with materials, even if they are simple, leads them to exploration and supports their contact with scientific knowledge and reasoning. This is evident in the fact that, as students were experimenting with the materials at their disposal, they discovered new ways of using them and different experiments than the planned, both before and after the instructor challenged them to perform the planned experiments. Students' alternative experiments play an important role in the process of learning because they have their own initiatives, and they are able to test their own hypotheses, leading them to discovery despite the result. Students themselves considered their discoveries important and decided to depict them in their drawings investing them with meaning. Since the alternative experiments were not included in the research's objectives, they are further elaborated in the discussion section as an outcome that came up, but it was not planned.

A typical alternative experiment was students' attempt to inflate the balloon that was placed inside the bottle using the straw. When they realized that the bottle's spout was a lot wider than the straw's diameter, they tried to pull the balloons from the bottles in order to inflate them using the straw. The latter proved more effective but not effective enough since the balloon's opening was then wider than the straw's.

Another example was their attempt to use syringes as pumps to inflate balloons inside the bottles. As in the previous example, this attempt was not successful, so again they tried to inflate the balloons individually by removing them from the bottles in which they were placed. In this experiment they noticed that the balloon inflated a little by pushing the syringe's plunger but when they tried to pull the plunger to refill the syringe with air, it was pulling the same air with which the balloon was inflated. This caused the balloon to inflate as much as the air inside the syringe.

What the students also tried with the syringes was to pull its plunger while having its opening closed with their finger. They tried that after the instructor challenged them to push syringe's plunger while having its opening closed with their finger. That made them want to try the other way around. They concluded that when syringe's opening is closed, the plunger could neither be pushed down nor pulled up.

Students seem to consider their experiments very important and worthwhile to depict them in their drawings. This is confirmed by Danae's drawing (Picture 9.3) as well as Nefeli's example in the picture below (Picture 9.5).

Scientific method processes should be cultivated from an early age as an extension of students' everyday lives (Fragkiadaki & Ravanis, 2016, p. 312). Students themselves, during science education, recall their personal experiences to explain phenomena or contrast a situation with a familiar one to them (Gutwill & Allen, 2012). Such examples were observed in the students' dialogues. Due to the fact that

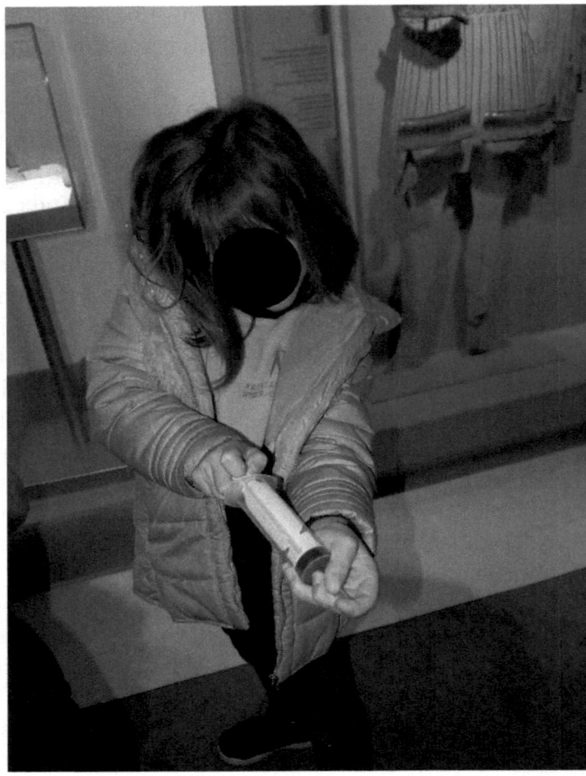

Picture 9.5 Nefeli's alternative experimentation

the latter was significant in the research but not considered as an objective of the research, this feature was further studied and included in the present section.

For this purpose, an additional node was created in NVivo, in which excerpts of students' dialogues were coded, where students connected concepts and phenomena of the program with their daily life (Fig. 9.9). Such examples are seen in the dialogues presented about the operational definitions where students describe the concept of air.

According to Fig. 9.9 students recall their experiences mostly in the third action where they formulate hypotheses and predictions. When the instructor asks students if they knew what the oracle is, in a class of first-grade students some answers were given:

- They tell us the future? (Vasilis)
- A church. (Odyssey)

In the above excerpt, Odyssey compares the oracle with a church, observing the mural on museum's wall. This probably shows that he identified some characteristics he sees in the church on museum's mural.

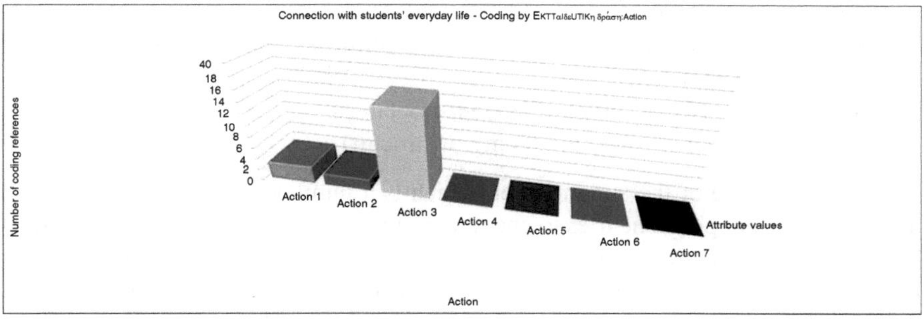

Fig. 9.9 Connection with students' everyday life in relation to the number of coding references in each action

When the instructor challenges students to propose ways of capturing air, George, a second-grade student, shares with his classmates a personal experience:

- Shall I tell you what I did? The smoke was going, I closed it in a bottle and left it and when I entered the house, I left it inside the house. (George)
- Really? So you could trap smoke huh? (instructor)
- Yes! (George)
- You trapped smoke in the bottle. (instructor)
- Yes (George)
- Hmm interesting. (instructor)
- I left him at home. (George)

The above excerpt supports the idea that experiment is part of students' lives, therefore it is not an unknown process to them.

Another example that students' education is directly connected with their everyday life and experiences which bring in the learning community socio-cultural characteristics, is the one below:

- I know how we will send it to Olympus. We will say a prayer to Christ and Christ will take the thunderbolt and send it to Zeus. (Vangelis)

Vangelis, a first-grade student, seems to invoke religion to help Zeus take his lost thunderbolt back.

The connection of concepts with students' everyday lives is a feature used not only by the students but also by the instructor, when she wanted to engage students in the educational program and its actions (Mujtaba et al., 2018). At some points she used such connections to break the ice, make students laugh and make them feel comfortable. When for example she was asking students where they could find the object depicted in their cards, if the students were hesitating, she was using the following example:

- What are these? Depicted in your cards, can you imagine? Where can we find them? (instructor)

- Look at your cards. (teacher)
- What do you think they are, can we find them in the supermarket, can we find them... (instructor)
- No! (laughing). No no. (a lot of students together)
- ... in the sea, where do we find them? (instructor)

Using the example of the supermarket, which is a familiar place for students from their parents' shopping, she tries to encourage students to give their answers without being afraid that their answer might be wrong. In the same way, she helps students formulate their own hypotheses and predictions about the ways they can capture air. After students have expressed their first ideas, she would give as an example, a students' practise from the school yard. This practise includes empty plastic bottles. They close their caps, and they are squeezing them – when they are squeezed very well, they open the cap which is skyrocketed.

Summing up, it appears that the cultivation of scientific method processes, particularly at young ages, and the formation of concepts from the world around them is achieved through the interaction among them and with their socio-cultural background, which is shaped by both the learning community and students' everyday reality.

9.6 Conclusion

Given the extensive research that shows that cultural, educational and cognitive factors influence students' understanding of science, more attention should be paid to students' prior knowledge and the resulting interpretive positions in the design of collections and related learning experiences (Mujtaba et al., 2018). It is important to carefully articulate the learning objectives of museum activities. Similarly, securing an appropriate "connection" to learning that fuels students' intrinsic motivation would be valuable in increasing students' performance and further learning (Mujtaba et al., 2018). In order for students' learning experiences to be more effective, it is important to have the opportunity to explore the exhibits interactively with others as well as on their own (Andre et al., 2017) and the opportunity to discuss what they learn with their peers, museum educators and their teachers. In addition, it is suggested that they will be encouraged to form links with their prior knowledge about the concepts of science education.

In this direction, and especially in institutions that do not focus on the dissemination of scientific knowledge, it is proposed to conduct a deliberated design of educational programs that will take into account many different factors for students' education (Delegkos & Koliopoulos, 2018). The design framework proposed in this research, SciEPIMGI, considers factors related to science education for young students and the advantages of collective and interactive learning through the principles of CHAT. On the other hand, the design framework recommends the utilization of the above via students' contact with their cultural heritage, such as the exhibits of museums of general interest. Of course, further research is required in order to test its limits and prospects, but the so far signs are encouraging.

References

Ambrose, T., & Paine, C. (2018). *Museum basics: The international handbook* (4th ed.). https://doi.org/10.4324/9781315232898

Andre, L., Durksen, T., & Volman, M. L. (2017). Museums as avenues of learning for children: A decade of research. *Learning Environments Research, 20*, 47–76.

Blunden, A. (2013). *Cultural-historical activity theory glossary of terms*. https://www.ethicalpolitics.org/ablunden/pdfs/Glossary_of_Cultural_Historical_Activity.pdf

Delegkos, N., & Koliopoulos, D. (2018). Constructing the "energy" concept and its social use by students of primary education in Greece. *Research in Science Education, 2018*, 1–26. https://doi.org/10.1007/s11165-018-9694-y

Engeström, Y. (1999). Activity theory and individual and social transformation. In Y. Engeström, R. Miettinen, & R.-L. Punamääki (Eds.), *Perspectives on activity theory* (pp. 19–38). Cambridge University Press.

European Commission. (2015). *Science education for responsible citizenship*. Luxembourg: Publications Office of the European Union, 2015. https://op.europa.eu/en/publication-detail/-/publication/a1d14fa0-8dbe-11e5-b8b7-01aa75ed71a1

Falk, J., Needham, M., Dierking, L., & Prendergast, L. (2014). *International science centre impact study: Final report*. John H. Falk Research.

Foot, K. (2014). Cultural-historical activity theory: Exploring a theory to inform practice and research. *Journal of Human Behavior in the Social Environment, 12*(3), 329–347. https://doi.org/10.1080/10911359.2013.831011

Fragkiadaki, G., & Ravanis, K. (2016). Genetic research methodology meets early childhood science education research: A cultural-historical study of child's scientific thinking development. *Cultural-Historical Psychology, 12*(3), 310–330.

Georgopoulou, P., Koliopoulos, D., & Meunier, A. (2021). The dissemination of elements of scientific knowledge in archaeological museums in Greece: Socio-cultural, epistemological and communicational/educational aspects. *Scientific Culture, 7*, 31–44.

Gutwill, J. P., & Allen, S. (2012). Deepening students' scientific inquiry skills during a science museum field trip. *Journal of the Learning Sciences, 21*(1), 130–181. https://doi.org/10.1080/10508406.2011.555938

Hooper-Greenhill, E. (2000). *Museums and the interpretation of visual culture*. Routledge.

Institute of Educational Policy. (2014). *New school (21st century school) – New curriculum, second part*. Retrieved from http://ebooks.edu.gr/info/newps/%CE%A0%CF%81%CE%BF%CF%83%CF%87%CE%BF%CE%BB%CE%B9%CE%BA%CE%AE%20-%20%CE%A0%CF%81%CF%8E%CF%84%CE%B7%20%CE%A3%CF%87%CE%BF%CE%BB%CE%B9%CE%BA%CE%AE%20%CE%97%CE%BB%CE%B9%CE%BA%CE%AF%CE%B1/2%CE%BF%20%CE%9C%CE%AD%CF%81%CE%BF%CF%82.pdf

Kolokouri, E., & Kornelaki, A. C. (2019). Introducing a new socio-cultural tool for science education in first grades: SCOPES. In K. Plakitsi, E. Kolokouri, & A. C. Kornelaki (Eds.), *ISCAR 2019 e-proceedings* (pp. 88–102). University of Ioannina.

Kornelaki, A. C., & Plakitsi, K. (2018). Thunderbolt hunt. Educational program for students from 5 to 9 years old in the archaeological Museum of Ioannina. *World Journal of Education, 8*(4), 87–101.

Kornelaki, A. C., & Plakitsi, K. (2020). Educational program 'thunderbolt hunt': An analysis with the experimental-genetic method. *Cultural-Historical Psychology, 16*(3), 38–46.

Kotnrlaki, A. C., & Plakitsi, K. (2022). Γέφυρες μεταξύ θεμάτων φυσικών επιστημών και εκθεμάτων μουσείων γενικού ενδιαφέροντος υπο το πρίσμα της θεωρίας της δραστηριότητας και το πλαίσιο σχεδιασμού SCIEPIMGI [Bridges between topics of science education and exhibits in museums of general interest in the light of activity theory and the design framework SCIEPIMGI]. *Science Education: Research and Praxis, 2022*(82–83), 37–35.

Mujtaba, T., Lawrence, M., Oliver, M., & Reiss, M. J. (2018). Learning and engagement through natural history museums. *Studies in Science Education, 54*(1), 41–67. https://doi.org/10.1080/03057267.2018.1442820

Osborne, J. (2010). Arguing to learn in science: The role of collaborative, critical discourse. *Science, 328*(5977), 463–466. https://doi.org/10.1126/science.1183944

Plakitsi, K. (2008). *Διδακτική των φυσικών επιστημών στην προσχολική και στην πρώτη σχολική ηλικία. Σύγχρονες τάσεις και προοπτικές [Didactics of natural sciences in pre-school and early school age: Modern trends and perspectives]*. Patakis.

Plakitsi, K. (Ed.). (2013). *Activity theory in formal and informal science education*. Sense Publishers.

Radice, Σ. (2014). *Designing for participation within cultural heritage: Participatory practices and audience engagement in heritage experience processes* [unpublished doctoral dissertation]. Politecnico di Milano, Design Department.

Reiser, B. J., & Tabak, I. (2014). Scaffolding. In R. Sawyer (Ed.), *The cambridge handbook of the learning sciences* (*Cambridge handbooks in psychology*) (pp. 44–62). Cambridge University Press. https://doi.org/10.1017/CBO9781139519526.005

Sannino, A. (2015). The principle of double stimulation: A path to volitional action. *Learning, Culture and Social Interaction, 6*, 1–15.

Science Centre World Summit (SCWS). (2017). *Tokyo protocol. On the role of science centres and science museums worldwide: In support of the United Nations sustainable development goals.* Retrieved from https://scws2017.org/tokyo_protocol/

Suryanti, I. M., & Lede, N. S. (2018). Process skills approach to develop primary students' scientific literacy: A case study with low achieving students on water cycle. *IOP Conference Series: Materials Science and Engineering, 296*, 1–6. https://doi.org/10.1088/1757-899X/296/1/012030

United Nations (UN). (2015). *Transforming our world: The 2030 agenda for sustainable development.* Retrieved form https://www.unfpa.org/sites/default/files/resource-pdf/Resolution_A_RES_70_1_EN.pdf

Veresov, N. N. (2017). ZBR and ZPD: Is there a difference? *Cultural-Historical Psychology, 13*(1), 23–36.

Vygotsky, L. S. (1962). *Thought and language*. MIT Press.

Athina-Christina Kornelaki is an Assistant Professor in the Department of Early Childhood Education at the University of Ioannina. She teaches courses in the field of Science and Museum Education in early grades. Her post-doctoral research project focused on the connection between formal and non-formal education toward designing science education programs in museums of general interest using SciEPIMGI framework. The research project was fully funded by the Greek State Scholarship Foundation (IKY). She defended her PhD "Educational program design in the field of science education for Informal settings under the prism of activity theory" with a scholarship from the Greek State Scholarship Foundation (IKY). Her research interests are science and museum education in a cultural-historical perspective, and she has collaborated with HEUREKA, the Finnish Science Center, during an internship and the Archaeological Museum of Ioannina during her PhD on the design of educational programs. She is a member of the international research group @Formal Informal Science Education Group (@FISE group) as well as in some Greek and international research associations such as ESERA and ISCAR. She has participated in national and international conferences and she has published articles in Greek and international journals. She has long experience managing and working as a researcher in several European-funded projects (Epoque, @Mindset, Coop4Edu, 21st TS).

Katerina Plakitsi is a member of the Governing Council of the University of Ioannina. She is a full professor of Science Education with two bachelor's degrees in Physics and Pedagogy, a master's diploma, and a PhD in Science Education. Her main researching interests are Science Education, Formal and Informal Science Education, and Cultural Historical Activity Theory applied in Science Education. She has written many books in Greek and English while she has published in different international academic journals. She is scientific coordinator in many European Projects and supervises some PhD scholars in Science Education. She has written many schools environmental science textbooks, and she coordinated a science curriculum reform in Greece for the contemporary Education. She co-founded the interdisciplinary master's program "Environmental Sciences and Education for Sustainability" in collaboration with the Faculty of Medicine and the Department of Biological Applications and Technology at the University of Ioannina. She is also the co-founder of the ISCAR-STEM Thematic Section and the principal investigator of the @formal and informal science education group (@fise group). She is the editor-in-chief of the international bilingual journal *Science Education: Research and Praxis*. She had been Head of the Early Childhood Department at the University of Ioannina. Katerina Plakitsi is President of the International Society for Cultural Historical Activity and Research (https://www.iscar.org/).

Part IV
Science Teachers Education Informed by Cultural Historical Activity Research

Chapter 10
Graph Analysis of an Expanded Co-teaching Activity in the Context of Physics Teacher Education

Glauco S. F. Silva, Gabriel Gomes dos Santos, Juliana Monteiro Rodrigues, Thiago Brañas de Melo, and Cristiano Rodrigues de Mattos

10.1 Introduction

One of the most severe educational problems that the world faces is adult literacy (UNESCO, 2013). Many countries have large educational deficits and require educational policies to reduce this important social gap.

Despite Brazil's economic and social problems over the past few years, the country has a large tradition of literacy for young people and adults. Especially from the 1960s onwards, Paulo Freire's work left profound marks when proposing educational activities that considered the acquisition of the writing system through the culture of the community to develop a critical view of its situation in the world (Mayo, 2008). Thus, apprentices could develop their emancipation for political life.

Historically, literacy has been considered the teaching and learning of the alphabetical writing system, especially for children. The main objective of literacy was to enable subjects to encode and decode the alphabetical system through writing and reading (Xavier, 2019). However, a significant difference established in the literature on literacy and the different theoretical and methodological assumptions is the age range of students. Thus, the difference between the literacy of young people and adults and that carried out with children is the different learning processes (Maciel & Santos, 2020).

G. S. F. Silva (✉) · G. G. dos Santos · J. M. Rodrigues · T. B. de Melo
Federal Center of Technology Education (CEFET/RJ), Rio de Janeiro, Brazil
e-mail: glauco.silva@cefet-rj.br; gabriel.santos.2@aluno.cefet-rj.br; juliana.rodrigues@aluno.cefet-rj.br; thiago.melo@cefet-rj.br

C. R. de Mattos
Department of Applied Physics, Institute of Physics, University of São Paulo, São Paulo, SP, Brazil
e-mail: crmattos@usp.br

K. Plakitsi, S. Barma (eds.), *Sociocultural Approaches to STEM Education*, Sociocultural Explorations of Science Education 21,
https://doi.org/10.1007/978-3-031-44377-0_10

However, young people's and adults' literacy was developed in the same way as children's literacy was proposed for a long time. In general, teaching strategies, methodologies, and teaching materials were addressed to children (Ferreira & Ricoy, 2019). The infantilized treatment of texts and activities was common, disregarding the life experience of young people and adults, their knowledge, expectations, desires, and needs about learning how to read and write (Barbeito, 2019).

The educational perspective we have adopted has three pillars. The first, related to the Freirean perspective, is based on the idea that the contents are developed based on the students' relations with their cultural environment, enabling the overcoming of immediate and daily visions, developing new mediations, and thus, identifying general aspects in a particular situation in which these students live - awareness process. In addition, in this perspective, Freire's pedagogical thinking brings a broader understanding of literacy processes, not reducing itself to the ability of decoding and coding the writing system but emphasizing that the literacy student understands and experiences the real functions of writing (or science) in their society and that their learning is an instrument for the production of citizenship.

The second pillar is related to Science, Technology, Engineering, and Mathematics (STEM). From these fields that make up STEM, it is possible to list examples and activities for physics teaching, which corroborate its integrative potential for understanding physical and social phenomena. Despite the polysemy of the term STEM and the different directions it has been developed, given its multiple perspectives, in this work, we focused on a scientific and technological education that allows adults to understand science, particularly physics, as an instrument of understanding their world and history.

The third pillar is related to initial and continuing teacher education. This dimension completes the classroom as a complex activity that expresses the multiple levels of teaching-learning processes. Whether initial or continuing, teacher education is a problem studied worldwide and requires joint efforts by the university and the school. In addition to the classic problem of initial and continuing training, we face a very complex problem: the relation of schoolteachers as student-teacher's mentors. Only recently the issue related to mentor's training has been investigated in the Brazilian educational literature. In the student-teacher's practicum, this situation is common and expresses one of the most significant contradictions between school and university: who is responsible for teacher education? Accepting the school as a place for teacher training implies political decision-making, which would include: (i) recognition of the mentor's work in mentoring student-teachers at schools; (ii) recognition of the student-teacher as part of the educational body of the school.

This chapter approaches education in its practice, where the three pillars that support our investigation converge. Our object is a physics class for young people and adults delivered in co-teaching by a mentor and a student-teacher. The class was based on a Freirean perspective, with a physics content contextualized in the students' society history and based on the perspective of understanding the role of science and technology in the mentioned society.

Finally, based on the methodology of analysis of verbal interactions using graphs, it is possible to suggest that the students' speech and to whom it was addressed allowed us to verify the students' engagement with the student-teacher as a teacher. This study also suggest that the convergence of these pillars made that class a promising example of educational success by engaging students with the subject-matter and the student-teacher. Furthermore, the co-teaching and the pedagogical approach to the content contributed to the development of the class and the legitimacy of the student-teacher as a teacher.

10.2 Historical Conditioning: Teachers' Education in Brazil

Villani et al. (2009) present a history of the science teachers' education in Brazil between 1950 and 2000, indicating the changes and transformations of the Brazilian socioeconomic and political conjunctures and the consequences for the Science teachers' education. The authors state three central agents constantly present in teacher education: public administrators, universities, and schools. Different needs and interests drive the articulation between them, thus causing tensions in the area. Those tensions result from contradictions that characterize research on teacher education and teacher training itself.

The traditional model of the relationship between university and school is the most dominant in Brazil today (Silva & Villani, 2021). Based on the separation of university and school activities, it seems that, at university, future teachers must learn theories, and, in schools, they learn to apply these theories learned at university (Zeichner, 2010). Thus, this dichotomy leads universities and schools to a type of relationship characterized much more by divergence and mutual ignorance than an effective collaboration (Vaillant & Marcelo, 2012).

In the Brazilian scenario, Teacher Education Programs (*licenciatura*, in Portuguese) take place in undergraduate level with an average duration of 4 years. These Programs are offered by higher education institutions, public or private, and by area of knowledge - their basic curricular structure includes both theoretical and practical training. Particularly in physics, in most Brazilian Programs, courses are organized into three sets: (i) specific content (basic physics, calculus, mathematical methods, etc.); (ii) pedagogical content (general didactics, educational policies, philosophy, and sociology of education); (iii) specific physics teaching contents (science education contents). Practical training takes place in a course whose activities are centered on the practicum. These activities are expected to occur from the second half of the Teacher Education Program, with a workload of 400 h, which must be developed in the school, under the supervision of a mentor.

In Brazil, practicum has been accomplished through courses called Teaching Practices. The objective of these courses, enacted in the 1960s through Decrees of

Law of the Brazilian government, is to promote the training of future teachers with the development of didactic activities in the school itself (Goulart, 2002). In general, we can identify three main agents related to the practicum: (i) the university professor, responsible for the Teaching Practice course at the university; (ii) the student-teacher, an undergraduate student enrolled in the Teaching Practice course; (iii) the mentor, schoolteacher responsible for mentoring the student-teacher during the practicum.

Currently, in this kind of course, the actions developed by the student-teachers are participation in the planning of classes, monitoring of students during classes at school, teaching, participation in other administrative activities of the school. All these actions must be continuously supervised by a school mentor teacher. Despite this rule, this is not what happens in most practicums in Brazilian schools (Rodrigues & Mattos, 2018).

The movement to professionalize teaching activity appears in several countries (Labaree, 1992), with a strong presence in the United States (Cochran-Smith & Lytle, 1990) and, to a large extent, caused a change in initial teacher education, with a more critical look at the university curriculum and the role of the school (Goodwin, 2012). In Brazil, this debate is introduced, stimulating investigations that sought to review the relationship between university and school (Rodrigues, 2013), mainly concerning the practicum (Silva, 2013).

In initial and continuing teachers' education, the practicum is one of the biggest challenges. In the last two decades, legislation for a new design of Teaching Practice in Brazil were edited. The legal documents offer higher education institutions only general guidance on which boundaries must be respected for implementing the practicum, leaving the specificities to the localities.

Thus, in most parts of the country, a type of traditional relationship model prevails in which "prospective teachers are supposed to learn theories at the university and then go to schools to practice or apply what they learned on campus" (Zeichner, 2010, p. 89). Consequently, student-teachers have difficulties accomplishing their practicum at schools since there are no institutionalized mediation elements between the theoretical reflections developed at the university and the concrete practices of the classroom. In other words, in the university's activity, the reflections for a concrete school practice are not developed, and the school is not properly prepared to receive the student-teachers and support the mentors, the latter without specific training for this function (Zeichner, 2010). Rodrigues and Mattos (2018) point out this situation as one of Brazil's main contradictions in practical teacher education.

It is currently possible to identify practicum projects that are carried out in schools to overcome the hegemonic power of universities over teacher education. Most of these projects include the empowerment of the mentor teacher as a teacher trainer. Among these projects are those that develop in co-teaching contexts as an important practical training strategy for student-teachers within the scope of the practicum. The main objective of this type of project is to overcome isolation, both for student-teachers and mentors, through joint practices.

10.3 Theoretical-Methodological Framework

To support the case study we are presenting, we are based on Cultural Historical Activity Theory (CHAT), on the theoretical-methodological perspectives of co-teaching, and the Graph Theory to social networks analysis.

10.3.1 *Cultural-Historical Activity Theory*

To understand the dynamic processes established in the practicum, we chose Cultural-Historical Activity Theory (CHAT) based on Vygotsky, Leontiev, and Engeström (Engeström & Sannino, 2021). From this perspective, the development of human systems is understood as activity systems – a complex system with different hierarchical levels that are fed back by one another. The structure of the system is dynamic, as the activity that composes it, at different levels, are coordinated actions, each one with its specific goal and with specific conditions to be carried out – operations. The elements to be investigated are the mediations between the subjects of the goal-oriented activity, their community, and the object of the activity. They determine how the subjects coordinate their actions to achieve their activity's objective. Thus, the genesis and the development of the activity takes place through different dynamics and chronotopes (time-space scales), in a myriad of mediations that give the activity system a complex character.

Taking CHAT as a basis implies adopting a theoretical-methodological perspective based on an onto-epistemology (Mattos, 2019). This means that the object does not pre-exist the investigation; the ontology of the studied object is developed simultaneously with the epistemology to know it is developed (Mattos et al., 2021).

A teacher education activity system has several hierarchical levels, from educational policies developed in the country, determining certain modes of teacher education at universities. More local policies are determined by the Departments of Education of states and municipalities until it reaches schools and classrooms, where local activities for the teacher education at the university and the elementary school intersect.

The object of investigation in CHAT is a unit of analysis – an activity – that expresses the complexity of the entire activity system. In the specific case of this chapter, we took the classroom as the unit of analysis, in which we have observed how the mentor, the student-teacher, and the high school students interact. Our research group has been investigating the structural contradictions of the teacher education system as understanding them will allow future actions to possibly overcome them (Ribeiro, 2016).

Thus, the practicum is understood as a social system marked by the complexity of dynamic relationships between agents belonging to the university as well as to the school. From a complex perspective, we understand that the subjects engaged in the activity cannot be taken separately from the network of relationships they are part of.

10.3.2 Co-teaching

The practice of co-teaching is widely used in different educational contexts in different countries (Murphy & Scantlebury, 2010). However, as Pitanga (2019) shows, in science education, the most common educational context is that of the teacher education, especially the practicum, in which the student-teacher, the university professor, and the mentor cooperate in joint teaching practices in the classroom at school. In some cases, it is possible to identify the complexity and the degree of commitment to co-teaching in the practicum, as is the case presented by Roth and Boyd (1999), in which a student-teacher in elementary school was closely accompanied for 4 months (the practicum's duration), in addition to the mentor himself, the university professor and the researcher.

Thus, in line with Roth and Boyd (1999), we consider co-teaching, in general, as a shared and co-participatory teaching practice in the classroom, as "co-teaching foregrounds the individual who co-participates in a collective activity, whereas team teaching highlights the collective work of a group which [...]. Co-teaching is consistent with the symmetry between Self and Another, whereas the notion of team teaching has a strong social over individual slant" (Roth & Boyd, 1999, p. 53).

Other authors, such as Roth and Tobin (2001), present the idea of co-teaching as a way of teacher education, in the professional environment itself, which minimizes the dichotomy between theory and practice, providing a type of training based on the experience of becoming a teacher in the classroom. Thus, Roht et al. (2002) seek to define co-teaching as a "method to deal with the problems created by a separation of theory and research from teaching" (p. 3). Even more complexly, Tobin (2006) states that co-teaching "involves two or more teachers who teach and learn together in an activity in which all co-teachers share the responsibility for the learning of the students" (p. 133). Finally, the work by Guise et al. (2017) presents a synthesis of some forms of experience with co-teaching, helping to determine its meanings, such as the practice of two or more teachers working together during a practicum class. In particular, the notion of co-teaching is noted in situations involving the initiation to teaching, an emphasis that is also given to our research work (Silva & Mattos, 2019).

10.3.3 Graphs and Social Networks

The use of networks to model social phenomena has been developed since the 1930s (Borgatti et al., 2009). The analysis through networks allows the graphic visualization of the investigated phenomenon and its relation with the mathematical modeling of its composition. The structure of a network is a methodological proposal in several investigations on social phenomena, as it highlights the role of social actors and their relationships. Social networks are representations of complex systems such as political, economic, cultural, and informational systems, which can take on different formats and levels of formality, following the development of the systems over time (Souza & Quandt, 2008).

Networks are represented graphically, typically composed of sets of connected points, that is, a set of edges in which social actors and their relationships are represented. We call **nodes** or **vertices** the points representing the subjects, or social actors, while the **edges** express the mediations, or connections, between the subjects, that is, how the flow of information between subjects takes place (Castells, 2009). Thus, knowing the variety of possible types of social networks, we used graph techniques to analyze a specific case of practicum in co-teaching.

Graph theory allows modeling a network through different metrics, however, a determining concept to characterize the network is the measure of **centrality**. According to specific criteria, this measure expresses the importance of the vertices in the network. If we take the arbitrary graph shown in Fig. 10.1 as an example, we can describe it as a star graph with five nodes (P1 to P5). Freeman (1979) uses the intuitive idea that a node with more interactions with other nodes has greater network centrality. Thus, it is possible to notice in the figure that the vertex P5, the most relevant is in the center of the graph since it connects with all other vertices.

However, not all relationships are simple enough to be represented by a star-type graph (Fig. 10.1), and, consequently, the centrality of a node is not always as evident and intuitive as the one presented in the example. There are several measures of centrality and each one of them, depending on the considered parameters, determines the relevance of each node. In this work, we used the **degree centrality** measure, defined as "a measure that reflects the direct relational activity of an actor. It measures the number of direct connections of each actor in a graph" (Lemieux & Ouimet, 2012, p. 26). In Fig. 10.1, vertex P5 is the most central of the graph because it has direct relationships with the other four vertices. These are adjacent to P5. Two vertices, when directly connected by an edge, are called adjacent. Moreover, the number of vertices to which it is adjacent is called the **degree** (Freeman, 1979). Thus, the more direct relationships a vertex has, the greater its degree. The degree centrality measure determines which vertices have the greatest number of adjacencies, i.e., the greatest number of relationships with other vertices. For this exemple, the number of interactions is not an important information, however, it will be considered in the case analised in this chapter.

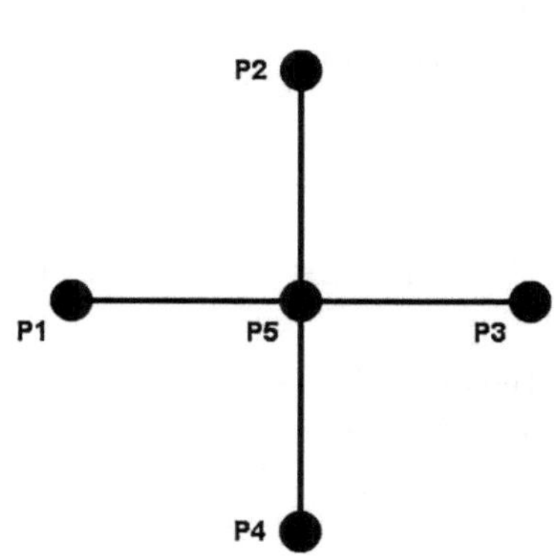

Fig. 10.1 Star graph with centrality in P5

10.3.4 Data Gathering and Processing

Using graphs to investigate the training and practice of science teachers in co-teaching, we performed data gathering through an analysis of verbal interactions between the students, the mentor, and the student-teacher captured through video recording of the class, whose communication between them was later transcribed and then the number of these verbal interactions and addresses were counted. From the analysis of the interaction between these subjects of the activity, we determined their degree centralities. This allowed us to identify the main actors in the classroom dynamics. This analysis makes it possible to understand, more clearly, who are the subjects who most verbally interact with others. The video recording allowed us to identify other dialogical elements than verbal utterances (change of position in the room, gestures, facial expressions, silences, external noises, etc.), which helped us to understand the complexity of the class. However, only the verbal interactions were included as data to produce the graph of the class presented and analyzed in this chapter.

In our analysis, we used a directional network in which the edges are arrows that indicate directionality, showing to whom the speech is being addressed. Thus, in addition to investigating the centrality, we identified which subjects were interacting during the class and the directionality of their verbal interactions (see Table 10.2), which allow us to characterize the type of class held.

In a directional network such as ours, we divided the degree centrality into two subtypes: **indegree** and **outdegree**. The indegree counts the number of addresses (arrows) directed to a subject (node), and the outdegree, the number of addresses the subject makes to another. In this case, the indegree determines how many people verbally addressed a subject, and the outdegree is the number of people the subject directed his speech to. From these measures, it is possible to identify how the addresses occur in the classroom, to know who the main interlocutors in the dialogic process are, whether between students, mentor, and students, student-teacher and students, or between teacher and student-teacher.

Finally, the verbal interactions'data, obtained from the video recording, were treated with *Pajek* software, specific for the construction of directed graphs. The software supports building and partitioning an extensive network into several smaller networks. This allows for the realization of varied statistical treatments and adequate to the size of the samples considered and provides more advanced visualization tools. The centrality measures were calculated with the software, as well as the graphical representations of the networks.

10.4 Empirical Framework

The research was developed in the context of a Physics Teacher Education Program (PTEP) at a federal university in the course called Teaching Practice (see Sect. 10.2). In the case presented in this chapter, the course is delivered to student-teachers in the final year who must fulfill 120 hours of practicum at the school.

Briefly, the methodology used in the course was carried out as follows: (i) the physics classes at the school were videotaped when the student-teachers developed some co-teaching activity with the mentor; (ii) the student-teachers and the university professor watched the videos of the classes; (iii) the student-teachers selected excerpts from the videos to analyze their classes and plan the following activities. The mentor did not participate directly in the activities developed at the university, but all didactic proposals developed at the university were discussed and planned with his participation. All involved agreed to participate in the investigation, being previously informed about the restricted use of the data and the research objectives.

The investigation took place in the second semester of 2015[1] (August to November), and the discipline had three student-teachers enrolled doing their practicum. Throughout the semester, the three student-teachers went together to classes at the school, which always took place with the presence of the university professor and the mentor. The mentor used to work with student-teachers from the perspective of co-teaching and, in between classes, he planned with the university professor and student-teachers the procedures for joint classes.

At the beginning of that semester, the mentor received the student-teachers and the university professor at the school. He explained to them his approach to physics contents, once the students were included in the Young People and Adults Education. In other words, the mentor used to relate the content taught to the students' daily experiences, whether their work situations or their life stories. After this explanation, the mentor introduced the students to the student-teachers on the first day, explaining to them the reason for their presence in the classroom.

The practicum in that semester took place, as we said, in a group of Young People and Adult Education, in the evening period (7 pm to 10 pm), with 30 students enrolled. The physics syllabus contents under study were energy conservation, linear momentum, thermometry, and thermodynamics. Physics classes were all concentrated in a single evening, while classes for student-teachers at the university took place the next day. At the university, planning and videos of classes held at the school were analyzed.

The practicum classes were developed throughout the semester with the participation of the student-teachers and the university professor, either with assistance in group work or in the practice of co-teaching between a student-teacher and the mentor. The investigated case occurred in a class of Student-Teacher E (ST-E) in a co-teaching class with the mentor.

---

[1]In Brazil, there are two academic semesters in the academic calendar: the first starts in mid-February and the second in August. They are usually called the first and second semesters, unlike other countries that indicate spring and fall semesters.

10.5 The Co-teaching Class

The class we are presenting is our unit of analysis, an activity that is not isolated from the system of activities to which it belongs. This complex system has hierarchically superior structures, for example, school and university activities, the activities of the Department of Education, and, even at a much higher hierarchical level, the activities of the Ministry of Education. All levels are mediated through rules, division of labor, and cultural instruments. Thus, the theoretical-methodological framework of the Cultural Historical Activity Theory:

> [. . .] offers a three-pronged analytical distinction between molar collective activity, modular individual actions, and their component automatic operations [...]. Activity is here seen as a collective, systemic formation that has a complex mediational structure. Activities are not short-lived events or actions that have a temporally clear-cut beginning and end. They are systems that produce events and actions and evolve over lengthy periods of sociohistorical time. (Engeström, 2008, p. 26)

The class we analyzed was the last of the academic semester (held in November) and lasted about 30 minutes, and 16 students were attending. According to the syllabus, the content of the class was thermodynamics, more specifically, the study of thermal machines. On that day, the three student-teachers were present. However, the class took place in co-teaching with ST-E and the mentor. The other two student-teachers were responsible for the video recording and helping to project the class's slides.

The content on thermal machines was taught from examples of steam locomotives, contextualized explicitly in the history of the city of *Petrópolis,*[2] where the school and the university are located. The first railroad built in South America passed through the city, connecting the coast of Rio de Janeiro to the countryside Over time, this railroad would become one of the main passenger transport routes between Rio de Janeiro and the State's countryside. This railway network was deactivated entirely in the 1960s, reducing the commercial and economic importance of several cities in the countryside of the State of Rio de Janeiro. Thus, approaching the physics content from the city's history allowed students to establish new meanings for the city. We suggest that their awareness provided the engagement and interest in participating in the class. In this sense, the entire development of the class, from its planning to its realization in co-teaching, has Freirean inspiration. The framing of thermodynamics

[2] *Petrópolis* is in the mountainous region of the state of Rio de Janeiro, 60 km from the capital, the city of Rio de Janeiro, and at an altitude of 860 m. During the nineteenth century, it was the summer seat of the Brazilian imperial family. The city is named in honor of Emperor Dom Pedro II. The railroad that passed through the city had historical importance, both because of the originality of having been built in the mountains and because it constituted the path to gold. The gold was extracted from the interior of the country to the coast, whose railroad was the only fast and safe route. The gold was taken directly to England due to trade agreements between the Brazilian and British crowns. With the end of the Empire, the railroad was used for passenger transport, but it was deactivated by the federal government in the 1960s due to the incentive of the pneumatic industry and the construction of highways (for more details, see BARMAN, R.J. (1999) Citizen Emperor: Pedro II and the Making of Brazil, 1825–1891. Stanford University Press).

in a topic dear to the students can be understood as the proposition of generative themes, which intends to mediate, through praxis, the more immediate contexts of the subjects with the more general contexts coming from historically stabilized scientific knowledge. This movement assumes that "for apart from inquiry, apart from the praxis, individuals cannot be truly human" (Freire, 2005, p. 72).

10.5.1 Class Description

At the beginning of this class, the mentor started presenting the content and explaining the purpose of the lesson to the students. As we can see in the transcript of his speech below, he was happy with how the ST-E had decided to address the content. The mentor showed in his speech that the choice of contextualization and problematization based on the city's history pleased him most. He then said that he was going to "copy the ST-E's idea".

> ***Mentor:*** *And then today the ST-E prepared some material, which I read at home beforehand to get an idea, and it is very good. And then, he's going to talk a little bit about this industrial revolution and this evolution of thermal machines, specifically, in Petrópolis, which I thought was a fantastic idea, right... I'll start copying his idea from next year. Very good idea. Why don't we talk about our reality? What was Petrópolis like in the past? And then I'm going to ask the Student-teacher to speak.*

So, right after his initial speech, the mentor moved a little further away from the students, allowing the ST-E to be closer to the class, in such a way that the two remained all the time ahead in the classroom. Right after that, the speech of ST-E was characterized by an invitation to the students: "I really need your participation!", indicating to them how he intended to maintain the interaction with them during the class.

> ***ST-E****: Well, good evening! I think... the Mentor made a great introduction, right. He presented very well what I'm going to talk about today. Well then, the title of the class is what will guide our class. Is "what the industrial revolution has to do with Petrópolis and what does all this have to do with physics? Before we start, I need to make a deal with you. Today I really need your participation! I don't like monologues; I don't like talking to myself. I feel crazy talking to the walls, so today is your day to talk to me, okay?*

In Table 10.1, we present the moments of participation of the mentor in a general and synthetic way when talking to the students, especially to complement some explanation by the ST-E.

The mentor's interactions mostly were related to the historical part due to his experience in the city. Moreover, every time he would speak, the mentor would then approach the students, and the student-teacher would take a step back. The mentor's coming and going dynamics remained throughout the class, as shown in Fig. 10.4, one of these moments in the class.

At the end of the class, the mentor interacts with the class to complement ST-E's explanation of the content.

Table 10.1 Mentor interventions during the co-teaching class with student-teacher 1

Time	Mentor's participation	Ways of intervention
08: 08	Mentor: "And the only way the ST-E spoke was to use these resources to be able to take letters, news [...] letters were used a lot, but also the spoken news!"	Interaction with the students
09: 16	The mentor moves a little further towards the students and the ST-E backs away a little. He talks about *16 de Março* street, located in the center of the city, saying that it is a tribute to the city.	Interaction with the class - brings elements of the city's history
10: 31	The mentor then responds to the student saying that the founder of the city, Mr. Köeller, would not have the vision of the city that the student currently has or that the city for other reasons has started to grow wildly.	Interaction with the class complemented with elements from the city's history
14: 19	The mentor intervenes: He moves towards the students, standing next to the ST-E. The emphasis of the mentor's questions was on the means of locomotion for people at that time.	Interaction with the class
23: 23	ST-E calls the mentor to complement his explanation. Then the mentor approaches the class, standing next to the ST-E, and explains about the telegraph. He says the telegraph and the train were lined up. When he finishes speaking, he returns to his position, a little further away from the students.	Interaction with the student-teacher, who is very comfortable asking the mentor for helping. He wasn't so sure about the telegraph explanation, as this was the mentor's request.
25: 42	The mentor and the ST-E take turns to talk about the train. The mentor provides explanations about the locomotives, saying they were small and had a lot of traction. The ST-E complements the explanation about the train hitches. The class at this moment is very participative.	Interaction with the class and the ST-E.
33: 06	The mentor consults the ST-E, whose speech is not possible to understand, and takes up the word again; he calls for attention from the class, who have become a little agitated after the quick discussion about cars and trains.	Interaction with the ST-E. The mentor seems to be aware of the schedule and dynamics of the planned class

Source: Silva and Mattos (2019)

ST-E: *[...] Vapor Horse, but it is slightly different from the horsepower unit. And then it's just a little bit, 220 horse-vapor is 223 horsepower units. And then, as the mentor talked about the units that we use the most, and he talked about power, right, last week, do you remember what he said? You can measure it in W or KW, so I transformed it there to W so you can get a sense of how much force this number can do.*

Mentor: *These units come from a comparison they barely made, Newton did it for the first time when he spoke of Horsepower. But later, with the thermal machines, they... the* *thing drifted to horsepower. These units are different, horsepower, ... which is Hp, ... are the same. But today you commonly speak of an engine with so many horses. You say an engine ... look at a car document, there is an engine with so many Hp's it is not written an engine with so many W, which is the official power unit, but it will appear in Hp in the official document of a car. And the horse that is no longer spoken as horsepower, we just abbreviate it to horse, the engine of so many horses, still widely used today, ok? For example, you take a... a... it's usually the engine of a boat that we use a horse, so I'm remembering here now.*

After the last slide, the Mentor closes this part of the class and invites the students to pay attention to the video about the content covered and organizes the class for the projection of the video.

Mentor: *So, everything that the ST-E said is summarized, hey, hey [asks for silence]. All of this that the ST-E said is summarized in a half-hour movie, the evolution from animal traction, passing through machines and reaching electricity. We're going to watch this movie, it's a super interesting movie, whoever wants to go there on YouTube and get this movie, can get it. It's called the History of Science. It is a film made by the BBC London. This is episode 4. There are six episodes. I got the number 4 [...]*

10.6 Co-teaching Graph Analysis

For the creation of the network, we used as a parameter the verbal interaction between the participants in the class analyzed here, based on video recording. Then, through this recording, intervals every 5 s were analyzed, in which it was observed who was talking to whom, that is, who was sending and who was receiving information. A new point (vertex) was formed on the network every 5 s. The parameter of 5 s was chosen because it seems to be the ideal time, as the speech addresses could be lost in a longer time interval, and with a smaller interval, it would make data collection difficult. Finally, because this class is based on an information receiver and sender, the network is directional, whose graph[3] we present in Fig. 10.3.

The direction of the arrows in Fig. 10.2 represents the occurrence of a directed sppech; for example, student 1 speaks directly to the teacher, but does not receive any specific speech from the teacher. The interaction between student 9 and ST-E, on the other hand, indicates that there was an communication in both directions of the dialogue.

The metric used is degree centrality, whose measures of the indegree and outdegree of each subject in the network are presented in Table 10.2. The analysis of the Table 10.2 shows that ST-E is the one with the highest degree centrality; he

[3]In this network, built from the verbal interactions between students, mentor and ST-E, it is important to highlight that the interactions of students with teachers are those that were more precisely analyzed, as they depend on students' behavior and quality from the video, it is difficult to know if it is the same student who is talking to the teachers again. In the recording of the analyzed class, the students did not appear in the video, being only identified by their voices, but during the class some noises occurred, making it difficult to recognize if it was the same person speaking again or if it was the students talking to each other.

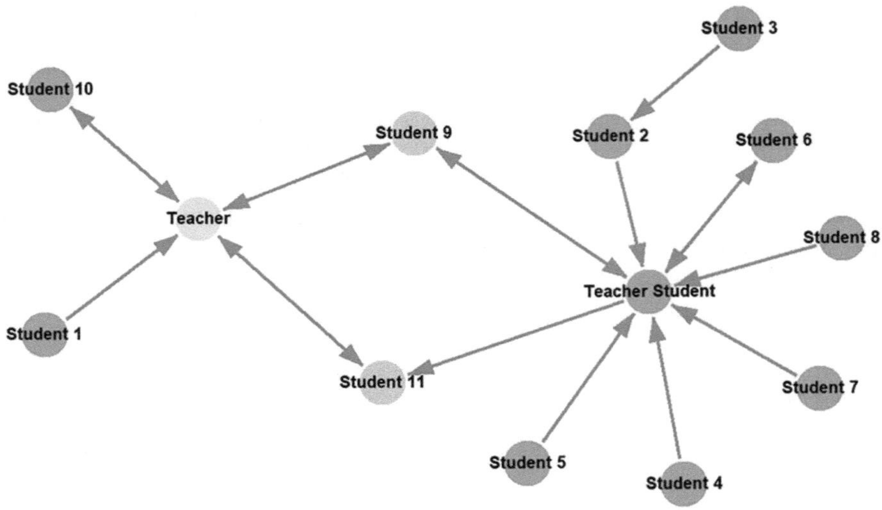

Fig. 10.2 Co-teaching class graph (colors legend: green, students iteracting either with the teacher or the studet-teacher; yellow, students interactiong with both teacher and student-teacher)

Table 10.2 Indegree and Outdegree Centralities of the subjects' verbal interactions

Rank	Vertex	Degree centrality	Indegree centrality	Outdegree centrality
1	Student-teacher E	10	7	3
2	Teacher	8	4	4
3	Student 8	3	2	1
	Student 9	3	1	2
	Student 11	3	2	1
6	Student 2	2	1	1
	Student 6	2	1	1
	Student 10	2	1	1
9	Student 1	1	0	1
	Student 3	1	0	1
	Student 4	1	0	1
	Student 5	1	0	1
	Student 7	1	0	1

Source: generated by the authors using the *Pajek* software

was the direct recipient of the speeches of 7 other people during the class, while he delivered a speech individually designed for three people. It is noteworthy that, for instance, for the construction of the network, only individualized speeches were considered, when there was a specific direction, from the student to the ST-E. The speeches of the ST-E and the teacher for the class as a whole were disregarded.

From the point of view of a teaching initiation process, it seems to us that in this situation occurred what was expected: the ST-E assumes the position of teacher, with recognition from the teacher and, especially, from the students. Unlike what happens in practicum situations commonly reported in Teaching Practice classes in which student-teachers are not properly incorporated into the routine of the school and the classroom, or situations in which the mentor teacher is usually "left to his own luck in front of student-teachers and little is known about how they effectively elaborate the work they are performing" (Sarti & Araújo, 2016, p. 176).

10.6.1 The Co-teaching Dance Subnet

To complement the analysis of the graphs, we proceeded with a qualitative analysis of a more specific moment in the class, whose verbal interactions between students, ST-E, and Mentor are represented in the sub-network in Fig. 10.3.

At the time of the class we are analising, ST-E presents a map of the city of Petrópolis after the train starts operating, contrasting with other maps of the city that he had shown previously. ST-E then pointed out on the map those places where the train passed and which still exist in the city. While presenting the map, it was possible to hear the students' reaction in the video, showing that they were surprised with the places where the train was passing at that time. However, these reactions are not included in the construction of the network because they are not speeches, the students react like "ohh" or "wow". Below are the transcribed speeches from this moment related to the subnet.

> ***ST-E:*** *Go to the next one [asks the other student-teacher]. That was Petrópolis after the train arrived [shows on the projected slide]. It has already changed its face incredibly, right? You see there [points to slide] That region there, in the first photo here, is where there is an old bus station. The bus station here in Centro, right, there it was a terminal for... There the train went around to return to Alto da Serra. That part there, look on the right, it's still there today, right? Have you seen it on the market's street? That's still there today.*

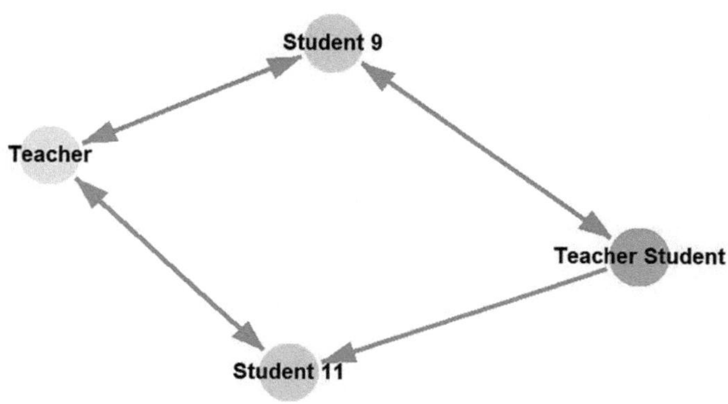

Fig. 10.3 Sub-network of the co-teaching class graph

Fig. 10.4 The co-teaching dance between the mentor and the ST-E, in the back and forth movement during class. (Silva & Mattos, 2019)

[students comment something, but it is not possible to understand]. And this one is a photo of the opening...

Student 11: *That was ugly, huh! [laughs].*

ST-E and Mentor: *[laughs].*

Student-teacher: *[...] it is a photo of the opening of Maria Fumaça, which I think was the second train of the ...of the...Petropolis railroad.*

Then a student speaks to the Student-teacher, who has stopped talking at that moment.

Student 11: *I've seen this train line here; I've seen this train in Petrópolis.*

Almost at the same time, another student asks a question:

Student 9: *They were too big!*

Mentor: *The trains?*

Student 9: *Yeah.*

Then the mentor steps forward and starts to explain:

Mentor: *No, the locomotives weren't big, they were small locomotives, but with very good traction. And that they managed to climb the mountain.*

Student 9: *But were there several trains, like wagons?*

Mentor: *Several wagons!*

ST-E: *[answers the student] It was one that pulled, and then the others were linked with ... with [makes the hand gesture] with a winch.*

Mentor: *One that pulled, one that stored coal, and the others are passengers or cargo, okay?*

ST-E: *So, there was one who pulled and one who pushed.*

Mentor: *In the middle... in the middle there was [pause because a student arrives late and interrupts the teacher's speech] in the middle of the train there was a gear, which was a special rail, that she could climb the mountain, ok?*

This excerpt from the transcription presents a significant moment in the class, as it explicitly shows the dynamics of verbal interactions between Student 9, Student 11, ST-E, and Mentor.

Student 9 initially asks the two teachers (ST-E and Mentor) the size of the locomotives, and then the mentor answers. Student 9 then returns with another question, directed at the mentor, from which he got the initial response, and then the mentor responds to him. However, ST-E also enters this scene. Right after the Mentor's answer, the ST-E complements, and there is an alternation of their speeches and positions (the mentor, when he speaks, takes a step forward and then returns).

Then ST-E asks the students a question, and only Student 11 answers. Initially, Student 11 gets the answer wrong, so ST-E corrects him, but the mentor sees the need to complement the answer. At this point, the mentor once again steps forward. Soon after, the student answers again, and ST-E goes back to explaining the answer given previously.

The sub-net of the graph in this class is indicative of good co-teaching practice. First, it shows a synchronized relationship between the two teachers, especially due to the come-and-go movements they performed during the alternation of their speeches. Second, it shows the participation of students with questions directed at the two teachers, indicating that the class did not differentiate one from the other. Thus, when the mentor felt the need to say something or complement the ST-E's speech, he would step forward and, with this gesture, the other would understand and retreat, giving the mentor, space to speak. Although ST-E and the mentor do not communicate verbally, i.e., they do not talk to each other, as shown in the network in Fig. 10.2, they communicate very well gesturally, giving rise to this movement. This movement of coming and going we call, by analogy, **the dance of co-teaching**.

In the description of the interaction between these four sub-network subjects (Fig. 10.3), it is possible to see the dance steps more clearly, as the ST-E and the mentor alternate between speeches to give a complete answer. Beyond that, it is possible to identify, at this moment that students also participate in the co-teaching dance between the two teachers. Students 9 and 11 participate with questions for both, with no distinction between ST-E or mentor. The content of the speeches indicates that the students recognized ST-E as a teacher, legitimizing him in this position during the class. Thus, we are interpreting that the back-and-forth movement between mentor and ST-E conveyed to the students the message that they were both teachers.

We also emphasize that the co-teaching dance comes from the link between the work of the mentor and the ST-E, since they complete each other by giving more precise information and helping each other, regardless of not having verbal communication, only gestural.

Thus, the mentor's movement of coming and going when he was talking to the class is indicative of a good partnership between the teachers, as there is an alternation of roles between them. Therefore, the ST-E assumes the mentor's role and, in turn, the mentor recognizes and legitimizes the other as the class teacher by taking a step back. From the perspective of the practicum, this seems to be an important movement in the process of becoming a teacher.

10.7 Final Considerations

The investigation we are carrying on about co-teaching involves the analysis of situations related to Teaching Practice and the development of the practicum at school. Thus, we relate the teaching training process as a process of structuring the activity of teacher education, through the coordination of actions between the ST-E and the mentor (Silva & Mattos, 2019).

In this specific context, the mentor must perform a double function: to be a teacher educator and teacher of the class for which he is responsible in his school; on the other hand, the student-teacher, in our case the ST-E, also takes on a dual role: as a student in the Physics Teacher Education Program and as a teacher in the class in which he places his practicum. Thus, this dual function of these two agents gives the activity and the object of that activity a hybrid characteristic.

Associated with this representation, we have the graph analysis, with which we identify the dynamics of the division of labor in the development of the co-teaching practicum activity. The degree centrality (Table 10.2) and the sub-network dialogues indicate that the teacher role is shared with ST-E, who emerges as a teacher during this class. Throughout the activity, the subject of the teaching actions is no longer the mentor, nor just the student-teacher, but a student-teacher-mentor dyad.

An object of the hybrid activity, the class that aims to teach thermodynamics to young people and adults also aims to teach and learn at school, in the context of the practicum represented in Fig. 10.5, based on Engeström (1987)'s activity model. In this sense, the instrument of the activity is co-teaching itself, which, as a methodology, mediates between specific content, historical context, and the students' experience, as well as teaching training.

To the left of the triangle shown in Fig. 10.5 are the rules of the Teaching Practice course (see Sect. 10.2) and those of the school activity. The community (in the center of the triangle's base) is composed of the school community and the participants of the practicum, particularly those who participate in the class in which the co-teaching activity is carried out. The division of labor (right at the triangle's base), initially established as being the ST-E responsible for co-teaching classes; the mentor responsible for teaching and accompanying the ST-E in co-teaching; the other student-teachers responsible for filming and helping to project the slides; and, finally, the students participating in the class.

However, we can verify the expansion of co-teaching as we identify the action of another member of the community as the subject of the activity: the students. The initial dyad "ST-E-mentor" becomes a triad: ST-E-mentor-students, as Silva and Mattos (2019) proposed. Thus, the co-teaching activity's expansion consists of the inclusion of the students as subjects, providing important changes in the organization of the activity, represented in Fig. 10.5 (the red lines).

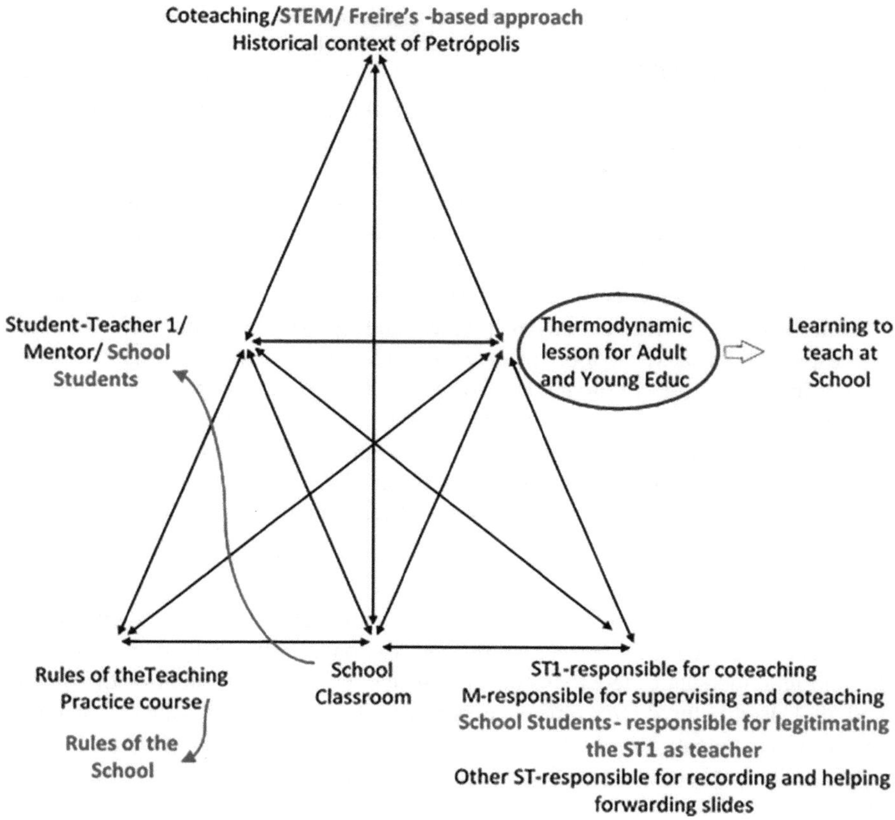

Fig. 10.5 Expanded co-teaching activity

In Fig. 10.5, we indicate the changes in the activity, placing students as subjects and not just as part of the community. In this way, the STEM treated based on a Freirean perspective also constitutes an instrument of the activity. The rules now include the rules of the school, and as expected in the activity transformation as a unit, the division of labor also develops, including now the students as subjects, since they start to have the role of legitimizing ST-E as co-teacher of the class (Silva & Mattos, 2019).

Thus, the transformation and expansion of the hybrid activity (Fig. 10.5) can be identified by two elements:

(i) the graph analysis shows the ST-E with the degree centrality of the network. The analysis of the class through the graph allows us to state that students also play a dual role, that of High School students and that of co-participants in the initiation of teaching by the ST-E. In other words, the triad of expanded co-teaching, in some way, expresses the mentor's and students' legitimation process of the ST-E as a teacher.

(ii) the report of ST-E in an interview presented in Silva and Mattos (2019). ST-E said that, at the beginning of the practicum, students perceived him as just an observer, but at the end, they considered him as a teacher. For ST-E, this is clear from the equitable treatment the students gave him and the mentor. Due to the contextualized theme (thermodynamics, thermal machines, and historical and life contexts), he expected the class to be more dialogic, with greater student participation. The dialogic perspective, foreseen in the Freirean-based methodology, was introduced so that, from the beginning, the ST-E established the dialogue by asking questions that demanded from the students the relationship between the elements of the STEM contents and their knowledge about the city's history. The students responded and contributed with their opinions, especially because the topic of the lesson allowed students to recognize aspects of their own lives in the content. Based on his report, it is possible to identify that he perceives that he has gained the respect and the consideration of students regarding his role as a teacher in that class. Thus, it seems evident that expanded co-teaching is related to the recognition that students have of ST-E as co-teacher, thus establishing a process of legitimizing the student-teacher as a teacher.

Therefore, the expansion of co-teaching brings a new division of work, as ST-E establishes new mediations in the classroom as a teacher. Likewise, students assume a new role as co-teaching partners. Silva and Mattos (2019) use the dance metaphor to help us understand the expansion of co-teaching. The movement of the mentor and ST-E is structured over time as a coordinated dance, but with the expansion of the activity, students start to mediate the dance, which makes them subjects of the activity. Students start to coordinate their actions with the initial dyad, which, when complexified, becomes a joint dance, now a triad – the new subject of the co-teaching activity – the expanded co-teaching.

As a conclusion of this case study, we suggest that the success of the co-teaching class is the result of a gradual process of co-construction of the activity. The practicum lasted a semester (August to November) in which the students,[4] the mentor, and the student-teachers who remain in constant contact with the mentor to plan the class activities were present in all physics classes at the school, for the most part, on a co-teaching basis. Thus, we conclude that the success of this class was due to the relationships established between content and methodology, but also due to the very historicity of the activity, developed during the relationships built between university and school, particularly through the Teaching Practice course. The students' participation also contributed to the success of the class, as they answered questions, complemented them with their personal experiences, and developed agency, determined by questions that related various aspects of the class to their lives.

[4] As highlighted by Sarti and Araújo (2016), Rodrigues and Mattos (2018), the most common practicum situation in Brazil is the disarticulation between the practicum agents, as well as between institutions.

A Freirean perspective for STEM content in this class plays a key role in students' engagement. The thermodynamics taught drawing from the historical elements of the city where these students were born, grew up, and raised their families, rescue their origins, in a perspective of the "ontological and historical vocation of *to be more* fully human" (Freire, 2005, p. 55). The experience described and analyzed in this chapter suggests that the expansion of educational activity is related to (and can contribute to) the debate on the role of school and science in the humanization process: the school as the place to establish conditions for overcoming experiential contradictions, expressed as limit-situations, and awareness, as the production of the ontology *to be more*, in the creation of actions that produce a viable unprecedented, in the direction of human emancipation (Freire, 2005).

Acknowledgements Glauco S F Silva thanks Rio de Janeiro Research Funding Support (FAPERJ) for the grant E-26/210.303/2019 (248327).

Gabriel Gomes thanks CEFET/RJ – Institutional Research Funding (PIBIC) for the scholarship.

Juliana Monteiro thanks Rio de Janeiro Research Funding Support (PIBIC/FAPERJ) for the scholarship.

Cristiano Mattos thanks Brazilian National Research Council (CNPq) for the financial support (grants 302100/2019-9 and 434918/2018-0).

References

Barbeito, M. H. Z. (2019). Teaching materials and adult education. In J. R. Rodríguez, T. B. Garcia, & E. Bruillard (Eds.), *IARTEM 1991-2016: 25 years developing textbook and educational media research*. International Association for Research on Textbooks and Educational Media (IARTEM).

Borgatti, S. P., et al. (2009). Network analysis in the social sciences. *Science, 323*(5916), 892–895.

Castells, M. D. (2009). *A sociedade em rede*: A era da informação: economia, sociedade e cultura; v. 1. rev. amp. Paz e Terra.

Cochran-Smith, M., & Lytle, S. (1990). Research on teaching and teacher research: The issues that divide. *Educational Researcher, 19*(2), 2–11. https://doi.org/10.2307/1176596

Engeström, Y. (1987). *Learning by expanding: An activity-theoretical approach to developmental research*. Orienta-Konsultit.

Engeström, Y. (2008). *From teams to knots: Activity-theoretical studies of collaboration and learning at work*. Cambridge University Press.

Engeström, Y., & Sannino, A. (2021). From mediated actions to heterogenous coalitions: Four generations of activity-theoretical studies of work and learning. *Mind, Culture, and Activity, 28*(1), 4–23. https://doi.org/10.1080/10749039.2020.1806328

Ferreira, V., & Ricoy, C. (2019). Multidimensionality of contents presented in textbooks. In J. R. Rodríguez, T. B. Garcia, & E. Bruillard (Eds.), *IARTEM 1991-2016: 25 years developing textbook and educational media research*. International Association for Research on textbooks and educational media (IARTEM).

Freire, P. (2005). *Pedagogy of the oppressed*. The Continuum International Publishing Group.

Goodwin, A. L. (2012). Teaching as a profession: Are we there yet? In C. Day (Ed.), *The Routledge international handbook of teacher and school development* (pp. 44–56). Taylor and Francis.

Goulart, S. M. (2002). A prática de ensino na formação de professores: uma questão (dês) conhecida, Revista da Universidade Rural, Série Ciências Humanas. UFFRJ, v. 24, n 1–2, p. 77–87.

Guise, M., et al. (2017). A; Continuum of co-teaching implementation: Moving from traditional student teaching to co-teaching. *Teaching and Teacher Education, 66*, 370–382.
Freeman, L. C. (1979). Centrality in social networks conceptual clarification. *Social Networks, 1*(3), 215–239.
Labaree, D. F. (1992). Power, knowledge, and the rationalization of teaching: A genealogy of the movement to professionalize teaching. *Harvard Educational Review, 62*, 123–154.
Lemieux, V., & Ouimet, M. (2012). *Análise estrutural das redes sociais*. Instituto Piaget.
Pitanga, C. F. (2019). Contextos e perspectivas internacionais do co-teaching no ensino de ciências. 89f. Trabalho de Conclusão de Curso (Curso de Licenciatura em Física). CEFET/RJ.
Maciel, F. I. P., & Santos, S. M. (2020). A história da alfabetização de adultos no ensino, na pesquisa e na extensão da UFU e da UFMG (1986–2019). *Cadernos de História da Educação, 19*(1), 24–41.
Mattos, C. R. (2019). *Racionalidades na Educação Científica* [Rationalities in Science Education]. Associate professorship thesis. University of Sao Paulo.
Mattos, C. R., Ortega, J. L., Rodrigues, A. M. (2021) Conceptual complexification as an onto-epistemological synthesis in science education activity. Submitted.
Mayo, P. (2008). Paulo Freire and adult education. In A. A. Abdi & D. Kapoor (Eds.), *Global perspectives on adult education*. Palgrave Macmillan. https://doi.org/10.1057/9780230617971_6
Murphy, C., & Scantlebury, K. (Eds.). (2010). *Coteaching in international contexts: Research and practice*. Springer. ISBN: 978-9048137060.
Ribeiro, D. F. B. (2016). *IFUSP, escola pública e formação de professores de física: contradição e alienação no movimento dialético do estágio (não) supervisionado*. 211f. Dissertação (Mestrado em Ensino de Ciências-modalidade Física). Instituto de Física/Faculdade de Educação- Universidade de São Paulo.
Rodrigues, M. A. (2013). *Movimento e Contradição*: a disciplina de Práticas em Ensino de Física e a formação inicial de professores de Física sob uma perspectiva histórico-cultural. 278f. (Doutorado em Ensino de Ciências-modalidade Física). Instituto de Física/Faculdade de Educação- Universidade de São Paulo.
Rodrigues, M. A., & Mattos, C. R. (2018). The contradictory nature of teacher education in the partnership between university and school. *Problems of Education in the 21st Century, 76*(1), 87–99.
Roth, W.-M., & Boyd, N. (1999). Coteaching, as colearning, is praxis. *Research in Science Education, 29*(1), 51–67.
Roth, W.-M., & Tobin, K. (2001). Learning to teach science as practice. *Teaching and Teacher Education, 17*, 741–762.
Roht, W.-M., Tobin, K., & Zimmermann, A. (2002). Coteaching/cogenerative dialoguing: Learning environments research as classroom praxis. *Learning Environment Research, 5*, 1–28.
Sarti, F. M., & Araújo, S. R. P. M. (2016). Acolhimento no estágio supervisionado: entre os modelos e possibilidades para a formação docente. *Educação, 39*(n2), 175–184.
Souza, Q. R. & Quandt, C. O. (2008). Metodologia de Análise de Redes Sociais. In F. Duarte, C. Quandt, & Q. Souza. (Org.). *O Tempo das Redes* (pp. 31–63). Perspectiva.
Silva, G. S. F. A. (2013). *Formação de Professores de Física na perspectiva da Teoria da Atividade*: a análise de uma disciplina de Práticas em Ensino e suas implicações para a Codocência. 327 f. Tese (Doutorado) – Universidade de São Paulo. Faculdade de Educação, Instituto de Física, Instituto de Química e Instituto de Biociências, São Paulo.
Silva, G. S. F., & Mattos, C. (2019). Análise da atividade de codocência na prática de ensino na formação inicial de professores de física. *Revista Brasileira da Pesquisa Sócio-Histórico-Cultural e da Atividade, 1*(2), 1–21.
Silva, G. S. F., & Villani, A. (2021). The physics teaching practice course and the student-teachers' activity in the benning ot the supervised practicum at schools. *Caderno Brasileiro de Ensino de Física, 38*(3), 1561–1588.
Tobin, K. (2006). Learning to teach through coteaching and cogenerative dialogue. *Teaching Education, 17*(2), 133–142.

UNESCO. (2013). *Adult and youth literacy*: National, regional and global trends, 1985–2015. UNESCO Institute for statistics. Available at: http://uis.unesco.org/sites/default/files/documents/adult-and-youth-literacy-national-regional-and-global-trends-1985-2015-en_0.pdf. Accessed 4 Aug 2021.

Vaillant, D., & Marcelo, C. (2012). *Ensinando a ensinar*: as quatro etapas de uma aprendizagem. Curitiba-PR: Editora UTFPR, 1ª Ed.

Villani, A., Pacca, J. L. A., & Freitas, D. (2009). Science teacher education in Brazil: 1950–2000. *Science and Education, 18*, 125–148.

Xavier, C. F. (2019). História e historiografia da Educação de Jovens e Adultos no Brasil - inteligibilidades, apagamentos, necessidades, possibilidades. *Revista Brasileira de História da Educação*, 19, e068. Access on 11 Feb. 2021. Epub July 29. https://doi.org/10.4025/rbhe.v19.2019.e068

Zeichner, K. (2010). Rethinking the connections between campus courses and field experiences in college- and university-based teacher education. *Journal of Teacher Education, 61*(1–2), 89–99. https://doi.org/10.1177/0022487109347671

Glauco S. F. Silva PhD, has been working as physics teacher in the Physics Teacher Education Undergraduate Program and in the Science, Technology and Education Graduate Program both at Federal Center of Technology and Education of Rio de Janeiro (CEFET/RJ), where he is an Associate Professor and held the chair position of the Physics Teacher Education Undergraduate Program (2013–2015). Graduated on Science Education (2013) at University of São Paulo, Glauco was high school physics teacher in Petrópolis (2000–2001), Juiz de Fora (2002–2004), and São Paulo (2007–2009). In 2012 he was a visiting scholar of the Urban Education Program (now Learning Sciences) at the Graduate Center of CUNY; a member of the International Organization for Science and Technology Education (IOSTE) and co-chair of the XX IOSTE Symposium (2022); a member and the current Vice-President (2023–2025) of the Brazilian Association of Science Education Research (ABRAPEC). Glauco has been carrying out research on science teacher education based on CHAT and focused on co-teaching, pre-service science teaching, and university-school partnership.

Gabriel Gomes dos Santos has been studying co-teaching on science teaching and students and teachers relationships through graphs as part of his scientific initiation since 2018. He is graduated from the Physics Teacher Education Undergraduate Program (2017–2021) and is currently taking his master's degree in the Science, Technology and Education Graduate Program (2022–2024), both at the Federal Education Center of Rio de Janeiro (CEFET/RJ). He is also taking an undergraduate degree in Computer Engineering (2022–2027) in CEFET/RJ.

Juliana Monteiro Rodrigues is graduated from the Physics Teacher Education Undergraduate Program (2023) at Federal Center of Technology and Education of Rio de Janeiro (CEFET/RJ). During her time at the university, she participated in scientific initiation programs in which she studied science education based on co-teaching using graphs. Currently, she is carrying out research about the Conceptual Profile of heat and physics textbooks.

Thiago Brañas de Melo PhD, has been working as mathematics teacher in the Mathematics Teacher Education Undergraduate Program and in the Science, Technology and Education Graduate Program, both at Federal Center of Technology and Education of Rio de Janeiro (CEFET/RJ), where he is Associate Professor and held the chair position of the Mathematics Teacher Education Undergraduate Program (2022–2023). Thiago has been carrying out research on mathematics teacher education focused STS teaching, socio-scientific questions, and Critical Mathematics Education.

Cristiano Rodrigues de Mattos has a PhD in Physics from the University of São Paulo investigating artificial cognitive systems. Currently, he is an Associate Professor at the Institute of Physics at the University of São Paulo and the leader of the Research Group in Science and Complexity Education (ECCo). He works on philosophical and psychological fundaments of teaching and learning processes, with emphasis on science education research to establish theoretical-methodological foundations based on the Cultural-Historical Activity Theory from a Freirean perspective. Through this framework, he investigated topics related to the teaching and learning processes of scientific and quotidian concepts, models of dialogic interaction, situated cognition, interdisciplinarity, and complexity, and developed practical educational activities aiming science as an instrument to develop citizenship and democratic education for social and economic equity.

Chapter 11
Expansive Resolution of Conflicts of Motives and Boundary Crossing Activity by Science Teachers

Sylvie Barma and Samantha Voyer

This chapter presents how two science teachers and a pedagogical counselor collaborated during 7 years to co-design and exchange technical and instructional artefacts to meet new curricular demands requiring the integration of technological design to science teaching. The colleagues gave new meanings to conflicting motives in a complex learning setting by means of learning actions. Epistemic developments happened as new ideas and forms of participation were materialized through technical objects and their instructional artefacts. Boundary crossing activity took place as the participants pooled in their respective expertise. The technical and instructional artefacts were exchanged with 170 teachers during training workshops also co-designed by the colleagues. Inspired by developmental work research and ethnomethodology, the chapter traces back 7 years doing CHAT research. Results present how the expansive resolution of conflicts of motives triggered transformative actions that resolved the inner contradictions identified in their activity systems. Boundary crossing was possible as new meaning, new roles and division of labor lead to the expansion of their practice.

11.1 About the Chapter

Between 2010 and 2017, Barma's research team engaged in a government school-university funded partnership with a school district gathering two science teachers and one pedagogical counselor. The participants acted as co-designers of technical objects and classroom and instructional artefacts with the goal to become facilitators for their peers facing challenges related to the integration of technology to science

S. Barma (✉) · S. Voyer
Department of Teaching and Learning Studies, Laval University, Quebec, QC, Canada
e-mail: sylvie.barma@fse.ulaval.ca; samantha.voyer@ctreq.qc.ca

K. Plakitsi, S. Barma (eds.), *Sociocultural Approaches to STEM Education*, Sociocultural Explorations of Science Education 21,
https://doi.org/10.1007/978-3-031-44377-0_11

education in their classrooms. These demands were requested by the ministry of Education in the form of a revised science and technology curriculum. More than 170 teachers, lab technicians and pedagogical counselors from 15 school districts benefited from six training workshops co-designed by the teachers and the pedagogical counselor. At the heart of the workshops was the intention, on the part of the two science teachers and the pedagogical counselor, to share the technical and instructional and concrete artefacts they co-designed. The technical objects and their instructional artefacts acted as boundary objects across all the tutorials undertaken by attending teachers coming from different school communities. This study documents how conflicts of motives were overcome as they shared their individual expertise and co-design eight teacher training workshops. Boundary crossing happened over the years as technical artefacts (prototypes) and their instructional artefacts were shared with 170 teachers and pedagogical counselors.

11.2 Science Education and Technological Education for the Development of Technoscientific Literacy

The twentieth century saw many rapid discoveries in science and technology. According to Fourez (1994), it is essential for a citizen to be scientifically and technically literate. He defines scientific and technical literacy (STL) as the capacity to build for oneself in a scientific-technical society a field of autonomy, communication and negotiation with one's environment. Fourez' definition also includes the collective dimension of scientific and technical literacy (Fourez, 2002). Other researchers such as Roth and Désautels (2002) invite us to take into account the experiential knowledge of groups of citizens with regard to techno-scientific issues and their capacity to engage collectively in the analysis of a question that concerns them (Roth & Désautels, 2002). Speaking of citizen collectives are engaged in carrying out certain research related to scientific activity themselves, Désautels (2002, free translation, p. 193) argues that "it can be said that they have demonstrated a high level of techno-scientific literacy, insofar as we conceive this competence not as an individual quality, lodged somewhere in the brain of a person, but rather as an emergent and distributed property resulting from the interactions between the actors engaged in the action" (2002, free translation, p. 193). Roth and Lee (2004) support that new aspects of scientific literacy emerge in unexpected and surprising ways. They also argue that teachers should avoid creating learning environments that funnel students into performance-based tracks and should instead offer students a broad variety of situations conducive to participation enabling students to make decisions in keeping with their own interests.

11.3 Challenging the Traditional Form of Science Education by Means of the Design of Technical Objects

From a research perspective, the importance of renewing teaching practices in science and technology education stems from an observed gap between current teaching methods and their relevance to students (Juuti et al., 2009). Hodson (1999) argues that teachers need to get students to recognize the areas where science and technology intersect, to articulate with each student's life experience. Latour (1989) agrees, explaining why science and technology are a creation of our imagination or, more precisely, the result of the attribution to a privileged few of the entire responsibility to produce facts. According to him, "science and technology are the cause that allows projects to succeed [and] as long as projects last, we do not know what is due to science and technology" (Latour, 1989, free translation, p. 283). He, thus, prefers to speak about techno-sciences and this allows him to include everything that is related to scientific contents (those who make science and the various social actors who surround scientific projects). This is expressed in particular by a disinterest among young people in science and, ultimately, in scientific and technical careers (Robitaille, 2010). According to Guchet (2011), science is no longer a contemplative activity seeking to describe an external reality independently of its activities, but rather an activity that produces its object through techniques. "Technique is no longer merely the instrument of scientific research, it now appears as a constitutive epistemological mediation on scientific reality" (Guchet, 2011, free translation, p.85). We follow these footsteps supported by several recent works to attest the relevance of presenting students with more meaningful activities since popular decontextualized didactic practices disinterest students (Osborne, 2003). For example, Roth (2001) focuses on the value of using technology to support science learning. Defining technological education, Impedovo et al. (2017), propose that: "In its primary meaning the word technology refers to the systematic study of processes, methods, instruments or tools to one or more technical fields, arts or trades" (p. 19). These authors also support the idea of the importance of an "educationally mediated intervention of design and technology education" (p. 24). Research sheds light on the importance of using pedagogical activities that guide students towards a process of designing or testing technical artefacts (Edelson, 2001). These activities promote a better critical analysis of these concrete objects and their integration with scientific concepts, which is reminiscent of Vygotsky (1962), who argues that learning depends on the activity, context and culture in which it takes place (Greeno, 1998). Sidawi (2009) emphasizes the importance of using a technological design approach to science education. According to him, a technological design constitutes a promising context for reinvesting the appropriation of scientific knowledge, which in turn feeds a design process.

> Through the pursuit of a solution to these problems, students experience the three aspects of developing, planning and communicating ideas; working with tools, equipment, aspects of developing, planning and communicating ideas; equipment, materials and components to

> make quality products, and evaluating processes and products (Qualification and Curriculum Authority, 2001, cited by Sidawi, 2009, p. 270).

Convert and Gugenheim (2005) have found that traditional models of science education that favour the transmission of content have led to a decline in student motivation and interest. Koballa and Glynn (2007), like Eick and Reed (2002), report that a majority of students would like to see a greater proportion of hands-on work in science classrooms. According to these researchers, teachers should not only focus on minds-on approaches, but also on hands-on approaches.

11.3.1 Merging Science and Technology Education: A Complex Relationship with Technical Objects

In 2006, Quebec high-school science teachers were asked to integrate technological design of concrete prototypes (technical objects) to support the appropriation of scientific concepts (GQ, 2006). The curricular prescriptions challenged the teachers' representation of 'Science' and 'Technology' especially since the two teaching subjects had been separated from one another in the past as they were trained either to be science teachers or technology teachers. After the implementation of the curriculum in 2006, there was a lack of teacher training that would explain the ministry's intention to integrate technology education to science teaching: tensions rose (Barma, 2011). The Ministry of Education missed out on providing teacher training sessions: from teaching ecology, biology or physical sciences, teacher's identity was redefined as they would be referred to as Science and Technology teachers (CES, 2013). From now on, science teachers were asked to integrate technological design process methods during class-workshops to support the appropriation of scientific concepts through the development of three disciplinary competencies one of which required hands-on activities (GQ, 2007). Most Quebec science teachers never had any training on machine tools nor had much knowledge of technological techniques like graphical language and manufacturing (CES, 2013). In the textbooks they were handed, they could read that technological processes were the means and method used to perform a task or obtain a concrete result like a technical object, a system, a product or a manufacturing process. The rapid and top-down implementation fostered resistance on the part of science teachers (CES, 2013).

According to Roth (2001), the value of using pedagogical activities that guide students towards a design or testing process for technical objects is important. These activities are conducive to the appropriation of the modeling approach and the construction of representations (mental or material). In addition, they promote a better critical analysis of the performance of these artefacts. In an article that raises questions about the possible interrelationships between science and technology, Roth (2001) refers to teachers who have questioned whether technology-focused activities can support science learning. According to Roth, several research studies highlight the value of using pedagogical activities that guide students towards a design or testing of technical objects. Still, according to Roth (2001), other researchers see technology as an application of science. They argue that students should not engage in the study of technology until they have appropriated the scientific principles necessary to understand technical objects or systems. But what do the pedagogical and didactic choices made by teachers tell us? Teaching science through technological design requires teachers to understand the connections between technology and science. This understanding can influence the teaching approach they adopt when planning their classroom lessons (Sidawi, 2009). For Roth (2001, p. 769), the way in which practices are organized in science and technology classrooms depends on how we see the links between these two fields of knowledge.

These reflections raise general research questions in the context of this study:

1. How do science teachers respond to a challenging curricular demand requiring the integration of technological design to science education?
2. What kind of actions do they take as they are recruited by a research team to co-design and exchange technical and instructional artefacts?

11.4 Theoretical Framework

11.4.1 Expansive Resolution of Conflicts of Motives to Trigger Teacher's Agency

An important body of studies focus on how teachers engage in transformative agency to envision and bring about new forms of curricular materials. Severance et al. (2016) have documented that democratic forms of participation are necessary for teachers to co-design curricular artefacts and demonstrate agency. Severance et al. (2016) investigated how artefacts their team designed, were helpful for teachers to break away from their current practice usually grounded in lecture-based teaching strategies. These artefacts were used as a first stimulus to trigger teachers' agency. For example, a teacher will employ an external conceptual artefact like a pedagogical strategy as a second stimulus, investing it with meaning and making a conscious decision to act on it (Barma et al., 2015). Reflecting on teachers' reasons to break out of a lecture-based practice, these authors document that science teachers often resolved double binds related to curricular prescriptions (conflicting stimuli such as addressing environmental issues versus focusing on assessing content) to increase their students' interest in their classroom. In other words, facing conflicting stimuli, science teachers become at the mercy of conflicting motives and will only resolve them by giving new meanings to the object of their teaching activity. To do so, they form and implement an external second stimulus to better cope with the evolving conflicts, which inevitably will overlap with other layers of the school community they belong to. Expansive resolution of conflicting motives unfolds as science teachers address new conflicts of stimuli and new conflicts of motives.

11.4.1.1 Expansive Learning

Expansive learning is a theoretical framework that focuses on new forms of learning and social practices developed beyond the activity of isolated individuals and considers the historicity of a collective process of transformation (Engeström and Sannino, 2011). Yrjö Engeström's framework is especially suitable for contradiction identification and resolution within and between activity systems overtly pursuing the same object and outcome. Engeström builds from the work of Davydov (2008) and conceptualizes four types of contradictions (i.e., primary, secondary, tertiary and quaternary), all in the context of an expansive cycle of learning. Following Il'enkov (1982), expansive learning sees contradictions as historically evolving tensions that can be detected and dealt with in real activity systems. Engeström characterizes contradictions as drivers of change that are inherent to all human activity. In capitalism, the pervasive primary contradiction between use value and exchange value is inherent to every commodity, and all spheres of life are subject to commoditization (Engeström, 2015). Most importantly, contradictions are the driving force of transformation. Expansive learning requires articulation and practical engagement with inner contradictions of a learners' activity system and, according to (Virkkunen & Ahonen, 2004).

A new form of activity is possible when an abstract germ cell is constructed by transforming a given situation through analysis and modelling to produce an idea that has the potential to transform the initial activity (Davydov, 1990). In dialectical terms, this corresponds to a continual back-and-forth movement and fosters a co-configuration of representations in keeping with the aim of constructing an innovative model of teaching practice. In the context of this study, where three participants work as a team to become creators of technical and curricular and instructional artefacts integrating technological design to science education, a possible initial germ cell would be to put in place conditions, so they share their respective expertise to co-model a technical object and its' instructional artefacts (Fig. 11.1).

When happening, expansive learning takes us beyond the limits of a single activity system and expands the unit of analysis to multiple activity systems interacting and potentially sharing a common object (Engeström, 2001). In Fig. 11.2, two activity systems have the potential to expand their initial objects (1) and transform it into objects (2) by means of sense making and dialogue, to create an object (3), a thirdness that is not yet there (Engeström, 2001). Miettinen (2006, p. 176) refers to expansive learning as "a process of shared construction of an object, a mobilization of mutual resources as well as a process of mutual learning".

Expansive learning occurs when teachers move away from an established form of dominant pedagogy centered on performance models where students are graded and compared, to competence models where pedagogical discourse is focused on themes

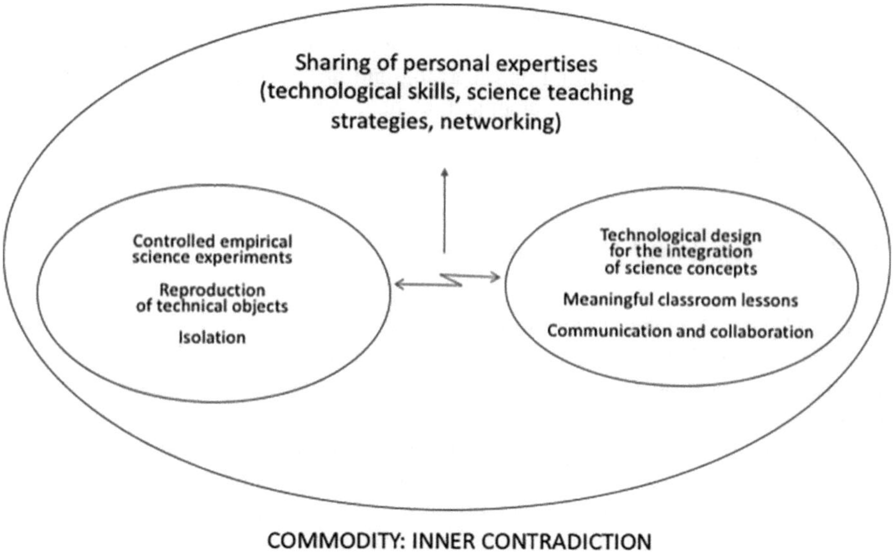

Fig. 11.1 Inner contradiction of teacher's activity: driving force of transformation

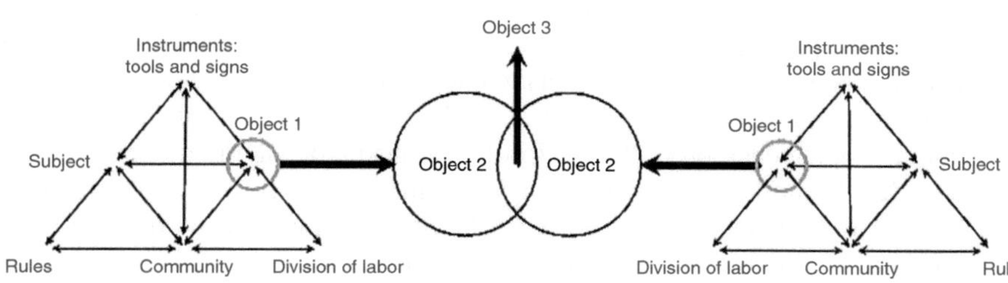

Fig. 11.2 Two interacting systems with a partially shared object as minimal model for third generation of activity theory. (Engeström, 2001, p.136)

or experience possessed by the learners (Yamazumi, 2006). In order to understand the activity under construction in this chapter (teachers' efforts to integrate technology education to science education), the concept of boundary is key. Kerosuo defines boundaries as "established distinctions and differences between and within activity systems that are created and agreed on by groups and individual actors during a long period of time while they are involved in those activities" (2006, p. 4). Gutiérrez and Penuel (2014) helps us ask new questions about "how students and teachers change and adapt interventions in interactions with each other in relation to their dynamic local contexts" (p. 19). Those ways take a variety of forms and shapes and involve boundary crossing as tensions arise and are overcome (Engeström, 2001). In this study, actions taken by three participants to create classroom and instructional artefacts have the potential of becoming instrumental to change the boundaries of their teaching activity. The funded partnership was also about establishing a strong collaboration between our university and school boards. Voogt et al. (2015) suggest that partnerships constitute a move from within, instrumental in exploring new ways of teaching and learning and how participant teachers created sustainable and transformative change that have an impact on their educational communities.

11.5 Method and Data Analysis

Inspired by developmental work research and ethnomethodology, the research team adopted a co-design and interventionist approach to document the integration of technology education to science education by teachers (Voogt et al., 2015). As shown in Table 11.1, 7 years of data, in the form of audio-video recordings of teacher training sessions, interviews and photographs during co-teaching sessions, and classroom and instructional artefacts produced by the team (teaching documents, YouTube videos, prototypes, online resources, administrative documents, speaking turns, etc.), were pooled in ('bricolage') and analyzed (Berry & Kincheloe, 2004).

Co-design of workshops and research activities unfolded in alternance giving form to the research project that consisted in co-modeling technical objects and implementing training workshops, where classroom and instructional artefacts were presented to the participant teachers and guidance counselors. Originally addressed to teachers in one school district but finally ending up reaching 15 districts over the years. Online questionnaires were sent and used as mirror data to inform the research team on the implementation of the classroom and instructional artefacts by teachers who attended the workshops. Six online questionnaires were completed by a total of 777 high school students. Following each teacher training workshops, an online questionnaire was sent to the 99 high-school teachers: 39 responded. The questions targeted information about:

- their opinion of the new curricular prescriptions asking them to make us of technological design to teach science;
- the impact of the training workshops on their teaching strategies;
- their need to collaborate with colleagues;
- the time spent to prepare their lessons in the science and technology class given the new curricular prescriptions.

At least one member of the research team attended all the workshops and collected ethnographic notes (Fig. 11.3).

In total, 20 sessions were conducted:

- seven with 52 high-school students who had experimented the technological design of the technical objects (prototypes) (co-modelled by the two participant teachers, the pedagogical counselor and the research team);
- three sessions with eight pre-service teachers;

Table 11.1 Data analyzed and used

	School years				
Type of data	2010–2011	2011–2012	2012–2013	2013–2014	2016–2017
Questionnaires	Online questionnaires: teacher (14) Questionnaires: students (77)	Online questionnaires: teacher (19)		Online questionnaires: teacher (27) Online questionnaires: students (299) Online questionnaires related to training sessions: students Microscope (132) Electronic (78) Colorimeter (35) Wind turbine (73)	Online questionnaires: teacher (39) Online questionnaires: students (304) Online questionnaires related to training sessions: students Microscope (148) Electronic (79) Colorimeter (84) Wind turbine (149) iPad (13)
Sessions (audio and videotaped)	Researchers – teachers (7) Researchers – students (7)	Researchers – teachers – pedagogical counselor (1)	Researchers – teachers (2) Researchers – pedagogical counselor (1) Researchers – professor (1)		Researchers – teachers (2) Researchers – pedagogical counselor (2) Researchers – students (4) Researchers – university students (3)
Ethnographic notes	22	6	14	18	10
Photos and videos	52		120		10 audio, video recordings 30 photos
Classroom and instructional artefacts co-produced	Wind turbine: 27 Electronic: 55	Microscope: 35	Colorimeter: 34		iPad: 5
Websites	Wind turbine on Pistes.org	Microscope on Pistes.org	Colorimeter on Pistes.org	5 LES on netsciences.ca	22 LES on netsciences.ca
Administrative documents	Wind turbine: 13 Electronic: 7 E-mails ≈ 15/week	Microscope: 4 Meetings reports: 2 E-mails ≈ 15/week	Colorimeter: 5 Meetings reports: 4 E-mails ≈ 15/week	Follow up documents: 2 Consent forms: 6 Recruitments letters: 2 E-mails ≈ 15/week	Meetings reports: 2

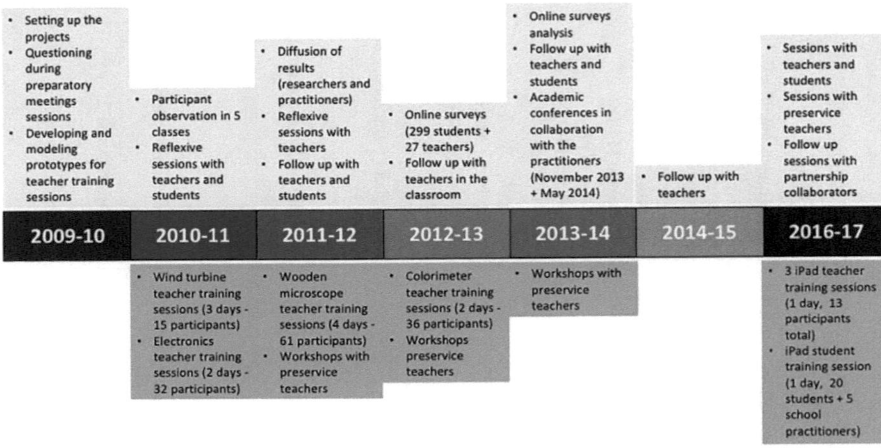

Fig. 11.3 (2009–2017) The co-design of the intervention project: workshops (dark gray) and research activities (light gray)

- 10 sessions with the two participant teachers and the pedagogical counselor, the main researcher and five graduate students.

All sessions were transcribed and analysed. The number of speaking turns was respectively: 667 (teachers), 266 (pedagogical counselor), 237 (students), 68 (school principal) and 53 (lab technician). They were key to understanding the way the participant teachers collaborated to design the prototypes and their classroom and instructional artefacts designated to be shared during the training workshops.

11.5.1 Data Analysis

Three levels of analysis were adopted to follow the development of the activity. Firstly, the identification of conflicts of motives that triggered the two teacher's and pedagogical counselor's agency to engage in co-modelling concrete artefacts. Secondly, conflicts of motives, in the form of recurrent tensions, reveal inner contradictions at the poles of emerging activity systems and help understand how they are eventually resolved through the use and exchange of the modelled concrete artefacts (prototypes) between activity systems. Thirdly, boundary zone activities are identified as the object of different activity systems converging to integrate technological education to science education.

11.6 Identification of Conflicts of Motives Through Discursive Manifestations of Contradictions

Contradictions are not directly accessible to a researcher: they become visible when recurrent tensions in the discourse are brought to the surface by way of a dialectical analysis. A dialectical contradiction refers to a unity of opposites, opposing forces, or trends within a given evolving system, a point of view that has informed the way verbatim transcripts were examined. The criteria to select the units of meaning is the presence of opposing forces potentially revealing conflicts of motives in the speaking turn: struggles, obstacles, tensions and clashes (Engeström & Sannino, 2011). To identify the emotional content in the narration, wherein individuals may express their doubts and hesitations, sentences such as 'I realize that', 'I must', 'It had to', etc. give good indications of a discursive manifestation of contradiction. The identification of tensions allowed to highlight how the teachers'conflicting motives were resolved as a second stimulus was built toward their resolution (Barma et al., 2015).

11.7 Use of Contradictions in Activity Systems to Understand the Development of the Activity

Using contradictions in empirical research pinpoints how their resolution demands the forming of new instruments and new form of division of labor (Miettinen, 2009). The activity system is a representation of human activity and when an inner contradiction is identified, the lens of the triangular representation helps understand the systemic dimension of the individual collective level of organization (Engeström & Sannino, 2011). The contradictory unity identified at the heart of the object of an object-oriented activity system (the commodity), "penetrates all corners of the triangular structure of activity" (Engeström, 1987, p. 112) by means of its dual nature; use and exchange value.

11.8 Expansion of the Activity: Boundary Zone Activity and Boundary Crossing

The third level of analysis is achieved by juxtaposing activity systems and by illustrating how they share certain boundaries and boundary objects (prototypes, classroom and instructional artefacts) (Barma & Bader, 2013). Two systems that share the same outcome will be pointed out, to the reconceptualization of the old activity to open up to a wider horizon of possibilities (Engeström, 2001). According to Edwards and Mutton (2007, p. 507), "Activity theory analyses have increasingly recognized that more than one system may be working on the same object of activity, but in doing so, they will emphasize different aspects of the problem they are working on". As the participants of the research projects learned to collaborate and shared their experiences, boundary zones are created and reflect how the sharing of the three collaborators' expertise led to the production of new teaching artefacts integrating technological design to science education.

11.9 Findings

11.9.1 Some Elements of Context

At the beginning of their collaboration with the research team in 2009, the two teachers and the pedagogical counselor adopted new roles on a daily basis: they became instructional and classroom artefacts developers up to 2017, not only for their own interests with their students, but also for their peers with whom they shared their work during teacher-training workshops once or twice a year. Fifty percent of the money granted to the researcher was used to unburden their teaching load. Approximately twenty percent of their teaching task was now dedicated to the design and the production of technical objects as well as classroom and instructional artefacts that would meet the curricular demands to integrate technological education to science education. '*A technical object* is a simple, practical object that has been manufactured, as opposed to an object found in nature (e.g. hammer, tweezers)' (GQ, 2007, p. 6). They also had the privilege of validating their work with their own students: something that would increase their credibility in the eyes of their peers. This section starts by presenting some of the conflicting motives they faced along the years and how they gave new meanings to their activities. The new motives they found resonated materially through technical objects and embodied practices reflected by the teaching strategies at the core of the classroom and instructional artefacts they produced. The selected excerpts illustrate what could have initially revealed a break to their agency but ended up resolved along the years. We present, in chronological order, the prototypes they co-designed (wind turbine, electronic devices, wooden microscope, colorimeter and an integration of the iPad). The analysis of the trajectory of their collaboration to co-design the technical, classroom and instructional artefacts makes it possible to identify inner contradictions in emerging activity systems as well as boundary crossing zones that were created. The expansive resolution of conflicts of motives starts with individual testimony from Henry, Helen and Stephen, with their own struggles and obstacles, but rapidly moves to a collective layer where they are addressed as a second stimulus is formed. The resolution makes possible the production of prototypes and their teaching artefacts. Via the teacher training workshops, the production is exchanged to reach another layer of the school community.

11.9.2 Resolving Conflicts of Motives and Engaging in the Design of Technical Objects, Classroom, and Instructional Artefacts

What drove the three participants to engage in creating new technical objects that would allow the appropriation of scientific concepts to the way they were designed along with their classroom and instructional artefacts varies according to the participants? For Helen, it was about a sense of incompetency related to the use of machine tools: the curriculum requirements demanded that teachers should now be skilled to work with them. Helen had never used them. She faced conflicting stimuli and the following excerpt illustrates why she decided to get involved in the project as she expressed some conflicting motives. "*In fact, I didn't understand anything about the reform when it happened. And then, I was a bit lost, so I got involved. I said to myself well...I'm going to go to the training that Stephen offered to all teachers. To find out exactly what a competence [related to practical skills] is*" (24-02-2011). Helen had taught science for 14 years but had never worked practically with saws or drill presses.

> Well, I was a bit panicked when I saw that I had to do the techno. That's one of the reasons why I decided to do the wind turbine with Stephen... I chose to do a techno project so that I could get help at the same time... it's all through this project that I learned. And now, I wouldn't do without techno (Helen, 24-02-2011).

For the other participant teacher and the pedagogical counselor, their expertise was different. Henry became a teacher by accident: he had previously been an electronic technician and had lost his job after the company he worked for reorganized its workforce (Barma et al., 2023). He and Stephen had been working together for 3 years at the same school and they had both been proactive in designing wind turbine prototypes on weekends and evenings at home. Stephen, on his part, was a true 'lobbyist'. He was good at recruiting colleagues, and his previous responsibilities made him go from schools to schools and connect with science teachers. Together, they enjoyed spending their free time teaching electrical concepts to present their students with more hands-on approaches. In the following excerpt, Henry reflects on his motives to collaborate with the research team and the two other participants: "*I am a good participant for the research group. I try to offer my help as much as I can. When it is time to prepare the workshops, I know I have a facility with the practical aspect of things. I joined the team because I see my role as a useful one. I have a sense of belonging to the group and the desire to contribute" (28-03-2011).* According to the three collaborators, at the beginning of the implementation of the new ST curriculum, their teacher colleagues leaned towards accepting to include the teaching of the design of technical objects to their students but were afraid to miss on the curricular benchmarks. For Helen, Henry and Stephen, it was a conflict of stimuli and they sensed there was a high degree of stress related to having the students perform at the ministerial exams.

Quoting a colleague in 2013, Henry says that they felt very strongly that it could be interesting [for the students] to engage in technological design. "*I think the kids would like it*". He goes on adding: "*It's not because they're not interested [the teachers]. It's really because they're afraid to take risks. It's the fear of missing out on their responsibility towards assessments*". During the research meeting sessions, the theme of assessment came back recurrently: "*The problem with assessment is that it kills our creativity and our will to improve our practice because we remain stuck in the present*".

Over the 7 years of the project, four prototypes were designed by the participants. The division of labor became implicit between them, even if each of them had strengths and weaknesses. Henry was responsible for the design of the prototypes, but alone, the technical object had no use in the classroom to meet the curricular requirements. Helen's role was to find theoretical anchors to link the technical objects to the scientific concepts, create classroom documents to accompany the teachers and have them improved by her two colleagues. Stephen and Henry focused on technical drawings and specifications. Stephen ensured that the ideas selected for the training would meet the needs of the greatest number of teachers by sending numerous emails to school boards. All prototypes and their classroom artefacts were developed with a turnkey vision in order to facilitate the teaching strategies in the classroom or with machine tools. Henry pinpoints an important aspect of their collaboration:

> Basically, the big challenge of our co-design activities is not to limit ourselves to making a mystery box that makes electricity. It is to succeed in closing the loop and to bring the student further. So, there's a big risk of slipping, because if you just make it work and you don't question yourself, you've missed the point. So, it can be a monumental waste of time to do a project like that or something that is going to be firmly contextualized, that is going to be meaningful for the student (Henry, 28-03-2011).

The wind turbine was the first prototype to be designed (Fig. 11.4). During the preparatory meetings in 2010, after a brainstorming session, the wind turbine was chosen because wind energy is a controversial subject in Quebec and therefore conducive to discussions with the students. Being a current topic, it would be a source of motivation for a greater number of teachers and students. In addition, in tenth grade, the curriculum focused on environmental issues, including energy challenges. That path seemed promising to the team: merging technical elements

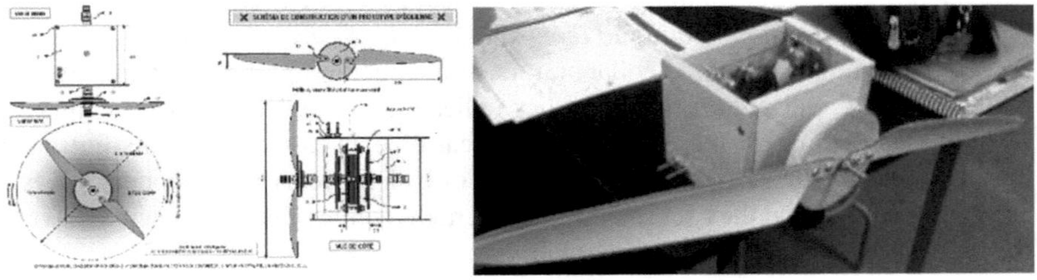

Fig. 11.4 The windmill prototype: technical object and classroom artefacts

but nevertheless achieving broader science education goals like the environmental one. The wooden wind turbine was composed of a system of rotor and magnets that induces an electric current in solenoids to light a bulb. Moreover, it had to respect the various technical, human, economic and environmental constraints of the specifications. Photos of all the steps of the realization were taken to better support the teachers during the reinvestment in class. Many concepts prescribed in the curriculum were touched upon, both in science (wind, air mass, electrical circuit, magnetic field...) and in technology (transformation of motion, function of conduction, insulation and energy transformation, etc.).

The 3 days of teacher training workshops dedicated to the windmill were held with tenth grade teachers. The days were divided into three parts: hands-on work to build the prototype, performance tests, and evaluation of their appropriation of scientific and technical skills. Reflecting on her first experience as a teacher trainer, Helen questions her capacity to act as an 'expert' when it came to woodwork and technical skills:

> I like to share my ideas with some teachers who want to get information on how to build the windmill. I'm happy to help them out. The only thing is that, since the reform was not well perceived at the beginning, I'm afraid that the teachers will say: Who is she to tell us what to do? She is a teacher like the rest of us. She's not better than us! So, I had to get over that. I felt insecure and I am afraid of the teachers' judgments. I am a teacher not an expert in electricity and wood work (Helen, 24-02-2011).

A few months later, the second two-day training workshop took place, with the goal of concrete applications of electronics concepts (Fig. 11.5). In 12 small tasks, the attending teachers were asked to familiarize themselves with direct and alternating current, resistors, diodes, the capacitor and the transistor. They were also asked to use measuring instruments such as the digital multimeter. The excerpt from an interview with Helen illustrates why this topic was chosen. *"Electronics, it's because it's a taboo there. You know, in the new curriculum, they have to do it and nobody has any electronics training, it was really a need" (24-02-2011).*

From 2012 to 2014, the participants focused on sharing and following up the implementation of the two prototypes in the science class of volunteers who had attended the training workshops. After having put their energy on tenth grade

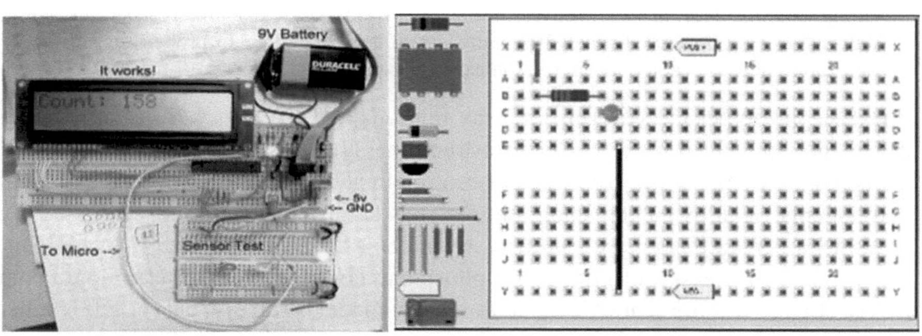

Fig. 11.5 Electronic component of one task

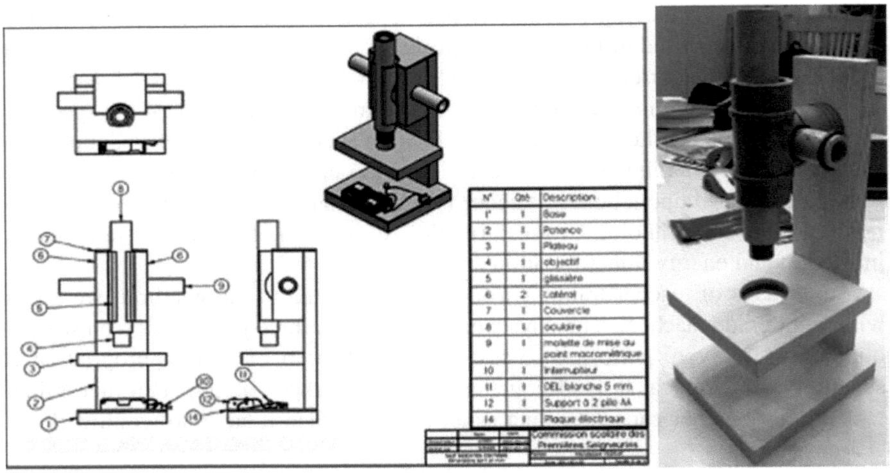

Fig. 11.6 The wooden microscope prototype: technical object and classroom artefacts

teachers, the team now wanted to reach out to ninth grade teachers whose curriculum was centered around the human body. Most teachers had taught biology for years and had never worked on machine tools in their formative years or teaching activities. Therefore, they developed a project to build and retro-analyze a wooden microscope as an alternative object to the optical microscope used in most schools (Fig. 11.6). The colorimeter required that you had to be able to work with a voltmeter and a battery, cut their pieces with the machine tools, calibrate a potentiometer and activate a switch. After analyzing the instructional artefacts, we noted that the teaching goals were related to the development of technical literacy but also health literacy, so adolescents came to understand the impact of energy drinks on their body. The two-day training session was a great success and was given a second time.

Reflecting on the workshops, Henry comments that:

> When you want to do things differently... There are a lot of problems that come with that. So, there's a lot of openness [on the part of the teachers attending the training workshops], as long as there's a sense of acceptable mastery or comfort that comes along. There's a very strong sense of, "Hey, this could be interesting, I think kids would like this. Teachers are not refusing to adopt a closed mind. It's really out of fear of failing to properly teach the integration of technical aspects in the context of a science class. (Henry, 17-05-2013)

The last prototype, the colorimeter, was first adapted by Henry from an existing project and enhanced by Helen with instructional artefacts (Fig. 11.7). The participants liked the idea of the colorimeter because, in addition to touching the electrical components, it allowed the integration of the environmental issue addressed by drinking water. In this project, students would be asked to build their own colorimeter capable of measuring nitrate and phosphate concentrations in a sample taken from a stream near the school using a previously made calibration curve. The teacher training workshops were demanding, this technical object was difficult to produce and required a good dose of knowledge in electricity. After the workshops, Henry

Fig. 11.7 Colorimeter built and used by students

Fig. 11.8 Students using iPad as a lab instrument

relates that: *"It's fair to say that not everyone was in their comfort zone when it came to electricity. It was also the case for the technological aspects of the tasks. Because everyone was saying that technology was not their department and it never would be" (17-05-2013).* Because of the strong resistance of attending teachers, Henry worked on seven versions of the colorimeter. *"It was too much for many [teachers], the gap was too big between the theoretical model that was proposed and what was required to be comfortable with" (Henry, 17-05-2013).*

The years 2015 to 2017 aimed at continuity, reinvestment and sharing of the best practices developed by the participants in order to decrease professional isolation and better support science and technology teachers. Four years later, the projects took another turn and had shifted from woodwork to integrating more computer-based applications electronics, robotics and 3D printing. The focus was on the use of the iPad in the science classroom. The first part of the training workshops was the presentation of applications to use the iPad as a measuring tool in the lab, to do interactive simulations or to create online quizzes. The second part was the sharing of labs using the iPad, such as for measuring magnetism and energy efficiency of different types of light bulbs where it is used to measure the brightness as shown in Fig. 11.8.

Table 11.2 Teacher training workshops: 170 peers from 15 school districts (2010–2017)

Participants attending	Wind turbine 2010 October (25–26-27)	Electronics 2011 January (26–27)	Microscope 2011–2012 November (3–4) January (26–27)	Colorimeter 2012 December (6–7)	iPad 2016 September (30) October (12 & 18)	Total
High-school science teachers	9	19	35	20	16	99
University pre-service teachers	2	0	1	0	5	8
Pedagogical counselors	2	4	7	2	3	18
Civil servant from the Ministry of Education	0	0	0	1	0	1
Lab technicians	2	9	18	13	0	42
Student researchers	0	0	0	0	2	2
Total	15	32	61	36	26	170

Table 11.2 presents an overview of the workshops developed over the years and reached more than 170 peers coming from 15 school districts (2010–2017).

Reflecting on conflicts that were overcome along the years highlights some trends in the type of conflicts that were resolved. For Helen, it was the sentiment of incompetency with practical work at the beginning and the necessity to link the technical object to the scientific concepts and develop environmental consciousness among the students (windmill, colorimeter). She gradually felt more competence when leading teacher training workshops and producing a substantial amount of online instructional artefacts. Henry feared remaining too focused on the technical object itself. With the help of Helen, he often redesigned drawings to make the production of the prototypes more accessible. He moved from a narrow vision of technical skills to a wider one that would merge more science concepts: "*Last year, I had pushed hard, and at one point in the year I was told by students that I should do more theory. It's not often that students ask their teacher to do more theory. Because in my opinion, in the project, they realized that, let's say, I hadn't mastered it" (Henry, 17-05-2013).*

As for Stephen, the number of emails he sent and had to manage to meet the demand from teachers ended up incalculable over the years, but he strongly felt content with the way the team responded.

> Student researcher: ...you're talking about support, because you're more from a certain school board..., for example in the other school boards, do the teachers benefit from support?
>
> Stephen: We are the ones who provide the support... it's calming down a little bit, but I would say, ... there was not a week when we did not answer a question coming from other school boards in the province. And there isn't a month in the year that we haven't answered questions. I don't answer technical questions about the projects we've done [when we were

> not directly on site] ... whether it's the wind turbine, the microscope, the electronics training, or most recently the colorimeter. (15-05-2013)

Finally, a general sense of satisfaction was expressed by Stephen at the end of the funding.

> With the projects we did, I find that teachers visit the websites and try more hands-on projects to integrate the theory. So, I feel we did not waste our time, we worked differently, we managed to make sense of theoretical notions through technological design approach. Some of the participating teachers at the professional development training workshops even changed our projects for their own purposes with their students (Stephen, 07-06-2017).

Helen's conflicting motives were mostly related to: (1) choosing to remain unskilled when it came to practical work, versus sticking to theoretical lecture based classes; and (2) adapt Henry and Simon's prototypes to transpose them into classroom and instructional artefacts coherent with the ministerial guidelines, versus using the instructional documents she had been working on alone on a theoretical basis.

Henry's conflicting motives were mostly related to:(1) designing technological artefacts that would remain mystery boxes to other teachers and students, versus linking the technological artefacts to meaningful problems or concepts in the lives of his colleagues and own students; and (2) forgetting about the end of the year ministerial assessments when designing the artefacts, versus creating the artefacts on the base of the scientific and technological contents present in the curriculum. Stephen' struggles were related to remaining to act on a local basis and not being overwhelmed with all the email management, versus networking with other school boards to expand the impact of the project.

11.9.3 The Contradictory Struggles for Co-designing the Technical and Instructional Artefacts

Analyzing how the sharing of the three collaborators' respective teaching expertise takes us beyond their individual activity. Each of them was acting in a specific context (own school, students, relations with the school principal, lab technician and other teachers). When we pool in their discursive contributions and analyze them with relation to an object-oriented, artefact-mediated collective activity system, it is possible to identify emerging tensions as areas of rupture, possible change and innovation to produce the new technical and instructional artefacts. The delineation of the activity system helps us to pinpoint the competing teaching strategies in the form of a double bind. The resolution of the double binds by the three participants will eventually see them face the secondary contradictions. Tensions are identified at each pole of the activity system. Supported by excerpts, we discuss how some double binds were identified and, in some cases, resolved. Figure 11.9 presents the inner contradictions of the collaboration of the team members as they engage in co-designing technical and instructional artefacts integrating technological design to science.

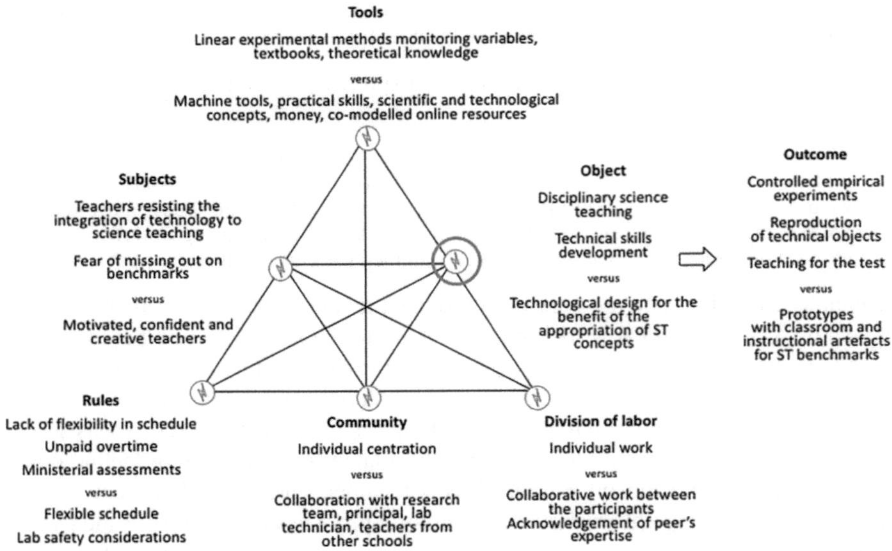

Fig. 11.9 Inner contradictions at each pole of the activity system

Zooming on conflicts related to the object of the activity, Henry expresses a conflict that could be a brake for engaging in agentive actions: "*The problem with assessment is that it kills our creativity and our will to improve our practice because we remain stuck in the present" (08-05-2012).* Two competing strategies are visible here: teaching for the test versus being creative and integrating technological design to science teaching. Reflecting on the outcome, he continues saying that when he started teaching science, he was not convinced that the students could benefit from the projects. *"So, I decided not to engage in projects and focus on the content of the new science program. The students hated it" (08-05-2012).* If we recall Helen's struggles in the first section, we better understand the insecurities she faced before collaborating with the other team members: "*We did our best to try to understand how to work in a workshop and how to integrate the evaluation aspects at the same time. I was working a little bit with my hands but not much, I was not competent with woodwork and zero with electronics, I liked teaching theory (Helen, 07-06-2017).* The following excerpt illustrates that this double bind was resolved over the years: *"It has changed so much over the years. Now it's more electronic, computer, 3D printer, it's beyond what we did at first. We pushed it so hard, we took a lot of different paths. We had to adapt" (Helen, 07-06-2017).*

The analysis of verbatim also highlights some important tensions regarding organizational problems, lack of flexibility in schedule, unpaid overtime, money issues, clashes when communicating with the lab technician and the school boards. There were also concerns about safety aspects when working on machine tools. At the beginning of the research project, Helen suffered from a lack of communication between her and Henry with regards to tight schedules: *"We could not find free time to work together. So, we each worked separately, but then we really tried to share our work (15-05-2013).* Helen also faced challenges when she was testing one prototype with her own students: it was too long to fit in the weekly schedule, so she *"coupled two lessons in order to help students finish their projects. It was an interesting accommodation" (08-05-2012).*

Other issues also had to be addressed when it came to managing collaboration with the lab technician, the testing of prototypes and the organization of the labs. Unpaid overtime became an issue: *"the relationship with the technicians, it is a little tense, because according to the union collective agreement. . .They must be paid just right" (Stephen, 15-05-2013).* Stephen recalls having meetings with the school board to address the issue of human resources as well as a lack of understanding between the administrative level of the university responsible to reimburse the expenses encountered by the school board:

> If we talk about constraints, we talk about irritants. Well, it's at the school board level. It's when you want to do business with a technician, when you need the service of a technician, well, the technician doesn't want to work because they're not paid the right way. . . as for the University, . . . they don't have the flexibility, they don't understand. (Stephen, 15-05-2013).

During the first years of the research project, having access to money quickly became an issue. The administrative aspects of managing the budget were not conducive to help Henry put in place ideas he has on the spot. He did not want to put his principal in an uncomfortable position: "*When you go to see your principal because you have an idea, you better make sure it will work because he is backing me up. I put 150$ on my credit card. I took the risk. . . . My principal can't do that" (08-05-2012).* He took risks to work on his idea.

When reflecting on some of the teacher training workshops, Stephen recalls what the team members did to integrate the theoretical notions of the curriculum to produce the technical objects and the instructional artefacts but expresses a great degree of frustration with regards to the attending teachers who faced a wall with electric theoretical and hands-on tasks.

> The theoretical level, it's too complicated for them [other teachers]. . . . it's not because on the theoretical level, the project goes too far. No. The project, basically, it's all notions that are in the curriculum. But to connect an RGB diode to the power supply is not a good idea. . . . If I add switches in parallel, in series, with diodes, they all get mixed up (Stephen, 15-05-2013).

This level of analysis is key to grab how the activity under construction occurs through dialogical interaction and meaning making, where the three participants modify their area of professional practice. The activity under construction grows multi-voiced as the expansive resolution of conflicting motives unfolds to meet the secondary contradictions and put in place innovative places of learning collectively.

11.9.4 Boundary Crossing Activity

The previous sections highlight the complexity of the situation the three collaborators were in, as they shared expertise to co-design and exchange the artefacts. It appears unlikely that alone, Henry, Helen and Stephen would manage to break away from a conflicting situation in a single attempt. Resolution of the conflict can unfold as a series of successive steps or "a mediational chain of agentive actions" (Engeström & Sannino, 2013). For Engeström et al. (1995): "practitioners must move across boundaries to seek and give help, to find information and tools wherever they happen to be available" (p. 332). This is how boundary-crossing unfolds as "horizontal expertise" is shared. Their capacity to reflect on their activity was strong and characteristic of professionals in practice (Wu et al., 2008).

Figure 11.10 presents boundaries that should be understood as dynamic and evolving. They were defined as the three participants negotiating new roles, new rules, new division of labor between their respective activity systems. This was possible since Henry, Helen and Stephen engaged in sharing their own expertise and supporting each other. They recognized their personal strengths but also knew that their collaboration would enhance their new form of practice. Stephen expressed that: "*What has changed is the attitude of some people. We put things together now. Many people in many schools have changed their attitudes and this is what is important to us... as concrete results" (Stephen, 2017).*

The initial germ cell that we presented was fruitful. As a boundary zone of activity emerged, boundary crossing unfolded, especially through the mediating actions of Henry, Helen and Stephen. Their new ideas were embodied in their

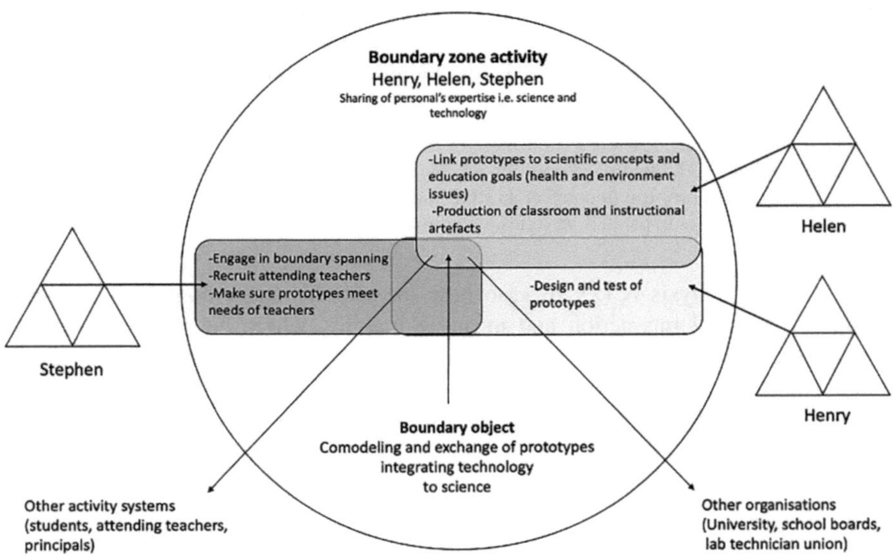

Fig. 11.10 Modelling the boundary zone activity of the collaboration of the three participants

practice in the form of material objects. The technical artefacts represented the material integration of technological design to science education. These artefacts became themselves a first stimulus as the need for producing curricular and instructional artefacts increased in anticipation of the forthcoming training workshops. Based on their fruitful collaboration, Henry, Helen and Stephen became facilitators reaching out to their 170 colleagues during eight workshops.

The boundary zone activity was formed and is a concrete example of the expansive resolution of the conflicting motives. Struggles decreased and a higher degree of satisfaction was palpable among the three colleagues. Henry expressed that he had: *"the impression that many of [his] students had changed their opinion about themselves. They now tell [him] they feel in control when engaging in workshops...that practical work gives them self-esteem and before they felt like second wheels"* (*28-03-2011*). The same feeling was expressed by Helen: *"The students like it too. We also had a lot of training at school, on machine tools, on CPR courses to make us safe and then there were even cabinet makers who came. There was a lot of training at the school to make the teachers safe" (Helen, 24-02-2011).* Henry was also content with the participation in the workshops of teachers coming from a fair distance and showed more interest in the prototypes he presented. *"I get the impression that the project is spreading by word of mouth and is doing its job of dissemination. It's not just a local anymore, it's scaled up" (Henry, 15-05-2013).*

As for Stephen, he was very much involved in boundary spanning:

> At the beginning, we sent the invitations to the school boards only to the Quebec City region. I always opened up to others, I always kept the pedagogical counselors that I'm in contact with across the province, informed. They always knew what we were doing. Even the first training session we did, some of them came from the Eastern Townships region [...] all those in my contact network were informed. (15-05-2013)

11.10 What Was Learned?

By expressing and overcoming conflicting motives in a complex learning setting, the three colleagues gave new meanings to the object of their collaborating activity. During the sessions with the research team, the participants expressed their own conflicting motives with the research team or with one another. Along the 7 years, new understanding of motives was attributed and found resolution through learning actions. In line with Hopwood and Gottschalk (2020) and Chaiklin (2012), epistemic developments happened as new motives reshaped new forms of collaborating practices. Helen, Henry and Stephen found new meanings to their practice (epistemological axis), more satisfaction (axiological axis), as well as usefulness to the integration of technological design to science education (praxeological axis). The pedagogical and didactic choices expressed in the curricular artefacts produced and exchanged tell us how they chose to integrate technological design to the new science and technology program. It was not solely about the technical object imbedding science and technical concepts. It was also about epistemological choices when it came to contextualize their approach in the classroom. In the footsteps of Hodson's (1999) proposal, they had the awareness of recognizing areas where science and technology intersected (environmental issues for the windmill and colorimeter) to link them to student's life experience. The double binds at the core of the inner contradictions due to isolation, disciplinary teaching and assessment were overcome when favorable conditions were put in place: flexibility with schedule, horizontal sharing of expertise, availability of money and a sense of safety with woodwork and electronics. Doing CHAT research revealed fruitful not only as a theoretical framework but also as a methodological and analytical grid to bring out the complexity of the individual collective activity.

The three colleagues demonstrated a high level of techno-scientific literacy. As we discussed previously, this competence shall not be considered as an individual quality but rather as emergent and distributed coming from the interactions between the actors engaged in the action. Those social interactions ended up being meaningful. The findings of the 7 years CHAT research project bring new light for the future of teacher's collaboration. Often, when a new curriculum is implemented teacher's agency is limited (Konopasky & Sheridan, 2016). Teachers do not always feel able to control, or even want to define or reshape, educational policies. The implementation of a new educational curriculum often means making sure members of a school community adhere to the proposed changes and are ready to invest time in developing new teaching practices. Focusing on individual conflicts of motives and joint agency allowed us to understand why and how participants engaged in expansive learning and came to collectively progress to create new tools that revealed boundary crossing objects and determined new roles to reconceptualize the object of their joint activity. The shaping of a boundary activity zone implied that different activity systems expanded their own activity to establish a zone of proximal development. The prototypes and their instructional artefacts were key players as they were exchanged between activity systems.

Expansive learning goes beyond the individual. Even if conflicts of motives start with the individual, they rapidly reach a collective layer to be resolved and then lead to the definition of a boundary zone where a new form of practice emerges. Science and technology teachers can be learners, but innovation does not occur without expansion of their respective activity systems. Challenging demands require time and require educational partners to interact and share the same problem space. The new form of activity is context based and tributary to specific needs of each educational setting.

References

Barma, S. (2011). A sociocultural reading of reform in science teaching in a secondary biology class. *Cultural Studies of Science Education, 6*(3), 635–661.

Barma, S., & Bader, B. (2013). How one science teacher redefines a science teaching practice around a theme: A case study in the context of educational reform in Quebec. *International Journal of Environmental and Science Education, 8*(1), 131–161.

Barma, S., Lacasse, M., & Massé-Morneau, J. (2015). Engaging discussion about climate change in a Quebec secondary school: A challenge for science teachers. *Learning, Culture and Social Interaction, 4*, 28–36.

Barma, S., Vincent, M. C., & Voyer, S. (2023). Decision-forming processes leading to peer mentorship. In Doing CHAT in the Wild (pp. 150–192). Brill.

Berry, K., & Kincheloe, J. (2004). *Rigour and complexity in educational research* (Conducting Educational Research). Open University Press.

Chaiklin, S. (2012). A conceptual perspective for investigating motive in cultural-historical theory. In *Motives in children's development: Cultural-historical approaches* (pp. 209–224). Cambridge University Press.

Conseil supérieur de l'éducation (CES). (2013). L'enseignement de la science et de la technologie au primaire et au premier cycle du secondaire. Québec. Retrieved from: https://www.cse.gouv.qc.ca/wp-content/uploads/2019/11/2013-08-lenseignement-de-la-science-et-de-la-technologie-au-primaire-et-au-premier-cycle-du-secondaire.pdf

Convert, B., & Gugenheim, F. (2005). Scientific vocations in crisis in France: Explanatory social developments and mechanisms. *European Journal of Education, 40*(4), 417–431.

Davydov, V. V. (1990). *Types of generalization in instruction: Logical and psychological problems in the structuring of school curricula*. National Council of Teachers of Mathematics.

Davydov, V. V. (2008). *Problems of development instruction: A theoritical and experimental psychological study*. Nova Science Publishers.

Désautels, J. (2002). L'alphabétisation technoscientifique et la démocratisation de la démocratie. *Revue canadienne de l'enseignement des sciences, des mathématiques et des technologies, 2*(2), 189–195.

Edelson, D. C. (2001). Learning-for-use: A framework for the design of technology-supported inquiry activities. *Journal of Research in Science Teaching, 38*(3), 355–385.

Edwards, A., & Mutton, T. (2007). Looking forward: Rethinking professional learning through partnership arrangements in initial teacher education. *Oxford Review of Education, 33*(4), 503–519.

Eick, C. J., & Reed, C. J. (2002). What makes an inquiry-oriented science teacher? The influence of learning histories on student teacher role identity and practice. *Science Education, 86*(3), 401–416.

Engeström, Y. (2001). Expansive learning at work: Toward an activity theoretical reconceptualization. *Journal of Education and Work, 14*(1), 133–156.

Engeström, Y. (1987/2015). *Learning by expanding*. Cambridge University Press.

Engeström, Y., & Sannino, A. (2011). Discursive manifestations of contradictions in organizational change efforts: A methodological framework. *Journal of Organizational Change Management, 24*(3), 368–387.

Engeström, Y., & Sannino, A. (2013). La volition et l'agentivité transformatrice: perspective théorique de l'activité. *Revue internationale du CRIRES: innover dans la tradition de Vygotsky, 1*(1), 4–19.

Engeström, Y., Engeström, R., & Kärkkäinen, M. (1995). Polycontextuality and boundary crossing in expert cognition: Learning and problem solving in complex work activities. *Learning and Instruction, 5*(4), 319–336.

Fourez, G. (dir.) (1994). *Alphabétisation scientifique et technique: essai sur les finalités de l'enseignement des sciences*. De Boeck Université.

Fourez, G. (2002). *La construction des sciences, 2ème version revue et augmentée*. De Boeck Université.

Gouvernement du Québec (GQ). (2007). *Québec education program: Secondary cycle two*. Ministère de l'Éducation du Loisir et du Sport. Retrieved from: http://www.education.gouv.qc.ca/fileadmin/site_web/documents/education/jeunes/pfeq/PFEQ_presentation-deuxieme-cycle-secondaire_EN.pdf

Government du Québec (GQ). (2006). *Quebec education program. Science and technology. Secondary school education. Cycle one*. Ministère de l'éducation. Retrieved from: http://www.mels.gouv.qc.ca/DGFJ/dp/programme_de_formation/secondaire/pdf/qep2004/chapter62.pdf

Greeno, J. G. (1998). The situativity of knowing, learning, and research. *American Psychologist, 53*(1), 5.

Guchet, X. (2011). Les technosciences: essai de définition. *Philonsorbonne, 5*, 83–95.

Gutiérrez, K. D., & Penuel, W. R. (2014). Relevance to practice as a criterion for rigor. *Educational Researcher, 43*(1), 19–23. https://doi.org/10.3102/0013189X13520289

Hodson, D. (1999). Going beyond cultural pluralism: Science education for socio-political action. *Science Education, 83*, 775–796.

Hopwood, N., & Gottschalk, B. (2020). From volitional action to transformative agency: Double stimulation in services for families with young children. *Pedagogy, Culture & Society, 30*(1), 35–52.

Il'enkov, E. V. (1982). *The dialectics of the abstract and the concrete in Marx's 'capital'*. Progress.

Impedovo, M. A., Andreucci, C., & Ginestié, J. (2017). Mediation of artefacts, tools and technical objects: An international and French perspective. *International Journal of Technology and Design Education, 27*(1), 19–30.

Juuti, K., Lavonen, J., Uitto, A., Byman, R., & Meisalo, V. (2009). Science teaching methods preferred by grade 9 students in Finland. *International Journal of Science and Mathematics Education, 8*(4), 611–632.

Kerosuo, H. (2006). *Boundaries in action: An activity-theoretical study of development, learning and change in health care for patients with multiple and chronic illnesses*. University of Helsinki, Department of Education.

Koballa, T., & Glynn, S. (2007). Attitudinal and motivational constructs. In *Handbook of research on science education*.

Konopasky, A. W., & Sheridan, K. M. (2016). Towards a diagnostic toolkit for the language of agency. *Mind, Culture, and Activity, 23*(2), 108–123.

Latour, B. (1989). *La science en action*. Folio.

Miettinen, R. (2006). The sources of novelty: A cultural and systemic view of distributed creativity. *Creativity and Innovation Management., 15*(2), 173–181.

Miettinen, R. (2009). Contradictions of high-technology capitalism and the emergence of new forms of work. In *Learning and expanding with activity theory* (pp. 160–175).

Osborne, J. (2003). Attitudes towards science: A review of literature and its implications. *International Journal of Science Education, 25*(9), 1040–1079.

Robitaille, J. P. (2010). *La relève en sciences et technologies au Québec: Un état des lieux*. Association francophone pour le savoir-Acfas.

Roth, W.-M. (2001). Learning science through technological design. *Journal of Research in Science Teaching, 38*(7), 768–790.

Roth, W.-M., & Désaultels, J. (2002). Science education as/for sociopolitical action: Charting the landscape. In W.-M. Roth et J. Désaultels (dir.), *Science education as/for sociopolitical action* (pp. 1–16). Peter Lang publishing Inc.

Roth, W.-M., & Lee, S. (2004). Science education as/for participation in the community. *Science Education, 88*, 263–291. Retrieved from: www.interscience.wiley.com

Severance, S., Penuel, W. R., Sumner, T., & Leary, H. (2016). Organizing for teacher agency in curricular co-design. *Journal of the Learning Sciences, 25*(4), 531–564.

Sidawi, M. M. (2009). Teaching science through designing technology. *International Journal of Technology and Design Education, 19*(3), 269–287.

Virkkunen, J., & Ahonen, H. (2004). Transforming learning and knowledge creation on the shop floor. *International Journal of Human Resources Development and Management, 4*(1), 57–72.

Voogt, J., Laferriere, T., Breuleux, A., Itow, R. C., Hickey, D. T., & McKenney, S. (2015). Collaborative design as a form of professional development. *Instructional Science, 43*(2), 259–282.

Vygotsky, L. S. (1962). *Thought and word* (F. Sève, Trans.). Messidor/Social Editions. (Original work published 1934).

Wu, W., Chang, H. P., & Guo, C. J. (2008). An empirical assessment of science teachers' intentions toward technology integration. *Journal of Computers in Mathematics and Science Teaching, 27*(4), 499–520.

Yamazumi, K. (2006). Activity theory and the transformation of pedagogic practice. *Educational Studies in Japan, 1*, 77–90.

Sylvie Barma is a full professor of Science Education at the Faculty of Education at Laval University since 2008. After teaching high school science for 20 years and contributing to the development of the new Quebec Science and Technology curriculum, she obtained a Ph.D. in Science Education.

Samantha Voyer taught science and technology at secondary level for 7 years. She then obtained a master's degree in science didactics under the supervision of Sylvie Barma. Her research focused on the treatment of socioscientific issues in the science classroom, from the perspective of activity theory. She currently works in a knowledge transfer and social innovation organization for educational success.

Chapter 12
Micro-Tensions from Students' Prototyping in a School Makerspace: Lessons from an Unfinished Work

Viktor Freiman and Jacob Lingley

12.1 Introduction

The study we discuss in this chapter explores several STEM-bound (Science, Technology, Engineering, Mathematics) trends that influence the K-12 education system in New Brunswick, Canada, over the past two decades. We explore these trends by analyzing the learning process of two students as they construct an invention in their school makerspace. One STEM-bound trend in New Brunswick reflects an increasing role of digital technology and digital skills in the modern society. To address this increase, the government installed high speed internet in all provincial schools and attempted to introduce several 1-to-1 laptop initiatives to support teaching and learning in some middle schools (Freiman et al., 2011). Another government supported initiative that was created to address the increased societal reliance on technology included the formation of the *Innovative Learning Agenda Funds program* (Blanchard et al., 2010). This fund was designed in part to promote teachers' innovative pedagogical ideas such as a robotics-based learning initiative and an experiment to explore the influence of a BYOD-based program (Bring Your Own Device) on student learning (Blanchard et al., 2010; Chiasson, 2013). While having brought some examples of successful integration of technology into schools, these initiatives have yet to succeed in creating a movement of transformation of teaching and learning to meet twenty-first century societal challenges (Chiasson, 2019).

V. Freiman (✉)
Université de Moncton, Moncton, Canada
e-mail: viktor.freiman@umoncton.ca

J. Lingley
Brilliant Labs, Fredericton, Canada

K. Plakitsi, S. Barma (eds.), *Sociocultural Approaches to STEM Education*, Sociocultural Explorations of Science Education 21,
https://doi.org/10.1007/978-3-031-44377-0_12

Nevertheless, the above mentioned and some other similar initiatives have appeared to inspire enthusiasm in many teachers, community leaders and educational stakeholders to appreciate the influence STE(A)M-based projects can have on student learning. Since 2014, several schools in New Brunswick have adopted the grassroots STE(A)M-based (where 'A' represents arts) philosophy whereby teachers have included open-ended projects for their students to explore cutting-edge technologies as part of their instruction (Freiman, 2020). Many of these classroom projects have been supported by *Brilliant Labs* (https://www.brilliantlabs.ca/), an Atlantic Canadian-based charity that supports students and teachers through its hands-on technology and experiential learning platform. One goal of *Brilliant Labs* is to create opportunities in which students can experience innovation, creativity, critical thinking, real-world problem solving and career building within a makerspace learning environment. Thus, this organization might have potential to contribute to the latest provincial 10-year Education Plan (GNB, 2016) by increasing students' interest in STEM-related careers while empowering them with modern technologies (e g. robotics, coding, 3D-printing, augmented reality, among many others). While the number of schools, teachers, and students involved in the movement has grown significantly, a clear understanding of what students learn and how schools can build a culture surrounding on this type of experience is still missing.

One of the Middle School makerspaces, located in an urban area central to the province, has created a design thinking class where all Grade 6 students were introduced to the process of technological invention during the fifteen 60-minute class sessions. Students were asked to develop a physical prototype for a solution to a problem that they identify in their community. In May–June 2019, one group of students and their teacher agreed to participate in a case study of makerspaces. This case study was from the research team of CompeTI.CA (In English: ICT competencies in the Atlantic Canada, www.competi.ca) partnership network and has been conducting similar research in some provincial schools since 2016 (Freiman, 2020).

As researchers to the maker phenomenon for quite some time, this case study permitted us to document amazingly rich K-12 learning experiences whereby students expressed feelings of pride and value over what they had made (e.g., Freiman & Kamba, 2020; Lingley, 2021; Robichaud & Freiman, 2020). When revisiting our data and discussing students' impressive creativity, we felt a need to dig deeper into these authentic moments of learning. While broadening our own perspective and zooming out on students' success, this data also did not always seem to reflect tensions regarding difficulties they experience when moving towards goal achievement. Indeed, challenges students face, even failures, as well as possible signs of students becoming emotionally exhausted and even disengaged also merit researchers' attention. These types of difficulties and obstacles are less frequently reported in the literature on makerspaces.

During our observation visits, we expected to see the refinement of skills between student project prototypes. We planned to listen and document the design-mediated student decisions that permitted them to arrive at some version of their prototype and not another one. We were also prepared to feel the different textures of materials students had curated for their projects. Under such circumstances, the context of our study made it easy to become positively enthralled in this sensory experience, particularly when a learning environment like the one that inspired this research had 15 different projects happening at the same time. Anything that could be considered as less than an ideal manifestation of "positive" experience could easily fade from the observers' attention. We have seen such signs before in more informal makerspace observations but have rarely given them much more thought than a fleeting moment as young makers seem to resolve these moments faster than it takes us to try to understand them.

It could be that as observers who only have a limited amount of time in a space of one lesson, we tend to redirect our attention towards another sensory rich activity that we understand in that moment, rather than continuing to observe a moment of disruption. We then could be missing out a wealth of student knowledge each time we allow our focus to be diverted away from what we impulsively decide to be a moment of tension. This left us curious about what we could understand from these tense moments and how it may influence our interpretation of what students are learning. We often asked ourselves questions like: *What happens when students do not actually achieve their intended goals and leave their work unfinished as they make the abrupt realization that their class has ended?* To better understand these moments, we decided to spend time zooming out and in on a single student project where two Grade 6 students, Bobby and Zack (names are fictive) collaborated on their invention during their experimental design thinking class over the period of one month. This helped us to understand follow-up questions like: *why did they do what they did?* and *were there moments that revealed some struggles or tensions and how did these students deal with these moments?* This process, while admittingly narrow in scope when there were plenty of student projects to include, finally permitted conjecture about the value of these learning experiences during the student making process. Our chapter represents a complex narrative journey through what we have learned about these tense moments that occurred as Bobby and Zack designed their invention: *Frankenstein Pencil.*

12.2 Theoretical Background

Our study explores a complex activity system where the ideas of teaching technological invention to middle school students interacts with increasing attention to pedagogical topics including STEM disciplines in twenty-first century education, design thinking and maker culture. To better understand this complex interaction, we use the Cultural-Historical Activity Theory (CHAT) grounded in Vygotskian insight about learning and post-Vygotskian perspectives developed, among others, by Alexei N. Leontiev (1981) and Yrjö Engeström (1987, 1999). Specific to our study, ideas of transformational change and the role of tensions explored by the CHAT literature enables us to grasp the potential of technology as a mediation tool across an activity system. Situating the use of tools in this context deepens our understanding of how students' expressions of skill, personality and consciousness can influence the evolution of practical social activities (Sannino et al., 2009). By introducing the process of creating innovative products by young learners in a collaborative learning environment, the school seems to have embraced the "social construction of knowledge rooted in different professional practices" (Wannenmacher & Antoine, 2016, p. 37).

We are exploring the tensions created by the dynamic nature of a complex activity system. While the bulk of this chapter focuses on the activities of Bobby and Zack, the influence of their activities, within the same classroom, may expand to, the entire class, the school, or even reaching a system level. For instance, the activities that occurred between Bobby and Zack described in this chapter occurred because of a teacher's decision, which was a result of a school decision, which at some point may have been inspired by a systemic trend valuing STEM activities and maker projects. When considering the influence of any one of these activities, we were interested in understanding the role of actors that comprised the activity system and the tensions that they may encounter as outcomes are produced. Like the findings discussed by Wannenmacher and Antoine (2016), whereby they suggest that an activity system is "fundamentally dynamic," we are curious to know the influence of any tension within a complex learning environment where at any time there are multiple concurrent activities between components of the activity system. The authors argue that "the dynamic is conveyed by the object of the system of activities and by the tension, opposition and even contradiction that appear within the system. The individuals (subjects) within teams (community), shaped by the division of labor and its regulation (rules), work to answer a need. This is what defines the object of a system of activities. The system is not immutable. On the contrary, its dynamic shows itself as an 'outcome', a trajectory that was not anticipated at the beginning" (Wannenmacher & Antoine, 2016, p. 38).

We consider this interpretation of activity and activity system useful to frame our analysis of students' learning within CHAT while focusing on what the authors call "moments of tension." Inspired by Wannenmacher and Antoine (2016) we use the term *micro-tensions* which we find to better reflect our effort to zoom out on challenges students face at different stages of the design process as they persevere towards prototype completion in their makerspace. In the following sub-sections, we zoom out from different aspects of learning in makerspaces, design thinking and activity theory that guided our investigation. It is this perspective that informed our ideas surrounding the seemingly disruptive micro-tensions from an activity system as potentially enriching student learning.

12.2.1 Learning in Making

Makerspaces are known as materially diverse learning spaces (digital fabrication labs, or 'fab labs'), introduced in early 2000s in the United States. (Gershenfeld, 2012). They present different layouts, in which students engage in a multitude of projects, during which they explore various technologies, create new artifacts, and share their results with others (Freiman, 2020). Makerspaces provide an environment in which students can design, experiment, construct and invent within a STE(A)M learning paradigm. Activities can range from cardboard construction to electronics, programming, robotics, and sewing. These informal activities, although previously known, took a new life in the early 2000s with the advent of Information and Communication Technology (ICT).

According to Oliver (2016), makerspaces take many forms but generally involve a physical (often non-classroom, e.g., in public libraries) space with shared resources to pursue technical projects of personal interest with the support of a maker community. Thus, "making is more commonly practiced in after-school camps and clubs", it might also "enrich the school-day curriculum and bridge formal and informal learning contexts" (Oliver, 2016, p. 160). In a K-12 context, according to Niederhauser and Schrum (2016), there is a "relationship between the maker movement and the effort to increase STEM-related curriculum and interest in STEM careers and to move beyond current career" (p. 329). According to Peppler and Bender (2013), the latter helps students to "make their own jobs and industries" (as cited in Niederhauser & Schrum, 2016, p. 329).

Besides establishing and promoting learning in a variety of technology-rich environments, Hagel et al. (2014) emphasize the involvement of "different players acting in the maker ecosystem – beginners, collaborators, and market innovators" (as cited in Oliver, 2016, p. 160) in the maker movement. Referring to MakerMedia (2013), Oliver (2016) also argues that makerspaces can be implemented across K-12 grade levels, to include easier electronic circuit projects and programming languages like Scratch for the elementary grades and more challenging 3D modeling and programming languages like Arduino for middle and high schools. In this rather informal context, learning is seen as both an autonomous and networking activity. Hence, it can be considered as self-directed, self-determined, problem-based while promoting collaboration with others to build together a path for creating and investigating many ideas. Makerspaces are a relatively new phenomenon in the K-12 education system, which has gained ground during the past decade in many countries, including Canada (Hughes, 2017; Sheridan et al., 2014).

In the maker pedagogy, the teacher is first and foremost a guide or a facilitator, with the main focus being to accompany students in a culture of collaboration and creativity (Gerstein, 2016). The role of a maker teacher has been envisioned in numerous NB ICT-related initiatives since 2000, but until recently, neither fully understood nor implemented in everyday practice (Freiman, 2020). While gaining traction in New Brunswick over the last two decades, elements of maker pedagogy are not new and can be retraced in history. The works of John Dewey (1916), Jean Piaget (1954), Lev Vygotsky and Michael Cole (1978), Seymour Papert (1980), and Jean Lave and Etienne Wenger (1991) have influenced instructional philosophies similar to making including STE(A)M, constructionism, project-based and problem-based learning (Lingley, 2021; Litts, 2015). Martin (2015) cites Montessori's (1912/1967) ideas of bringing children to learn by building with a variety of tools and materials. The tools have evolved, but the big idea is the same. In fact, Niederhauser and Schrum (2016) consider "making" as a pedagogical orientation with its main focus on "integrating creativity and imagination with design and encourages problem-finding in addition to problem-solving" (p. 359).

Several possible learning benefits of this pedagogy are mentioned in the literature. Interdisciplinarity is often found to be one of the main advantages of school makerspaces that have a particular focus on STEAM skills (Litts, 2015). Sheridan et al. (2014) argue that the work in makerspaces fosters students' autonomy and collaboration. Other elements mentioned in the literature are the development of critical thinking and argumentative skills (Litts, 2015), the increase in the capacity to do problem-solving, as well as the use of gaming in learning, which can make learning more cooperative (Williamson, 2015). Overall, the researchers seem to agree that when working on their projects, students explore different possibilities for their future career choices (Litts, 2015), while becoming active members of their learning community (Sheridan et al., 2014). Moreover, this experience seems to engage and motivate young learners because they do work according to their personal interests (Litts, 2015). Some researchers have also observed increased perseverance and self-esteem among the learners (Blikstein & Krannich, 2013).

12.2.2 Design Thinking Stages Within Making Activities

Fasso and Knight (2020) consider design thinking to be "at the heart of makerspaces", as an approach being "situated within the genuine practice of designers, technologists, engineers and STEM professionals" while promoting working in teams, social engagement, and collaboration. In the context of technology integration ("low-tech/high-tech"), makerspaces seem to increase capacity of designing, prototyping, and evaluating of artifacts (Fasso & Knight, 2020; Lingley, 2021). Moreover, according to Hatzigianni et al. (2021), design thinking, in the context of making, can "scaffold children's experimentation and problem-solving skills" while helping them to "explore and build contextual knowledge which is then applied in creating useful, practical objects directly connected to their everyday world." (p. 2).

There are several cyclical models to study design thinking. For instance, the IDEO-model consists of discovery, interpretation, ideation, experimentation and evolution stages (idem.). A similar approach was used by Forbes et al. (2021) to study STEM learning in 3D-technology enhanced makerspaces. A slightly different five-stages model has been used by Hughes et al. (2019) consisting of an ask-imagine-plan-create-improve-ask cycle (Fig. 12.1).

In our study, the model used by the teacher was inspired by the i*Cubed* framework *Investigate! Invent! Innovate!* (The Learning partnership, 2015, Fig. 12.2) which considers a more complex process beginning from identifying the problem, towards conducting research, designing, building, testing & evaluating, redesigning, (eventually marketing), and sharing the invention. This approach focuses on finding

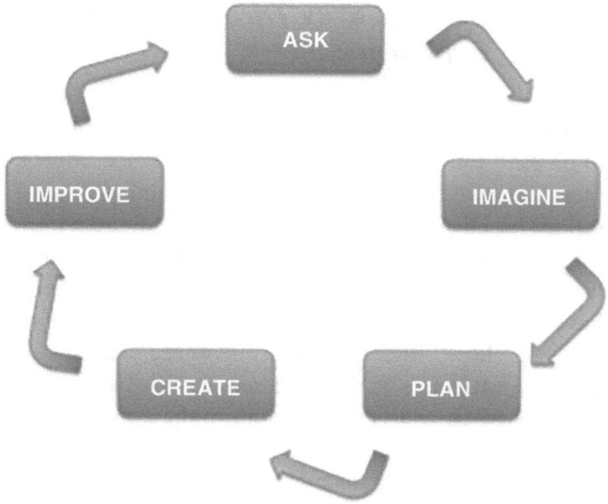

Fig. 12.1 A design-thinking cycle (Hughes et al., 2019, p. 347)

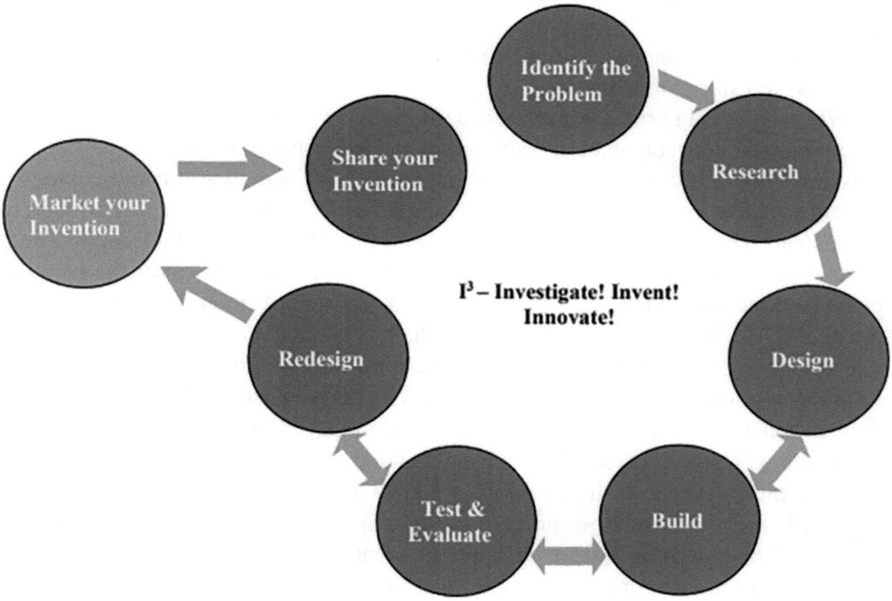

Fig. 12.2 I^3- *Investigate! Invent! Innovate!* Framework. (The Learning Partnership, 2015)

and solving a problem in our daily lives while reinforcing interest in STEM career fields, getting familiar with cutting-edge technologies (robotics, computer programming, apps development, microcontroller boards, 3D printing etc.), and exploring maker culture.

12.2.3 Using Activity Theory to Analyze the Making Experience

In a growing field of research of making and makerspaces, Activity Theory appears to be used for different purposes. For instance, Mersand (2020) used it to conduct a literature review by revealing, among other issues, the "underrepresentation of school-based makerspaces and maker-style activities" (p. 183). Walan (2021) has focused her analysis on stimulating interest of young girls in STEM and developing twenty-first-century skills when combining making and drama. Keune et al. (2015) have examined, from the socio-culture perspective and the construct of "boundary crossing" how "interdisciplinary learning and co-construction were communicated through artifacts" (p. 2). There were also several studies that focused on micro-tensions when analyzing learning in the context of makerspaces. A doctoral study conducted by Nelson (2019) has investigated promises and micro-tensions of implementing makerspaces in a high school. Chu et al. (2017) have revealed that implementing makerspaces in a formal school context could potentially lead to several micro-tensions related, among the other, to curricula pressure. Kajamaa and Kumpulainen (2019) however, seemed to acknowledge the positive influence to learning some micro-tensions from making may have on the emergence of students' transformative agency.

While reviewing how other authors used models of activity theory to understand the interaction between objects, community, and subjects in their studies, we became reassured that we could use a similar model to bring simplicity to the entangled system of activities we observed during a single visit. For instance, it was difficult to understand the influence of one micro-tension from Bobby and Zack across other student groups in the class. Many students were using the same tools, had similar conversations and since they all go to the same school, their inventions, while drastically different in concept, were to help a similar demographic. However, when zooming-in on how Bobby and Zack were directly influenced by a moment of micro-tension, even though we were able to better understand the direct influence a micro-tension had on the students' activity, we realized that these influences were entangled with activity systems within and beyond their learning environments. This observation created a desire to adopt a CHAT-based discourse that situates any observed micro-tension from an activity system along a continuum of entangled systems. This methodological decision to create our own interpretation of the activity theory framework led us to discover the influence of several micro-tensions observed within the makerspace – specifically their influence on Bobby and Zack's learning.

12.3 Method

12.3.1 Case Description: Design Thinking Class

The activity of each maker in the space, the materials they choose and the influence their construction will have on their community are incredibly diverse. For an observer of the activity of students in a makerspace it is difficult to identify the nature of how (built-in-the-process-of-making) young makers are mobilizing to enact the next iteration of their prototype. In the context of STEM related pedagogies, the boundaries of intra-, inter-, and transdisciplinary knowledge students create are difficult to identify and any attempt to try to classify them contains a risk of missing or misinterpreting some rich, original idea that inspired that stage of student activity.

The case we report was observed as part of a Grade 6 classroom dedicated to Design Thinking at a middle school in an urban center of New Brunswick, Canada. This Middle School (Grades 6–8) has approximately 750 students Grades 6–8 in three instructional programs; Early French Immersion (Grade 3 entry), Late French Immersion (Grade 6 entry) and English Prime (all classes except for Post Intensive French are instructed in English). Over the Grade 6 school year, four themed classes based on groups' rotation, have been organized namely, Art, Technology, Music, and Design Thinking.

This allowed each of the Grade 6 groups to spend fifteen 45-minute sessions in a learning environment that is located nearby the Brilliant Labs Facilities (BLF). A so-called Brilliant Classroom is in an area of a school that is being purposefully redesigned to become more a collaborative space for making. When students are in their Design Thinking class, they are encouraged to explore the materials offered both in the Brilliant Classroom and BLF. In the classroom itself, there is a wall of tools and consumables (glue guns, cardboard cutters, calipers, rulers, glue, tape, etc.), a set of 6 laptops, a station for gluing and a cart of cardboard. The classroom has been designed to encourage collaboration. Each of the tables accommodates 5 students, with one student taking turns sitting on a flip top desk on wheels. On average, there are between 24 and 29 students in the class at any one time with one teacher who has been teaching the evolving Design Thinking class for the last 5 years.

The Design Thinking class is unique to this school. This offering is in alignment with the School Improvement Goal of creating a culture of instruction that instills a sense of appreciation for inquiry, critical thinking, making, student voice, and community service. The administration and school shared leadership team has been focused on this goal for quite some time with similar instructional visions at each grade level. In Grade 6, all students cycle through this Design Thinking class, in Grade 7 they all produce a Science Fair project and in Grade 8 they are all part of an end-of-the-year Marketplace where each student is an entrepreneur, showcasing their final product. The Science Fair and Marketplace initiatives were in place far longer than the recent addition of the Design Thinking class. This addition completed the school leadership's goal of experiential learning initiatives at each Grade level. The achievement of this goal is the first of many aspirational goals whereby the instructional vision of the school's leadership is one of interdisciplinarity, making and global awareness.

Brilliant Labs has supported the Grade 6 initiative since it began 5 years ago under a different name of *iCubed. ICubed*, I^3 or *Investigate! Invent! Innovate!* is a program offered by the Canadian charity The Learning Partnership. Once registered with this program, teachers are provided Professional Learning and a package of instructional resources for 18 lessons. It is the goal of this program to have students invent a solution for a problem that they observe in everyday life. Initially, Ms. Larsen (a fictive name) ran the *iCubed* program in her classroom, in the Grade 6 wing of the school as an extracurricular addition to her teaching assignment. Ms. Larsen continued to offer *iCubed* in her homeroom for 2 years at which time the school leadership team realized the instructional opportunity to offer this class to all Grade 6 students and Brilliant Labs installed their BLF to an old automotive garage in the school. As BLF evolved, the school leadership began to collaborate with Brilliant Labs to enact their plan of transforming the adjacent learning environments to become more collaborative and flexible. It was in Ms. Larsen's third year of offering *iCubed* module that they moved the physical location of the class to the Brilliant Classroom, a shared space between the school and Brilliant Labs.

Between years 3 and 5, Ms. Larsen has significantly changed the structure of the *iCubed* class, towards design thinking approach by refining the lessons content in a such a way that students had more time to go through different stages of design of their invention, beginning with identifying and exploring their problem (analysis), producing ideas of solution (ideation), and prototyping and testing. All students are told during their first class that the goal of this course is to develop a physical prototype for a solution to a problem that they see in their community. During project ideation, students work extensively with Ms. Larsen to refine their ambitious initial ideas as they often include complex materials and processes that would take years to accomplish. Therefore, students are encouraged to consider problems that can be addressed in the limited time of the 15 one-hour periods. Student learning is primarily supported by Ms. Larsen's guidance, however, given the proximity of BLF, students can often get support provided by Brilliant Labs' staff members.

With regards to material, after students complete an invention utility survey with their peers, students are required to complete a detailed drawing of their invention before creating an initial prototype. Many initial prototypes use repurposed materials like cardboard to promote thoughts of sustainability and material accessibility. For those students who are interested in integrating other materials like 3D printing and physical computing, they are encouraged to remain mindful of their limited time and to first ensure that they complete a non-functioning model with repurposed materials. This often creates a micro-tension that evolves into opportunistic discourse between the role of fostering creativity while remaining mindful of the effect material choice has on time for students to complete an iteration of their idea. During the early sessions of the class, Ms. Larsen uses her experience not to dissuade students from creatively exploring the affordances one material may have over another in their project, but to inform students as to the entire process required to use a material like 3D printing, physical computing, and cardboard construction. When investigating a group of Grade 6 students going through a design thinking cycle while working on the prototypes of their invention, we focused on their learning experiences, efforts they were putting into producing their ideas, creating, and testing prototypes, and sharing their solutions with other members of their learning community.

12.3.2 Participants

The participants were 15 students who had their consent forms signed, according to the ethical requirements. The students were working in teams of 2 or individually in their projects. There were 9 projects simultaneously developed during the design thinking cycle by students who have permission to participate (at least one student per team). After making sure we do not collect data on projects conducted by students who did not have forms signed and briefly talking to students who had their teammates not participating in the research, we decided to focus on four teams trying to get a deeper knowledge about their projects and learning experience.

12.3.3 Data Collection

Over the fifteen, one-hour long sessions devoted to the design-thinking module, spread out across a 1.5-month period (May–June), we decided to use periods 5, 10, and 15 for data collection. The first observation period was used to obtain a general perspective on what was going on in the makerspace. We also were working on clarifying which of the projects we were permitted to film. Careful consideration was paid to the decision to use video-recordings as a primary method of data collection. There was some fear that the presence of recording devices would disrupt the students' process. To address issues of confidence produced by data collection via video-recordings, we relied on what we could hear and what we could see throughout the activity system of the classroom. This helped to alleviate some of the limitations presented by the invasive nature of recording equipment. During the second and third observation period, we used two cameras, one focusing on two teams (4 students) working around one table and another used for impromptu observations when the researcher was moving from one project to the other sometimes asking questions about what students were doing.

The last time we observed this class was an extra period where students were preparing to present their projects to their class. No video-recordings were taking place on that last day. In addition to video-observations, we decided to talk to the students from our focus groups before the observation period asking them about what was done so far and what were their plans for the coming period and overall, how the project was progressing. The last interview (with no observation) included general questions about students experience with making. We also did an interview with the teacher about the project, students' successes, and difficulties and about teaching and learning in the makerspace.

12.3.4 Analysis: Focus on Bobby and Zack

As we mentioned at the very beginning of our chapter, during our 4 visits, we realized that one of the groups was clearly struggling to get through different stages of the design process. During our fist observation, when we decided to choose this group as a focus group, we did not see any particular signs of merits telling us why this groups should be studied more deeply than the others, this was more of an intuition. Yet already the first interview prior to the second observation period provided some data of particular interest regarding this group of students. They shared with us that this was not their first idea to invent. At some point, they decided to switch towards a different idea; however, it came at a cost of some delays with their project. During the observation lesson, as we will see below, students spent a considerable time designing a project nametag, using *Tinkercad (www.tinkercad.com)*, on which they placed the name of their project which they called *Frankenstein Pencil*. This was also to us a crucial moment where the students and us as observers encountered micro-tensions. Micro-tensions included those between students and their project delays, students and their design choices and micro-tensions between what we thought students were learning and what they actually expressed during their design thinking cycle.

After spending some amount of time designing a 3D representation of their logo, Bobby and Zack explored a variety of shapes provided by the software. The accessibility provided by the click, drag-and-drop interface seemed to give permission for students to play in a risk-free environment whereby they were able to virtually construct an unlimited number of new shapes. As they became more familiar with the functionality of *Tinkercad*, the first iteration of their complex geometric construction began to take shape. Their design consisted of two concentric cylinders, one inside of the other. Once we fully understood the narrative behind idea, it was clear that the interior cylinder would become a hole in which to accept a solid object once brought into the physical world via 3D printing. This task was met with numerous cycles of ideation where Bobby and Zack attempted to ensure that their measurements were correct as to allow the diameter of a pencil to fit inside the cylinder. This complex process was also identified as critical, for our analysis as it revealed numerous micro-tensions related to constructing and measuring. For example, one micro-tension was observed just before our second observation period, where these students expressed their frustrations as they discovered a conflict between their initial measurements and what they saw on the *Tinkercad* interface.

At this time, in addition to needing to (re)design their virtual prototype, they were trying to add new functional components. Protruding into the interior cylinder of their construction were a series of cones that they referred to as *spikey thingies*. After watching these students confer between their hand drawn diagrams and the geometric shapes on their computer screen, we understood the *spikey thingies* to be the structures that would hold the parts of the broken pencil. Just as Dr. Frankenstein brought his creation to life in the classic story of *Frankenstein*, these students would use these *spikey thingies* to reanimate their broken pencils. As inspiring as it was to observe a fascination of literature become entangled with their invention, this moment was not without the observance of several micro-tensions. Perhaps mediated by the imminent project presentations, most of our conversations with these students before their presentation occurred through their own anguish. During the interview, these students acknowledged that the invention they were trying to prototype was too difficult. Even though they insinuated that they were close to making a physical prototype at some point during their project, to our knowledge, they never brought their ideas into a physical space via 3D printing. Regardless of their unresolved issues with their design, students persevered as they continued to reflect on the aspects of the project that made it too difficult to design – making incremental changes in an effort to finish their project. It was this emotional connection to a project that was never brought into a physical space that inspired us to further analyze trying to identify the influence of these moments of micro-tension had on their learning.

12.4 Results

To understand the results of this case, the question "*What were they trying to make?*" is important – as the most apparent answer to this question is that Bobby and Zack did not make anything in a sense of creation of some tangible artifact. This type of answer however is confined to a physical space: a physical construction, something you can hold, something that is a prototype of an idea. This is a rather obvious expectation as at the outset of the very first Design Thinking Class, it was made explicitly clear that the goal was to make something. After 15 classes, Bobby and Zack were however unable to make any physical device representing their invention. While it may appear to an external observer that these two students failed in the construction-based goal of the class, a deeper analysis of interview transcripts and a frame-by-frame analysis of video segments captured during classroom observation visits demonstrate that Bobby and Zack have indeed made something that makes their learning visible. The something that they made however, did not occupy physical space. Bobby and Zack's representation of their idea exists as an entangled material experience with artifacts that exist between virtual and physical planes of existence. For instance, as part of their required classwork, Bobby and Zack illustrated a detailed drawing using pencil and paper. This occurred in physical space. However, in the virtual space of *Tinker*cad, Bobby and Zack extended what they had illustrated on paper and constructed a three-dimensional representation of their ideas. It is this transdimensional space that we believe was a major source of tension, filled with multiple instances of micro-tension, that resulted in Bobby and Zach's frustration – an observation that ultimately inspired the detailed analysis below. Situating this observation within the CHAT, we argue that as these students attempted to resolve their design issues, they expressed new knowledge and transformed their learning experience because of the micro-tensions that emerged from their discourse.

As mentioned above, students had some issues with their initial project idea, so they decided to switch to a different idea of invention. Namely, during the first conversation (prior to the second observation visit, session #10), Bobby shared the following:

> *Bobby: Maybe we should tell them (Interviewer1; Interviewer2) about that first thing we started with. We wanted to do a pocket projector, and you need a lot of lights for it, and you just put a picture, and then all the light, if it's dark, will make the projector and just in your pocket.*
>
> Interviewer1: *Why did you decide to switch to a different idea.*
>
> *Bobby: Yeah, we realized we probably wouldn't be getting a light source that would be bright enough. We needed something that would help everybody with, but it was a bit hard. It wasn't really enough (light), so we switched to the pencil one.*

From this quote, we learn about Bobby's sensitivity and empathy to the task of making something "that would help everybody with." It appears Bobby's empathetic goal directed their imagination towards the act of creating some kind of device that will bring pencils back to life. This reanimation device ("*if you like break it, you have this little thing that puts it back together*"), that would restore functionality to broken pencils was affectionately named *Frankenstein Pencil*. While the above interview excerpt revealed that Bobby and Zack struggled with the idea of the influence of switching ideas would have on their project, the struggle produced this empathetic and personalized opportunity. *Frankenstein Pencil* seemed to combine their two passions, one for pencils and another one for their interpretation of Mary Shelley's 1818 story of *Frankenstein*:

> *Bobby: We just trying to figure out what kind of name, (pointing to his colleague): He started it, and then we finished it off.*
>
> ...
>
> *Zack: We thought of a name, the Frankenstein Pencil. Because Frankenstein comes back from the dead.*

Students also explained why such a device (students called it "little thing") would be useful (from the second pre-observation interview):

> Interviewer2*: What is the idea behind reassembling pencils?*
> Bobby*: Well, it would be useful.*
> Zack*: We have broken a lot of pencils.*
> Bobby*: So, you have to replace the pencils all the time.*
> Zack*: I like pencils.*

These excerpts imply that Bobby and Zack at least thought of two metaphors involving the idea of reanimation. The first related to their emotion of 'feeling sorry' for a 'broken pencil' that still had potential for use, and the second related to their passion for the story of *Frankenstein*. It was these metaphors that seemed to inspire their design trajectory:

Interviewer1: What was the title of your project to refresh my memory?
Bobby: *Frankenstein Pencil.*
Interviewer2: Why *Frankenstein Pencil*?
Zack: *Because it comes back from the dead.*
Bobby: *It comes back to life, so we're basically bringing a pencil back to life.*
Interviewer2: So, you had this idea from some experience?
Bobby: *I knew about Frankenstein for a long time, but I just heard about it again and I just thought about it for a few weeks.*

Bobby and Zack's special (even empathetic) attachment to the project was also identifiable during their work on the computer-aided design of their device. They had spent a considerable amount of time designing a *logo* for their project (Fig. 12.3), time that could be argued would have permitted them to bring a version of their virtual design into a physical world.

We learn from these fragments of data that beyond Bobby and Zack's concerns about the choice of their invention, their imagination is simultaneously fed by the task (inventing something that helps to solve a real problem) and their personal attachment. It is also grounded in their own real (pencils that get broken) and imaginary (inspired by a story) experience. Bobby and Zack continued to express of their personal connection to the project during the last interview, when they answered the question 'What was the source of your inspiration? Zack said again: *I like pencils*. In addition, at some point in our observations Bobby and Zack shared with us that even the name of their *Tinkercad* account was also associated with *Frankenstein*.

When digging deeper into our interview data and several videos that captured both students working on their *Tinkercad* design, we tried to get a sense of what they

Fig. 12.3 Students designing a logo for their project using *Tinkercad*

were actually designing beyond an idea. With the absence of physical artifact, it was interesting to attempt to glean any insight into how Bobby and Zack were intending to design functional elements that would ultimately become a testament to the *Frankenstein Pencil* name. Indeed, prior to designing their prototype, students had to do some preparatory work researching the topic and thinking about how to approach their invention (brainstorming). They were also asked to complete some pages in the booklet explaining their idea and making a rough draft of how they believed their idea could be actualized. This was followed by a detailed drawing from different perspectives. It was here that Bobby and Zack chose to include a magnified drawing of different components of their device, noting their necessary measurements.

By reflecting and refining the ideas that inspired their *Frankenstein Pencil* project, Bobby and Zack were able to complete the next stage of the design thinking process, prototyping, with intricate detail. Students were asked to provide a description of the project and make a detailed drawing. This is where the ingenuity of Bobby and Zack's imagination of their invention and the complexity of its technological components became entangled and resulted in us identifying several micro-tensions.

As we began to examine the evidence, we used Zack's comment about their design: "*We wanted to bring broken pencils back to life*" as the context in which we situated our analysis. Perhaps there would be evidence of a mechanism beyond what Zack implied to be the end goal of their device. Since Bobby and Zack have never 3D-printed their prototype, what still remains unclear is if their idea would function as intended. It was clear that they intended for their device to function in such a way that two unusable pieces of pencil would be fused together to live another life as a functioning pencil. Or their device would be a kind of prosthetic device helping to hold the broken pieces. Regrettably, none of our observations in isolation contain sufficient details on the entirety of the students' design process for their *Frankenstein Pencil* prosthesis. Similar to the *Frankenstein* literary metaphor throughout Bobby and Zack's project, we were like Dr. Frankenstein. Just as Dr. Frankenstein had to piece fragments of anatomical structures together, we too had to rely on fragments of data. Once assembled the data became a narrative interpretation for a construction that only functioned in Bobby and Zack's imagination. One such fragments of data was a sketch of their device in their project booklet (page on the left) and detailed drawing (page Detailed Drawing on the right) (Fig. 12.4).

While their sketch is not very clear (a screenshot from the video is rather blurred), it shows a cylindric shape. Their detailed drawing shows a more complex configuration of their device as being composed of two pieces; both look like cylinders. Different views of the first piece show that the cylinder is 101 mm long (front view) and 5.25 mm in diameter (top view). From the top and the bottom, it looks like there are two concentric circles. The way Bobby and Zack have used a dark circle in the top view led us to the interpretation that the outermost cylinder has a circular hollow at the center. The drawing in the right upper corner includes handwriting that says 'x-ray.' It appears that this shows the inside of the device and supports their detailed drawing of the *first* piece whereby their device consists of an interior cylinder with a

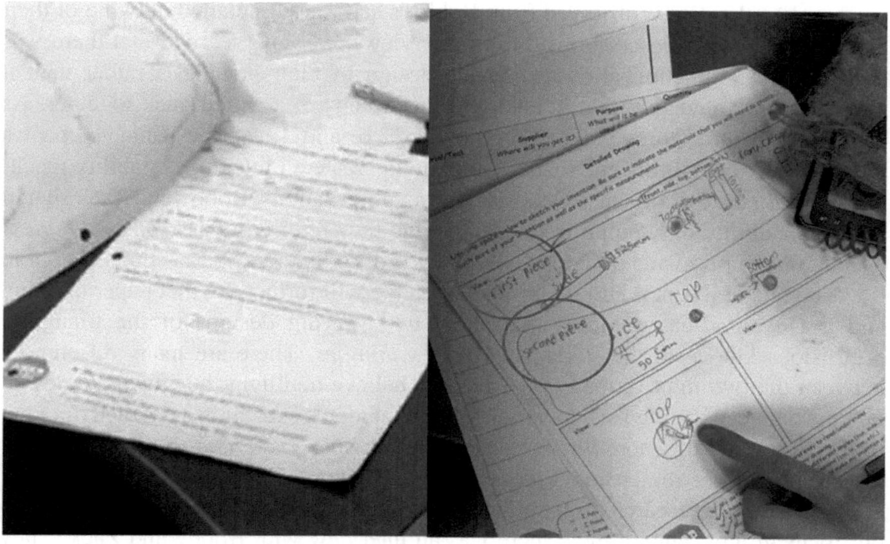

Fig. 12.4 Students' sketch (on the left) and detailed drawing (on the right) of their invention

length of half of the first piece. The 'x-ray' view also shows what appears to be a solid structure at the center which could be a point where two broken pieces of a pencil would meet once inserted into the device. We will see below that this construction was never clearly reproduced in *Tinkercad* as a digital prototype. Unfortunately, we were never able to seek more clarification as to the function of certain structures due to the limitations of our data collection method.

Working within the limited data that we were able to collect, it was plausible that the invisible cylinder of the first piece (shown on the 'x-ray') is actually the second piece which is also drawn in a shape of a cylinder but only half of the length of the first piece (50.5 mm). From the top view and the bottom view, it looks the same as the students' only use a dark circle on the first piece which we interpreted as the hollow. With this, it appears that the second piece is to be placed inside of the first piece. It is however unclear why the bottom view has additional details, namely a dimension of 9 mm long and an indication (arrow) of some dimension that is equal to 3.42 (mm). Moreover, there is another part of the drawing presented only as a top view that shows a circle with the diameter of the first piece, 5.25 mm, and three other 'shapes' which look like 'teeth'. Would this mean that inside of the second piece, there would be some *spikey thingies*? As said before we were not able to check it with Bobby and Zack as they were too busy during their work so we could not interrupt them particularly since these observations were occurring late in June towards the final days of the school year. Overall, the whole device (two pieces) has the appearance of a very complex composition of shapes which once combined may have functioned as intended, at least to some degree.

Based on the details observed from Bobby's and Zack's detailed drawing of their invention and their descriptions during interviews, we have learned that there was more complex function attributed to the design of their prosthesis rather than it simply being an adapter that holds two broken ends of a pencil back together. As illustrated in the figures below (Figs. 12.5, 12.6, and 12.7), the students spent a considerable amount of time discussing how to create an integrated stability mechanism inside the cylindrical prosthesis. During observations, Bobby and Zack shared the responsibility of attempting to design what they referred to as "*spikey thingies*" ('teeth' on their drawing above) to act as a type of rudimentary screw threads to hold the two ends of a pencil into their prosthesis. We will further see that paper drawing and a *Tinkercad*-based model that show two varying designs of the triangular protrusions known to the students as *spikey thingies*. There are many differences between the two figures; many of which we believe highlight the entangled design process between the physical experience of the students sketching a detailed drawing of their idea and the more virtual experience of using *Tinkercad* to transform their sketch into a three-dimensional model.

Following our research protocol, we began our second observation visit talking to our focus groups, among them there was an interview with Bobby and Zack. They shared with us the state of their process which they identified as "*we're still designing*." In speaking specifically about their work of a digital prototype the said:

Bobby: *[W]e're still designing, and we'll have to 3D print the first prototype, and hopefully that works because we're really on a tight schedule. If the first few prototypes don't work, we'll be mad.*
Iinterviewer3: Is it stressful?
Zack: *Yeah.*
Bobby: *So, we're going to have to work really hard on the design, so we don't have to redesign. I think we will make one or two (prototypes).*

From the video screenshots below, we witness students learning to make a cylinder with the hole which is plausibly connected to the first piece of their drawing. While the dialogue above demonstrates that Bobby and Zack were frustrated about their "tight schedule" and the need to produce their first prototypes, the video screenshots below demonstrate that there were areas in which the students should have been proud. For instance, Bobby and Zack were quite skilled at manipulating the software being able to draw, replicate and then scale the cylinder to transform it into a negative volume in which a pencil could be inserted. According to Bobby and Zack, it was this process that appeared to be difficult (Fig. 12.5).

While Fig. 12.5 demonstrates multiple iterations of one component of Bobby and Zack's design, there are still micro-tensions left unresolved with regards to the question: *What did they actually make?* Up until these moments captured at the end of our second of three observations, there was only limited evidence of an initial prototype.

There was a week in between the second and third visit. We witnessed more struggle as Bobby and Zack continued to make progress towards the completion of their *Frankenstein Pencil*. The excerpt below will present more moments of tension as Bobby and Zack began to see many of their ideas falling apart:

	Students first have drawn a cylinder, then made its diameter equals ‘five’.
	Students have replicated the first cylinder making the diameter of the second one equal to ‘four’ (supposed-to-be-inner cylinder).
	Students have tried to insert the second cylinder inside of the first one but bringing them closer to each other.
	By composing a new shape, students investigate how to make sure the cylinder inside the other one makes a hole. The task does not appear to be easy; they try to re-position their construction to get a better sense of what is going on. Their work draws attention of one of their peers.
	Few minutes later students still struggling with inner-outer cylinders as they make effort to align the shape; the situation draws teacher’s attention. So, the teacher takes control of students’ laptop showing some tricks on how to make an alignment (*“It’s awesome”*) while mentioning that *“there is still a couple of problems*” whereas it is necessary to make a right selection of parts to be aligned and axes for a proper alignment; she completes the work asking, *“is that better?”* Happy with the results, students started building up their logo.

Fig. 12.5 Students’ design of a cylinder with a hole

Fig. 12.6 Students work simultaneously on designing two shapes

Fig. 12.7 Students' tinkering process when trying to adjust measurements to their initial drawing

4:30	P2	I'm going to move it (referring to blue prism)
4:31	P1	Nnnnnnn nnnnnn no. I have to measure it.
4:34	P2	I put it back I put it back. There, is this good? Is this good? Look.
4:43	P1	(Leaning over P2's computer to consult the booklet – detailed drawing) We've got to do the...
4:45	P2	Side. but which side?
4:47	P1	Yeah. No. The bottom. 43.92 (spoken: forty three point ninety two)
4:48	P2	I did that. wait
4:52	P2	(With mouse cursor on bottom middle anchor tool - used to change base diameter of cylinder) 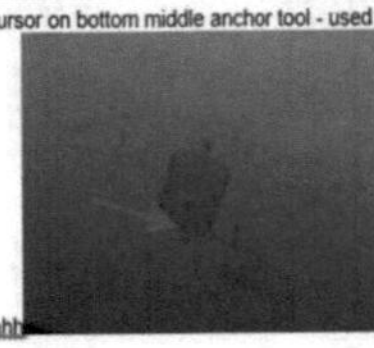 This one? Ohhhh
4:56	P1	It has to be 50.5 (spoken: fifty point five)
4:56	P2	I'm just going to erase it. (deletes cylinder)
5:01	P1	(difficult to hear) Ok whatever. Here, you do the triangle.
5:03	P2	Yes. (Stands up to switch computers so presuming that P1 wants him to work at his computer.
5:10	P1	No no no, you can do it on yours. (It appears that P1 was just standing up to get a better look at the detailed drawing that was on P2's side of the table. P2 interpreted this movement to mean that they were going to switch computers. Perhaps this shows that P2 doesn't know that their that their cloud-based design is updating across the computers in real-time. 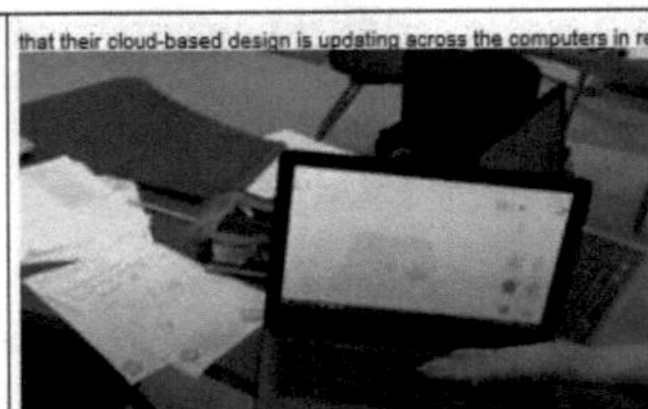
5:20	P1	Five point three five. (Pointing at one of the triangles in detailed (top) diagram that protrudes towards middle of circle representing the cylinder) Way over here then, no no, I can't read your writing (perhaps this means that P1 was the artist) (P2 does sit down at P1's computer. P1 continues to examine detailed drawing). 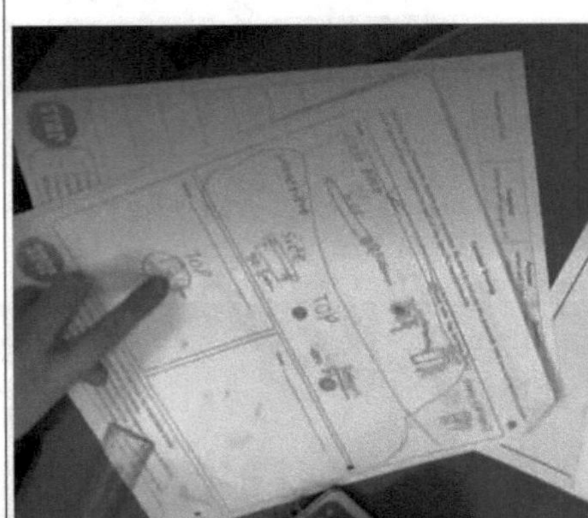 Also... at 5:28 - you see that students used text labels in their drawing to indicate First and Second piece.

Interviewer1: You are trying hard to get through this process, right? What do you do to make it happen, to make progress?

Zack: *Keep going, don't take big breaks or anything because that doesn't work well. We had time to work on it, if we don't design it, then that's not good. I just realized a second ago that we did the measurements all wrong.*

Interviewer2: Oh really? Just while you were sitting here? What's wrong with the measurements?

Zack: *Well, we thought that we were doing the first piece, which already has the grooves, and then you slip it in, but we couldn't put the spikey thingies to make the grooves on the piece that's supposed to put the pencil back together.*

In this fragment of the interview, students share their struggle with the design when trying to move towards a functioning mechanism to hold the pencil pieces in place. It appears that their struggle involved the process to "put *spikey thingies* to make the grooves" along the edge of the broken pencil. Their struggle is compounded by the fact that it appears as though they found some problems with measurements that they had taken towards the beginning of the project. We also learn some important details about their prototype which, we recall was composed of two pieces: the first piece was supposed to "already" have the grooves. This piece was planned to be slipped "in" (into the second piece) but students were having difficulties placing "the *spikey thingies*" (we called 'teeth' above) in the design. Moreover, no matter the piece to which they were referring in this interview, it seemed it was critical as it was supposed to "put the pencil back together". At the end of our conversation, we asked students any last-minute design changes they were they were going to attempt:

Interviewer2: So, talk to us about today? What do you hope to do this period?

Zack: *Finish it.*

Interviewer2: Finish it? What does that entail? What do you have to do to finish it?

Bobby: *First of all, either get rid of parts or just restart all the measurements. I think our project was a little too hard right now because we have to make it so that the spikes make the grooves, and the grooves have to fit with the part we want it to fit and then we have to put it inside of the thing. So that's probably a bit too hard and we don't know what we're going to do.*

While the exact meaning of the "first piece" remains a mystery for us, it was clear that Bobby and Zack's last attempt to finish their prototype was to fix the issue with the measurements and with "*spikey thingies*" to "make the grooves." Students seemed to confirm both intentions in the following part of the interview yet recognizing that the task appears to be harder than they thought and that they actually "don't know what we're going to do". In this last statement, there seems to be a micro-tension between their original plan (detailed drawing) and how to keep moving forward trough the design cycle and create a physical prototype.

Nevertheless, students seemingly began to restart their project trying to re-design their prototype by fixing both issues: adjusting measurements and inserting *spikey thingies*. They also decided to work simultaneously on two laptops using a shared collaborative space via cloud computing within *Tinkercad*. As a result, a new type of micro-tension appeared, this time with regards to how they would synchronize collaborative work. The first moment we noticed this micro-tension was as students began to work with two shapes, a cylinder (Zack) and a prism (Bobby) (Fig. 12.6).

At some point, Bobby also wanted to do something with the cylinder, yet, at this moment, perhaps due to a communication error between the server and user, Zack's session was terminated, and he had to start his design all over again. There was also a moment where Bobby and Zack started to tinker with measurements and shapes. We assume that this was an effort to resolve the issues they mentioned in the previous interview. In the video, they often referred to their paper drawing, pointing out details of their device including specific measurements. Even though this observation was very short, we were able to capture an intense discourse that illustrates the intricacy of their work (P1 = Zack; P2 = Bobby) (Fig. 12.7).

From this excerpt of their discourse, we see several references to the detailed drawing and the particular attention they paid to the measurements. As they conversed, Bobby and Zack changed places in an attempt to coordinate their actions both within their physical learning environment and the virtual *Tinkercad* design environment. This coordination seemed to be a response to their suggestion that achieving the functionality intended for the *Frankenstein* Pencil was a challenging task.

In the next video sequence, which was only 3 minutes and 41 seconds long, we see students continuing working with a triangular prism while simultaneously being seemingly unhappy with their progress. Perhaps out of frustration, at one point, Zack starts to make random movements with the mouse without any clear plan. Since they are working collaboratively across a shared instance of *Tinkercad* via the internet, Bobby sees Zack's random manipulation of what we believe was to become *spikey thingies* on his screen and looks somewhat unhappy, asking: "*What are you doing?*"

As result of this mounting tension, Bobby seemed to take the lead; the role he kept untill the end of this session. Zack was sitting aside. At some point, the situation seems to become more tense with Bobby becoming nervous (tapping with force on buttons of his laptop) and his teammate looking a bit exhausted. Nevertheless, perhaps mediated by their teacher, Ms. Larsen, the next segment showed students, at least to an external observer, deescalating and being able to reproduce they concentric cylinders correctly aligned on their screen – they had finally reproduced the same result we had seen in the last observation. Later in the observation, they also continued working on the triangular prism trying to make it sharper as to secure a piece of a pencil inside the cylinder. This is where their teacher, Ms. Larsen attempted to facilitate a resolution, first by giving some hints sitting with Bobby, then calling both students to an interactive board to discuss a way in which to combine the cylinder with the triangular prism-like spikes (Fig. 12.8).

While the conversation between the Ms. Larsen and the students is not fully audible, we believe that Ms. Larsen is explaining to Bobby that the triangular prisms can act as both a solid and a hole inside the cylinder. One of her remarks was heard on the video: "the hole will be sharp too" meaning that she was suggesting that the students consider the result if they use their *spikey thingies* as negative rather than positive volume.

In one of the final segments, we were able to film before Ms. Larsen told students to clean their workstations and wrap up their session, 24 sec long, we still see some signs of discouragement in Bobby and Zack. Bobby seemed to be randomly placing

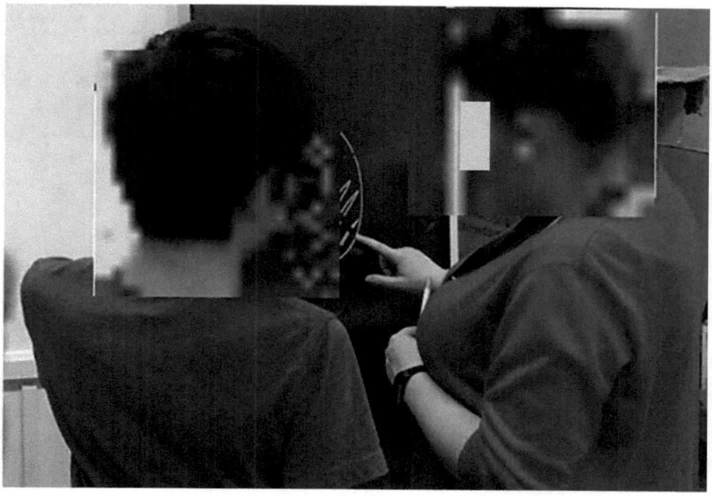

Fig. 12.8 Ms. Larsen gives some hints to the students about how to put spikes

Fig. 12.9 Students' efforts to design a prism to make spikes

black scribbles on the triangular prism to erase the shape, and Zack idly sitting to the side seemingly disconnected from his design work (Fig. 12.9).

It does not seem that Bobby and Zack finished their goal that day. Nevertheless, during the last interview, one of them mentioned "*We actually finished our prototype*" with a strong emphasis on '*actually*.' However, we have no evidence to suggest that their work was completed. There was no culminating event to suggest that their virtual prototype was complete on *Tinkercad* nor was there at least one physical object printed, even though several of their classmates were preparing to present an incomplete version of their own project. The following excerpt from their

Fig. 12.10 Students show the diameter of their cylinder

last interview, Bobby and Zack mentioned again the issue with their measurement which seemed to remain a major obstacle preventing completion (Fig. 12.10).

Interviewer1: Are you any closer? What did you achieve?
Bobby: *(incomprehensible)... we couldn't figure out what to do.*
Zack: *You know how pencils are 6 mm long?*
Interviewer1: Mmm hmmm.
Zack: *We had it measured as 5,*
Interviewer1: Ok. Sorry. Say that again.. So 6...
Bobby: *Student Right: It's actually 5.25.*
Zack: *yeah... (looking confused at student right).*
Interviewer1: And you said long (grabbing for a pen to use as an example).
Zack: *(gesturing to I1 holding a pencil with his fingers at both ends) Just like that.*
Interviewer1: Ok.
Bobby: *(interrupts) I think (gesturing his fingers as scissors around the diameter of the pen) for our pencil, this is 5.25.*
Zack: *(interrupts student right) but a full pencil is 6. (P1 is scratching his chin. There is a pause...).*
Interviewer1: So, the diameter is...
Bobby: *Well. we don't actually know (diameter of the) pencil.*

12.5 Discussion and Conclusions

The data were collected over the fifteen 1-hour long sessions during one and half months of the school year. The focus of our analysis in this chapter was one single project of invention chosen by a group of two students who was attempting to design a device (kind of prosthesis) to connect two parts of the broken pencil. Our findings, while limited to some video-recordings during students' work and a few interviews at different moments of their projects, reveal several micro-tensions we discuss in this section. Concerned with the problem of pencils that easily break down and need to be replaced, on one side, and inspired by the story if *Frankenstein*, students have imagined that building their device would help to give a second life to the pencil ("bringing it back to life"). Yet, students' ingenious idea was never materialized in some kind of tangible prototype while remaining at the stage of computer-aided design using *Tinkercad* digital platform.

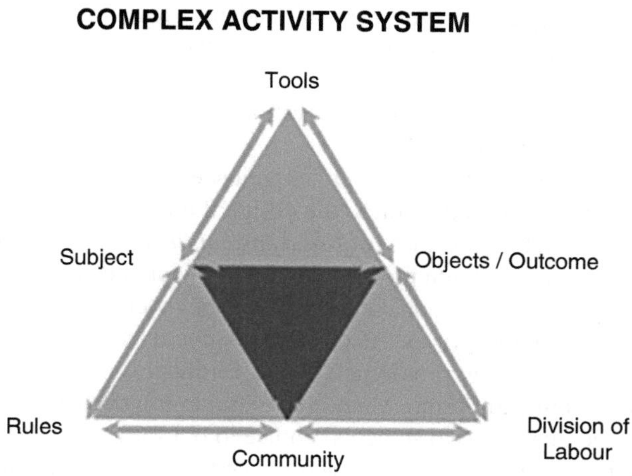

Fig. 12.11 Complex activity system as described by Engeström (1987)

While failure is considered as a "key concept in design thinking" by playing a central role in innovation projects" (von Thienen et al., 2017, p. 1), it is not easy to identify what are the potential gains in students' learning. This is why, among a multitude of projects that were running in parallel in the makerspace, many of them leading to creation of very promising prototypes we were impressed with the determination and intensity of two students' efforts to bring their 'broken pencil' to life, and decided to zoom out their unfinished venture, full of micro-tensions. The lenses of the CHAT turned to be helpful to feed-in our micro-level analysis of students' activity (Leontiev, 1981) within a larger and more complex *activity system* Engeström (1987, Fig. 12.11) which, in our case, consists of two students trying to achieve an *object-goal* of inventing a product/process (*Frankenstein Pencil*) using different types of mediating tools (sketch, detailed drawing, *Tinkercad*, 3D-printer, . . .).

Their work is regulated by *rules* (explicit and implicit regulation and norms, like stages of the design thinking process, safety rules, building something useful), supported by a (learning) *community* (students, their teacher, Brilliant Labs' staff members, . . .), *division of labor* (in our case, tasks, power and authority sharing among two students, or, on the class level, sharing tasks and resources among community members to achieve a common goal). The result of the activity within the activity system (*outcome*) is defined by a combination of intentions (building a capacity to innovate; socio-emotional growth; transdisciplinary learning outcomes; interest in pursuing study of STEM disciplines, . . .). Along with Jurdak (2006), we share an assumption that "our knowledge of the world is mediated by our interaction with it, and thus human behavior and thinking occur within meaningful contexts as people conduct purposeful goal-directed (socially organized, V. F. & J. L.) activities" (p. 286).

Further, a *developmental trajectory* of students' learning, on one side, and dynamic character of activity system consisting in the "construction and resolution of successively evolving tensions or contradictions" (Engeström, 1999) to eventually "reveal opportunities for creative innovations, for new ways of structuring and enacting the activity" (Foot, 2014).

To our knowledge, Grade 6 students' experience in producing a technological invention was their first introduction to the design thinking process. At some point, the students' switched to a new idea (of *Frankenstein Pencil*) and succeeded to move from the first steps of the design process (including producing a detailed drawing of their invention) to an attempt to make a prototype they were supposed to test and to present to the rest of the group. Also, students could choose from using some physical materials for their prototype (like a cardboard) or designing a virtual model that they could 3D-print. Many students have indeed opted for physical materials. The fact that our two students' choice to opt for a computer-aided design appears to add more complexity to their activity system with possible micro-tensions arising around the mediation tool (*Tinkercad*).

In this case-study, we use the term micro-tension to describe moments in which we questioned if what we had just observed had the potential to create conflict within the project, between the students' ideas or the learning environment in general. These opportunities for conflict could have easily gone unnoticed as they were small or micro and it is possible that what we characterized as even a fleeting moment of micro-tension was actually something else. Nonetheless, in a makerspace, amongst the flurry of activities, all interpretations of phenomena, regardless of their place on a continuum between micro- and macro-levels represent opportunities to reflect on observations to understand their influence on the learning environment or adjacent activity systems. From the perspective of the Activity Theory, this influence appeals to the transformative agency not only students have in this space but the agency an educator or an observer experience when they must look past what we may describe as micro-tensions. No matter the size, any observation of tension, struggle, or even disengagement is really an invitation to understand the influence these observations may have on learning across entangled systems of activity (Fig. 12.12).

Our findings suggest several types of micro-tensions which seemed to emerge as students persevered through the design process. Some of them reflect inner tensions within some elements of the activity system (for example, regarding the representation of the object, or division of labor), others appear to be connected to interactions between different elements of students' activity (for example, during mediation process towards a goal achievement). Indeed, the very idea of building a device which connects two parts of a broken pencil is a source in itself of a tension. From one perspective, students seem to be highly motivated and engaged in the design of their idea; from another, they might be overexcited and fixated on the novelty their idea will bring a community of users. No matter the origin of this micro-tension, it left Bobby and Zack struggling to design their device without ever assessing its feasibility, prior to beginning any required design work. It is possible that this micro-tension was exacerbated with their decision to change their project idea after they

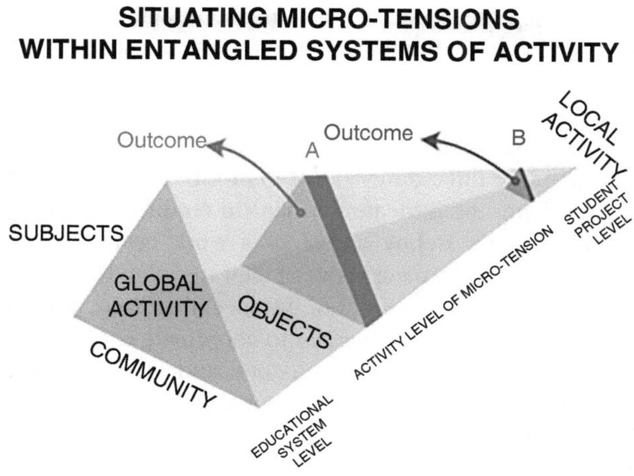

Fig. 12.12 Freiman & Lingley's interpretation of activity system theory to account for entangled activities across a complex system of multiple activities of different scale

had already explored their first. However, neither idea was fully explored before they began their virtual prototype.

Another source of significant tension in this study surrounded the detailed drawing. Even though this drawing contained very few details, Bobby and Zack continued to refer to it as a critical guide in their design. Despite it being an important design guide, it created numerous micro-tensions including questions about the precision of measurements and overall lack of coherent details. Nevertheless, as Bobby and Zack became aware of these tensions, it created an opportunity in which they could express their ideas using complex materials and knowledge. For instance, even with the lack of details, the detailed drawing was a visual reference: anchoring their virtual design within a physical realm. Even though we also found the lack of details of their drawing to limit our ability to interpret meaning, we found even their simple depiction of a triangular prism to be compelling enough to look closer at the video evidence. It took looking at the video evidence frame-by-frame to reveal the complexity of their invention. Upon reviewing the video at this level of analysis, we found that their invention would require lengthy experience with geometric design for their two-dimensional ideas to function correctly once transferred to a three-dimensional platform.

Beyond the technical difficulties Bobby and Zack encountered with *Tinkercad*, they also faced that the complex knowledge of geometry that they required to achieve success in their project was far beyond the scope of the Grade 6 mathematics curriculum. While *Tinkercad* does provide the necessary tools to make this level of knowledge more accessible, there were still several moments that required additional support from Ms. Larsen. This process has sometimes lead Bobby and Zack to become frustrated or at times disengaged because of the disparity between what they wanted to achieve, and their ability mobilize the required knowledge for achievement. While they succeeded in mastering some functions within *Tinkercad*, like the construction of nested cylinders, Bobby and Zack still were not able to resolve the issues of implementing the functional, triangular protrusions or *spikey thingies*.

With regards to the use of *Tinkercad* as a design tool, we believe that it was the collaborative nature of the three-dimensional platform that helped bring Bobby and Zack closer to actualizing the intricate mechanism required for the function of the *Frankenstein Pencil*. While Bobby and Zack's work became more efficient by working concurrently on their project between their two cloud connected computers, it was not without moments of tension. For instance, they were able to capitalize on the virtual design environment of *Tinkercad* to coordinate design tasks, however in doing so, it created a source of confusion. As one student manipulated an object, it not only effected the other's on-screen design but also disrupted what was thought to be a shared vision. As this tool-mediated asynchrony occurred within the activity, a power struggle was created between the two originally united students. This may have been because of many reasons, including an underlying disparity between one student's familiarity with the technology or direction of the project. Regardless of the origin of tension between students, it ultimately led to one student taking the lead and the other looking a bit discouraged, or at least indifferent.

Ms. Larsen's influence on Bobby and Zack's actions also seemed to contribute to their progress towards their design goal. Ms. Larsen, who had to deal with several projects running at the same period of time, in the same space, appeared to be aware of the micro-tension our two students experienced. This was certainly a monumental undertaking as we can only presume that there were a similar number of moments of tension amongst the other projects in the classroom. Nonetheless, Ms. Larsen attempted to direct the design process of Bobby and Zack on two occasions by providing (1) a written comment in their booklet asking students to reconcile the lack of details and issues of measurement in their drawing; (2) helping to find a way to put a hole through the entire length of the cylinder; and (3) providing some hints referring to how to make the *spikey thingies* sharper. Throughout the entirety of the observations, she shared her awareness of students' difficulties saying that students' ideas for inventions are often beyond their capacity to realize them. Yet, she believes that their struggles, or micro-tensions, help to build stronger transdisciplinary skills, including resilience, perseverance and autonomy while fostering their creativity and inventiveness.

To summarize our findings, we noticed that design thinking activities can engage students through a complex process of finding the significance of the incomplete. Even though it must have been difficult to admit, Bobby and Zack humbly acknowledged their defeat linking it to issues arising from their own actions: "our measurements were wrong". Perhaps in their defense, what we define as *measurements* in this virtual design process remains entangled in physical dimensions. Even as researchers who have an interest in this topic struggled with the correct vocabulary to describe ideas that are abstracted from the physical to the virtual. It is always the intention to actualize these ideas again into a physical space. Bobby and Zack are not the only ones with an incomplete project, we too have not been able to complete these thoughts and hope to be able to refine them further in future research where we have an opportunity to collect more evidence across materials within the activity system.

This type of abstraction activity where some physical object (prototype) was envisioned as a goal, also created difficulties for us as researchers to understand the state of mind of students who did not create anything tangible. However, when analyzing students' learning experiences on video, we found students to be quite comfortable with the idea of virtuality which was arguably based on their own meaning of what is considered as a prototype. As they plunged into a virtual world, they seemed at ease as they navigated representations of their ideas. While there were aspects of Bobby and Zack's ideas that transcended the plane of virtuality and held on to firm, tangible material characteristics of a physical world, our only data was based on the students' abstractions. Our data reflect the students' representations: abstractions of an idea that was always in a state of becoming. From this perspective, we consider their prototype to have *existed in discourse* - something that may have been lost had they created a physical prototype.

It was the realization of this complex process of abstraction that provided us with a rich answer to our first question: *What did Bobby and Zack actually create?* Upon first review of the data, we were perplexed. The only credible answer that we could provide, from data observed in a class where students only goal was to make *something,* was that they created *nothing.* Considering the goal of the class, this answer would be filled with implied tensions. At the surface level of interpretation, we would have had data that supported the fact that two students had spent 15 classes yet produced *nothing*. We now know that this micro-tension is misplaced. While the tension of *nothing* is powerful, it suggests that it may be an artifact of a time in which students of a traditional classroom were expected to produce constructions that occupied physical space. Our study reveals that the tension of nothing is an issue of perspective and presumption that may be linked to the process of design. Even though students and educators are adept at exploring tools that create virtual experiences, these tools may still be considered to merely be tools that bring ideas into our comfortable notion of physical reality. The more accurate tension surrounding the fact that Bobby and Zack created *nothing*, is that many of us are not yet comfortable with the *something* that was revealed to have been made to only occupy a virtual space.

Our study also revealed a series of micro-tensions that were observed as Bobby and Zack engaged in an activity, but also had implications on related systems of activities with outcomes more influential than those at the level of the students. For instance, one such micro-tension we observed related to students' *emotional experience*. Bobby and Zack seemed to be inexorably connected to the desire to complete their work despite realizing how difficult is to invent something new in a context of making (Kumpulainen & Kajamaa, 2020). In fact, during the project, students expressed a vast array of emotions. Their emotional state varied from being excited (when talking about their ideas) to exhausted. Only for them to then become exasperated with the scope of their idea saying, "it's much harder than what we thought". Sometimes, their work became intense: resulting in students to be more physically active with their gestures and movements throughout the makerspace. This created a micro-tension with classroom logistics that contributed to the overall climate of learning environment. This was the first kind of micro-tension that made us consider the nested influence of a localized activity, zoomed-in on Bobby and Zack, on a related, however more global activity where the outcome effects a system rather than individuals. Further research is required to look more closely at how materials of Bobby and Zack's activity system were influenced by or produced from actors beyond the local. Not only will this help to clarify the entangled and dynamic nature between activity systems but also help researchers to collect more evidence on what gives rise to micro-tensions.

We found the following observations provoked similar thought experiments inspired by the entangled nature of activity systems as supported by the CHAT literature:

Micro-Tension of a 3D Perspective in Virtual Environment One of the possible reasons why Bobby and Zack had difficulty completing a physical prototype may have been what other research has found to be a students' inexperience with software mediated geometric modelling (Bhaduri et al., 2021). Within in *Tinkercad*, Bobby and Zack had difficulties connecting shapes and navigating the virtual construction plane in all three dimensions. It appeared that some of this difficulty originated in Bobby and Zack's inexperience with the transition between their ideas that were written on paper and those that appeared on the screen. It was clear that they had an issue with comparing the measurements on their draft paper and those defined by *Tinkercad*. Also, some software features may have been difficult to navigate when implementing design changes to small objects. For instance, the diameter of the pencil Bobby and Zack were measuring was only 5.25 mm. Some of their frustration was likely a result of how tedious it was to manipulate the small *spikey thingies* inside a cylinder of only 5.25 mm. Finally, Bobby and Zack seemed fixated on the scribble function in order to produce the protrusions. There are no dimensions until an *uneditable* scribble is finished. While they were motivated to use this design tool, it was likely disorienting for an object that was meant to be significantly smaller than their pencil, to fill the screen.

Micro-Tension of Functionality While unsettling to observe Bobby and Zack struggle with a growing disparity between how they had hoped their invention would function and the reality of available resources, it seems as though one of the micro-tensions they experienced may have been less related to their mechanical design and more related to their fascination with the story of *Frankenstein*. Bobby and Zack referred to the classic story several times during our observation as it pertained to their desire to bring life back to the broken pencils. Even though their design process was similar to Mary Shelley's story, their invention requires scavenged, broken parts of pencils, what Bobby and Zack were designing was a prosthetic. Even once brought into being, their *Frankenstein Pencil* would have not had the original function restored. The reassembled pencil would merely have the ability to function as it once did because of the prosthetic. This distinction is nuanced but nonetheless significant. We found that the entangled micro-tension of pencils and *Frankenstein* was actually an opportunity by which the students created yet another prototype – their own version of *Frankenstein*. Much like their pencil however, the plot of Bobby and Zack's *Frankenstein* story remains unfinished.

Micro-Tension of Complexity While it took an intense, frame-by-frame analysis to feel comfortable making a decision with regards to what Bobby and Zack actually made, it is our interpretation that, along with the guidance of Ms. Larsen, they were designing a complex, grooved prosthesis. The grooves, or *spikey thingies* would have helped create resistance by making impressions into the wood of a broken pencil. Even if our interpretation is not entirely correct, it would require a complex design process to achieve. Many of these design-rich achievements we observed. However, perhaps more impressive than the design goal was the fact that Bobby and Zack were constructing the interior structure of their device as a solid, only to then transform it into a *hole.* Once aligned and grouped together the hole would create the grooves in the exterior cylinder. Part of this micro-tension of complexity was that both the pencil-and-paper task and the *Tinkercad* task included virtual ideas. In the pencil sketch, Bobby and Zack showed a "x-ray" view of their design where a structure was shown to accept the broken pencil. When attempting to recreate this idea in *Tinkercad*, the interior structure disappeared as the students did not expect it to be *inside* the exterior cylinder. This perplexed Bobby and Zack and likely contributed to the emotional intensity we observed. This unresolved micro-tension of complexity created an opportunity for Bobby and Zack to explore ideas and limitations of virtual representations that may have not been explored had they never used a virtual design software.

Micro-Tensions of Being Researchers Immersed in the Action As we constructed and reconstructed our interpretations, we discovered that students were doing work that was much more sophisticated than what we first thought. Every time that we were confused, it was because we were not looking close enough at the evidence we had captured. Every observation always contained richer and more complex actions by students than what we originally perceived. In such an immersive learning environment, it would have been easy to focus on another project that had made something tangible. However, we would have completely missed the value of Bobby and Zack's unfinished work. There was rich significance to their work despite the fact they never thought about bringing their idea into physical space. They never 3D printed even the most primitive of prototypes. Not one *spikey thingie*, or cylinder was ever made by Bobby and Zack. These students did not try to ever make their ideas physical even though their classmates were prototyping with cardboard all around them. Regardless of all of the micro-tensions related to their unfinished work, Bobby and Zack shared rich moments of learning with us as researchers.

Micro-Tension Related to the Process Versus Product Dialectics The expectation of what happens in a makerspace is defined well before a student enters. It is in the name: this is the space where things are made. Since many materials are accessible, the making that happens in a makerspace is often rapid. This speeds up the cyclical design process, producing a prototype with each new cycle. Bobby and Zack were seemingly unaware of this opportunity. Perhaps they did not have an opportunity to consider what the process of becoming into a physical space could afford their ideas. Rather they resorted to exhaustion. Through the exhaustion however, Bobby and Zack seemed to have moments of dialectical exploration – moments that inspired our own dialectical thoughts:

1. On the metaphysical level, Bobby and Zack's *Frankenstein* narrative can be imagined as a once broken pencil re-appearing within a system of learning that has been brought back to life.
2. On the plane of experience of navigating between physical and virtual world: just like their physical pencil, their virtual model was also broken. The students were struggling to bring it back to life. At the end of their journey, students nevertheless seemed to be proud of what they have accomplished claiming that "*We actually made a prototype*", according to their own representation of this part of the design process.

This final remark made by one of the students underscores, in our view, a richness of students' experience within a largely virtual space. While leaving their project at its initial steps, students nevertheless have left a wealth of knowledge about what the real process of invention is and how difficult it is to move through its different stages. At every point in their unique process, even though micro-tensions arise, opportunities are revealed as students work on resolving issues that create tension. The willingness for Bobby and Zack to attempt to reach resolution of the issues that caused tension within their activity system is inspiring, even if they were in this case unsuccessful leaving their work unfinished. However, isn't this what is happening to all kinds of inventions? Moreover, could this experience be precursor of further accomplishments within similar or larger activity systems? No one can answer this question. But we hope that our chapter has provided at least some insight for continuing investigation.

References

Bhaduri, S., Biddy, Q. L., Bush, J., Suresh, A., & Sumner, T. (2021). 3DnST: A framework towards understanding Children's interaction with Tinkercad and enhancing spatial thinking skills. In *Interaction design and children* (pp. 257–267).

Blanchard, S., Freiman, V., & Lirrete-Pitre, N. (2010). Strategies used by elementary schoolchildren solving robotics-based complex tasks: Innovative potential of technology. *Procedia-Social and Behavioral Sciences, 2*(2), 2851–2857.

Blikstein, P., & Krannich, D. (2013). The makers' movement and FabLabs in education: Experiences, technologies, and research. In *Proceedings of the 12th international conference on interaction design and children* (pp. 613–616).

Chiasson, M. (2013). *Bring your own device: Action research report*. Govt of New Brunswick.

Chiasson, M. (2019). Investigating teachers' transition towards innovative practices in a modern learning space. In *Proceedings of the conference Transitions19: One journey many pathways* (pp. 27–36). Melburn Graduate School of Education.

Chu, S. L., Angello, G., Saenz, M., & Quek, F. (2017). Fun in making: Understanding the experience of fun and learning through curriculum-based Making in the elementary school classroom. *Entertainment Computing, 18*, 31–40.

Dewey, J. (1916). *Democracy and education: An introduction to the philosophy of education*. The Free Press.

Engeström, Y. (1987). *Learning by expanding: An activity theoretical approach to developmental research*. Orienta-Konsultit.

Engeström, Y. (1999). Activity theory and individual and social transformation. *Perspectives on Activity Theory, 19*(38), 19–30.

Fasso, W., & Knight, B. A. (2020). Identity development in school makerspaces: Intentional design. *International Journal of Technology and Design Education, 30*(2), 275–294.

Foot, K. A. (2014). Cultural-historical activity theory: Exploring a theory to inform practice and research. *Journal of Human Behavior in the Social Environment, 24*(3), 329–347.

Forbes, A., Falloon, G., Stevenson, M., Hatzigianni, M., & Bower, M. (2021). An analysis of the nature of young students' STEM learning in 3D technology-enhanced makerspaces. *Early Education and Development, 32*(1), 172–187.

Freiman, V. (2020). Issues of teaching in a new technology-rich environment: Investigating the case of New Brunswick (Canada) school makerspaces. In *STEM teachers and teaching in the digital era* (pp. 273–292). Springer.

Freiman, V., Beauchamp, J. B., Blain, S., Lirette-Pitre, N. L. P., & Fournier, H. (2011). Problem-based scenarios with laptops: An effective combination for cross-curricular learning in mathematics, science and language. *World Journal on Educational Technology, 3*(3), 136–152.

Freiman, V., & Kamba, J. (2020). 3D modelling and printing to support students' STEM explorations in school makerspaces: Lessons from one case study from New Brunswick, Canada. In A. Savard & R. Pierce (Eds.), *MACAS in the digital era: Proceedings of the 2019 MACAS symposium* (pp. 96–110). Montreal. https://www.mcgill.ca/macas2019/proceedings

Gershenfeld, N. (2012). How to make almost anything: The digital fabrication revolution. *Foreign Affairs, 91*(43).

Government of New Brunswick (GNB). (2016). *10-year education plan: Everyone at their best*. GNB.

Hagel, J., Brown, J. S., & Kulasooriya, D. (2014). *A movement in the making*. Deloitte Press. Retrieved from: http://dupress.com/articles/a-movement-in-the-making

Hatzigianni, M., Stevenson, M., Falloon, G., Bower, M., & Forbes, A. (2021). Young children's design thinking skills in makerspaces. *International Journal of Child-Computer Interaction, 27*, 100216.

Hughes, J. M. (2017). Digital making with "at-risk" youth. *The International Journal of Information and Learning Technology*.

Hughes, J., Morrison, L., Kajamaa, A., & Kumpulainen, K. (2019). Makerspaces promoting students' design thinking and collective knowledge creation: Examples from Canada and Finland. In *Interactivity, game creation, design, learning, and innovation* (pp. 343–352). Springer.

Jurdak, M. E. (2006). Contrasting perspectives and performance of high school students on problem solving in real world, situated, and school contexts. *Educational Studies in Mathematics, 63*(3), 283–301. https://doi.org/10.1007/s10649-005-9008-y

Kajamaa, A., & Kumpulainen, K. (2019). Agency in the making: Analyzing students' transformative agency in a school-based makerspace. *Mind, Culture, and Activity, 26*(3), 266–281.

Keune, A., Gomoll, A., & Peppler, K. (2015). Flexibility to learn: Material artifacts in makerspaces. In *Fifth annual FabLearn conference: Equity and diversity in making*. Stanford University.

Kumpulainen, K., & Kajamaa, A. (2020). The sociomaterial ecology of emotions in a school's makerspace. In *Maker literacies and maker identities in the digital age* (pp. 149–166). Routledge.

Lave, J., & Wenger, E. (1991). *Situated learning: Legitimate peripheral participation*. Cambridge University Press.

Leontiev, A. N. (1981). *The problems of the development of the mind*. Progress Publishers.
Lingley, J. (2021). Mathematics, making, and materialism [Master's thesis, University of new Brunswick]. *UNB Scholar Research Repository.*. https://unbscholar.lib.unb.ca/islandora/object/unbscholar%3A10513
Litts, B. K. (2015). *Making learning: Makerspaces as learning environments*. Doctoral dissertation. The University of Wisconsin-Madison.
Maker Media. (2013). Maker faire: A bit of history. Maker faire Website. http://makerfaire.com/makerfairehistory/
Martin, L. (2015). The promise of the maker movement for education. *Journal of Pre-College Engineering Education Research (J-PEER), 5*(1), 4.
Mersand, S. (2020). The state of makerspace research: A review of the literature. *TechTrends*, 1–13.
Montessori, M. (1912/1967). *The Montessori method: Scientific pedagogy as applied to child education in "the Children's houses" with additions and revisions by the author*. Robert Bentley.
Nelson, J. B. (2019). *Schooling makerspaces: On the promises and tensions of implementing this much-touted innovation in a high school* (Doctoral dissertation).
Niederhauser, D. S., & Schrum, L. (2016). *Enacting STEM education for digital age learners: The "maker" movement goes to school*. International Association for Development of the Information Society.
Oliver, K. M. (2016). Professional development considerations for makerspace leaders, part one: Addressing "what?" and "why?". *TechTrends, 60*(2), 160–166.
Papert, S. (1980). Mindstorms. Children. Computers and powerful ideas.
Peppler, K., & Bender, S. (2013). Maker movement spreads innovation one project at a time. *Phi Delta Kappan, 95*(3), 22–27.
Piaget, J. (1954). *The construction of reality in the child*. Basic Books.
Robichaud, X., & Freiman, V. (2020). Creativity as a learning factor in an interdisciplinary environment including mathematics, music and technology. In A. Savard & R. Pierce (Eds.), *MACAS in the digital era: Proceedings of the 2019 MACAS symposium, Montreal, Quebec* (pp. 79–95). McGill Faculty of Education.
Sannino, A. E., Daniels, H. E., & Gutiérrez, K. D. (2009). *Learning and expanding with activity theory*. Cambridge University Press.
Sheridan, K., Halverson, E. R., Litts, B., Brahms, L., Jacobs-Priebe, L., & Owens, T. (2014). Learning in the making: A comparative case study of three makerspaces. *Harvard Educational Review, 84*(4), 505–531.
von Thienen, J. P. A., Meinel, C., & Corazza, G. E. (2017). A short theory of failure. In *Electronic colloquium on design thinking research* (Vol. 17, pp. 1–5).
Vygotsky, L. S., & Cole, M. (1978). *Mind in society: Development of higher psychological processes*. Harvard University Press.
Walan, S. (2021). The dream performance–a case study of young girls' development of interest in STEM and 21st century skills, when activities in a makerspace were combined with drama. *Research in Science & Technological Education, 39*(1), 23–43.
Wannenmacher, D., & Antoine, A. (2016). Management of innovative collaborative projects: Moments of tension and the peer-mediation process—A case-study approach. *Knowledge Management Research & Practice, 14*(1), 35–45.
Williamson, B. (2015). Hackerspaces and homeschooling: Making 'startup schools'. Hackerspaces and Homeschooling: Making 'Startup Schools' - Connected Learning Alliance (clalliance.org)

Viktor Freiman PhD, full professor in mathematics education at the Université de Moncton, Canada, and director of the CompeTI.CA (ICT competences in the Atlantic Canada) Partnership Network. His research interests and publications are in the field of teaching and learning algebra, problem-solving, use of technology, STEAM education, history of mathematics education, and mathematical creativity and giftedness. He is a president of the International Group for Mathematical Creativity and Giftedness as well as co-organizer of Mathematics and its Connections to the Arts and Science symposia. He is co-editor of the book series Mathematics Education in the Digital Era.

Jacob Lingley (MEd from University of New Brunswick, Canada) is a Program Director with Brilliant Labs, an organization that promotes STEM education throughout. His research interests include mathematics, making, and materialism. For more information about his work and the work of Brilliant Labs, visit www.brilliantlabs.ca or contact Jacob directly at jacob.lingley@nbed.nb.ca.

Index

K. Plakitsi, S. Barma (eds.), *Sociocultural Approaches to STEM Education*, Sociocultural Explorations of Science Education 21,
https://doi.org/10.1007/978-3-031-44377-0

MIX
Papier aus verantwortungsvollen Quellen
Paper from responsible sources
FSC® C105338

If you have any concerns about our products,
you can contact us on
ProductSafety@springernature.com

In case Publisher is established outside the EU,
the EU authorized representative is:
Springer Nature Customer Service Center GmbH
Europaplatz 3, 69115 Heidelberg, Germany

Printed by Libri Plureos GmbH
in Hamburg, Germany